Contributions to Statistics

For further volumes:
http.//www.springer.com/series/2912

Alexander Krämer • Md. Mobarak Hossain Khan •
Frauke Kraas
Editors

Health in Megacities and Urban Areas

Physica-Verlag
A Springer Company

Editors
Alexander Krämer
Department of Public Health Medicine
School of Public Health
Bielefeld University
Universitätsstr. 25
33615 Bielefeld
Germany
alexander.kraemer@uni-bielefeld.de

Md. Mobarak Hossain Khan
Department of Public Health Medicine
School of Public Health
Bielefeld University
Universitätsstr. 25
33615 Bielefeld
Germany
mobarak.khan@uni-bielefeld.de

Frauke Kraas
Institute of Geography
Cologne University
Albertus-Magnus-Platz
50923 Cologne
Germany
f.kraas@uni-koeln.de

ISSN 1431-1968
ISBN 978-3-7908-2732-3 e-ISBN 978-3-7908-2733-0
DOI 10.1007/978-3-7908-2733-0
Springer Heidelberg Dordrecht London New York

Library of Congress Control Number: 2011929497

Cover design: eStudio Calamar S.L.

Printed on acid-free paper

Physica-Verlag is a brand of Springer-Verlag Berlin Heidelberg
Springer-Verlag is a part of Springer Science+Business Media (www.springer.com)

Foreword

Highly diverse driving forces, processes and actors are responsible for different trends in the development of megacities. Under the dynamics of global change, megacities are themselves changing on the one hand they are prone to increasing socio-economic vulnerability because of pronounced poverty, socio-spatial and political fragmentation, sometimes with extreme forms of segregation, disparities and conflicts. The juxtaposition of vastly different local life worlds, life-forms and lifestyles plays a significant differentiating role. On the other hand – and an often neglected aspect – megacities offer positive potential for global transformation, e.g. minimisation of space consumption, high effective use of resources applied, efficient disaster prevention and health care options – if good strategies are developed.

In many megacities of the developing world and the emerging economies the quality of life is eroding. Most of the megacities have grown to unprecedented size, and the pace of urbanisation has far exceeded the growth of the necessary infrastructure and services. As a result, an increasing number of urban dwellers are left without access to basic amenities and face appalling living conditions. Already, existing symptoms of economic, ecological, infrastructural and socio-economic overload are increasing, producing emerging urban security risks at a local, regional and international level. With regard to the environment, water and health, problems of emission reduction, the provision of clean drinking water, and sewage and rubbish disposal are the most important issues.

The inadequate environmental situation is already directly responsible for avoidable health problems. Despite a long history of urban sanitary reform and healthy-city movements, inhabitants of rapidly growing urban agglomerations in the developing world and emerging economies are increasingly confronted with severe environmental health risks. Additionally, social inequalities lead to subsequent and significant intra-urban health inequalities. Land-use changes often create changes in environmental conditions and the habitat for a number of species, which can trigger the outbreak of diseases; overcrowding in urban agglomerations provides an easy pathways for the spread of communicable diseases; large-scale migration to urban areas and integration into a global market where borders are frequently crossed and large distances easily travelled by a growing number of people allow the rapid movement of infected individuals into previously unexposed populations. Since the mid 1970s, the World Health Organization (WHO) has identified 30 new diseases. In addition, there has been a significant resurgence and redistribution of old diseases carried by mosquitoes, such as malaria and dengue fever, which can now affect regions and urban areas where they were not prevalent before.

Against this background, the aim of our book is to contribute to the important cross-sectional multidisciplinary topic of health. Several chapters are based on research conducted at two different locations, namely the Pearl River Delta (PRD) in China and Dhaka, the capital of Bangladesh, under the priority programme of the German Research Foundation "Megacities – Megachallenge: Informal Dynamics of Global Change" (SPP 1233).

The book is divided into six parts. In the introductory part, "Challenges, Theories and Concepts", a conceptual framework from the perspective of the health sciences is presented. The second part, "Case Studies and Examples", addresses the situation in both developed and developing countries. The third part focuses on "Environmental Health Risks", which includes chapters about the health effects of air pollution, thermal stress and the effects of climate change on the epidemiology of infectious diseases in South Asia. The fourth part, "Informality and Health", highlights issues like informal working conditions, the informalisation of health care, rural-urban migration, the health of migrant populations and effects of mega-urbanisation on water quality and health. Then, we examine aspects of "Spatial Dimension and Health", hereby addressing spectral surface reflectance fields, remote sensing and Geographical Information Systems (GIS) in public health, and health economics considerations. The last part of the book provides insights into "Urban Livelihoods, Urban Food and Health".

As such, the guiding idea our book lies in a multi- and interdisciplinary approach to the complex topic of health in megacities and urban areas, which can only be adequately understood, when different disciplines share their knowledge and methodological tools to work together. We hope that our book will allow readers to deepen their understanding of the complex dynamics of urban and megacity populations through the lens of public health, geographical and other research perspectives.

Alexander Krämer, Md. Mobarak Hossain Khan, Frauke Kraas

Contents

Contributors

Rolf Annuss NRW Institute of Health and Work (LIGA.NRW), Department of Prevention and Innovation, Ulenbergstr. 127-131, 40225 Düsseldorf, Germany

Rafig Azzam Department of Engineering Geology and Hydrogeology, RWTH Aachen University, Aachen, Germany

Klaus Baier Department of Engineering Geology and Hydrogeology, RWTH Aachen University, Aachen, Germany

Sabine Baumgart Department of Urban and Regional Planning, Faculty of Spatial Planning, TU Dortmund University, Dortmund, Germany

Hans-Georg Bohle Geography Department, University of Bonn, Bonn, Germany

Tabea Bork Institute of Geography Cologne University, Cologne, Germany

Irene Brüske Helmholtz Zentrum München – German Research Center for Environmental Health, Institute of Epidemiology II, Munich, Germany

Susanne Breitner Helmholtz Zentrum München – German Research Center for Environmental Health, Institute of Epidemiology II, Munich, Germany

Katrin Burkart Department of Geography, Climatological Section, Humboldt-Universität zu Berlin, Berlin, Germany

Carsten Butsch Institute of Geography, Cologne University, Cologne, Germany

Wilfried Endlicher Department of Geography, Climatological Section, Humboldt-Universität zu Berlin, Berlin, Germany

Benjamin Etzold Geography Department, University of Bonn, Bonn, Germany

Rainer Fehr NRW Institute of Health and Work (LIGA.NRW), Department of Prevention and Innovation, Ulenbergstr. 127-131, 40225 Düsseldorf, Germany

Bettina Gransow Seminar of East Asian Studies, Free University Berlin, Berlin, Germany; School of Sociology and Anthropology, Sun Yat-sen University, Guangzhou, China

Oliver Gruebner Geomatics Lab, Department of Geography, Humboldt-Universität zu Berlin, Berlin, Germany

Kirsten Hackenbroch Department of Urban and Regional Planning, Faculty of Spatial Planning, TU Dortmund University, Dortmund, Germany

Lu Han Department of Social Medicine and Health Management, School of Public Health, Sun Yat-sen University, Guangzhou, China

Shahadat Hossain Department of Urban and Regional Planning, Faculty of Spatial Planning, TU Dortmund University, Dortmund, Germany

Patrick Hostert Geomatics Lab, Department of Geography, Humboldt-Universität zu Berlin, Berlin, Germany

Heiko J. Jahn Department of Public Health Medicine, School of Public Health, Bielefeld University, Bielefeld, Germany

Markus Keck South Asia Institute, University of Heidelberg, Heidelberg, Germany

Md. Mobarak Hossain Khan Department of Public Health Medicine, School of Public Health, Bielefeld University, Bielefeld, Germany

Aklimunnesa Khan Department of Public Health, Sapporo Medical University School of Medicine, Sapporo, Japan

Alexander Krämer Department of Public Health Medicine, School of Public Health, Bielefeld University, Bielefeld, Germany

Frauke Kraas Institute of Geography, Cologne University, Cologne, Germany

Volker Kreibich Faculty of Spatial Planning, TU Dortmund University, Dortmund, Germany

Mareike Kroll Institute of Geography, Cologne University, Cologne, Germany

Elmar Kulke Department of Geography, Humboldt-Universität zu Berlin, Berlin, Germany

Li Ling Department of Medical Statistics and Epidemiology, School of Public Health, Sun Yat-sen University, Guangzhou, China

Britta Mey Leipzig Institute for Meteorology (LIM), University of Leipzig, Stephanstr. 3, D-04103, Leipzig, Germany

Mitsuru Mori Department of Public Health, Sapporo Medical University School of Medicine, Sapporo, Japan

Hirofumi Ohnishi Department of Public Health, Sapporo Medical University School of Medicine, Sapporo, Japan

Asae Oura Department of Public Health, Sapporo Medical University School of Medicine, Sapporo, Japan

Annette Peters Helmholtz Zentrum München – German Research Center for Environmental Health, Institute of Epidemiology II, Munich, Germany

Paulo Pinheiro Department of Public Health Medicine, School of Public Health, Bielefeld University, Bielefeld, Germany

Dietrich Plaß Department of Public Health Medicine, School of Public Health, Bielefeld University, Bielefeld, Germany

Luise Prüfer-Krämer Travel Clinic, Bielefeld, Germany

Sanjeevi Prasad French Institute of Pondicherry, Pondicherry, India

Regina Rückerl Helmholtz Zentrum München – German Research Center for Environmental Health, Institute of Epidemiology II, Munich, Germany

Fumio Sakauchi Department of Public Health, Sapporo Medical University School of Medicine, Sapporo, Japan

Patrick Sakdapolrak Department of Geography, University of Bonn, Bonn, Germany

Alexandra Schneider Helmholtz Zentrum München – German Research Center for Environmental Health, Institute of Epidemiology II, Munich, Germany

Thomas Seyler French Institute of Pondicherry, Pondicherry, India

Erhua Shang Department of Public Health, Sapporo Medical University School of Medicine, Sapporo, Japan

Ronny Staffeld Department of Geography, Humboldt-Universität zu Berlin, Berlin, Germany

Ramona Strohschön Department of Engineering Geology and Hydrogeology, RWTH Aachen University, Aachen, Germany

Claudia Terschüren NRW Institute of Health and Work, Düsseldorf, Germany

Manfred Wendisch Leipzig Institute for Meteorology (LIM), University of Leipzig, Stephanstr. 3, D-04103, Leipzig, Germany

Kathrin Wolf Helmholtz Zentrum München – German Research Center for Environmental Health, Institute of Epidemiology II, Munich, Germany

Yinghua Xia Department of Medical Statistics and Epidemiology, School of Public Health, Sun Yat-sen University, Guangzhou, China

Yuan Yuan Seminar of East Asian Studies, Free University Berlin, Berlin, Germany

Arina Zanuzdana Department of Public Health Medicine, School of Public Health, Bielefeld University, Bielefeld, Germany

Wolfgang-Peter Zingel South Asia Institute, University of Heidelberg, Heidelberg, Germany

Part I
Challenges, Theories, Concepts

Chapter 1
Public Health in Megacities and Urban Areas: A Conceptual Framework

Alexander Krämer, Md. Mobarak Hossain Khan, and Heiko J Jahn

1.1 Introduction

In this chapter, first, we will briefly discuss worldwide urbanisation processes with major dimensions of public health challenges in megacities and urban areas. Second, we present some empirical findings from public health surveys conducted in the megacity of Dhaka, Bangladesh. Third, a conceptual framework is proposed based on our research on megacities within the framework of the German Research Foundation's priority programme "Megacities – Megachallenge: Informal Dynamics of Global Change" and forth, a description of the burden of disease – classified as group I, II and III diseases – in urban areas including their determinants are presented. Lastly, strategies to improve the quality of life in megacities and urban areas are discussed.

1.2 Urbanisation and Megacity Development

Urbanisation is a worldwide phenomenon mostly occurring in developing countries. Over the last 20 years many urban areas have experienced dramatic growth, which is the result of a combination of factors such as geographical location, natural population growth, rural–urban migration, national policies, continued global economic integration and globalization (Cohen 2004; Cohen 2006; UN-HABITAT 2008). Urban areas in developing countries absorb about 5 million new residents every month (UN-HABITAT 2008). In the near future, the pace of urbanisation will be even faster than in the past. Recent data show that worldwide the urban population will reach 4.58 billion by 2025 from 3.29 billion in

A. Krämer (✉) • M.M.H. Khan • H.J. Jahn
Department of Public Health Medicine, School of Public Health, Bielefeld University, Bielefeld, Germany
e-mail: alexander.kraemer@uni-bielefeld.de

A. Krämer et al. (eds.), *Health in Megacities and Urban Areas*,
Contributions to Statistics, DOI 10.1007/978-3-7908-2733-0_1,

2007, in contrast the rural population will be 3.43 billion by 2025 from 3.38 billion in 2007 (United Nations 2008). Thus, virtually all population growth (over 96%) over the next two decades will be concentrated in urban areas and most of urban growth will be concentrated in less developed regions (1.21 billion out of 1.29 billion), mostly in Asian cities (United Nations 2008).

Although 74% of the total population lived in urban areas in more developed regions in 2007 as compared to 44% in less developed regions, most of the megacities (with a population of at least 10 million) are located and will develop in developing countries mainly in Asia (Table 1.1) (UN-HABITAT 2008; United Nations 2008). In 1950, there were only 2 megacities in the world located in developed regions (New York-Newark, USA and Tokyo, Japan), which increased to 3 megacities in 1975, and 19 megacities in 2007 (United Nations 2008). The major contributing factors for megacity development are increasing globalization and industrialisation and subsequently rapid urbanisation by rural–urban migration. Presently, there are no megacities (with 10 million or more inhabitants) in eastern and southern Africa, northern and southern Europe and the Caribbean. These areas

Table 1.1 Distribution of worldwide megacities, 2000–2025

Areas	2000	2010	2020	2025	New megacities 2000–2025
Africa	1	2	3	3	2
Eastern Africa	0	0	0	0	0
Middle Africa	0	0	1	1	1
Northern Africa	1	1	1	1	0
Southern Africa	0	0	0	0	0
Western Africa	0	1	1	1	1
Asia	8	11	13	16	8
Eastern Asia	3	4	5	6	3
South-central Asia	5	5	5	7	2
South-eastern Asia	0	1	2	2	2
Western Asia	0	1	1	1	1
Europe	1	1	2	2	1
Eastern Europe	1	1	1	1	0
Northern Europe	0	0	0	0	0
Southern Europe	0	0	0	0	0
Western Europe	0	0	1	1	1
Latin America and the Caribbean	4	4	4	4	0
Caribbean	0	0	0	0	0
Central America	1	1	1	1	0
South America	3	3	3	3	0
Northern America	2	2	2	2	0
Oceania	0	0	0	0	0
Developing countries[a]	11	15	18	21	10
Developed countries	5	5	6	6	1
Total	16	20	24	27	11

Source: UN-HABITAT (2008); United Nations (2008)
[a]Including China and Turkey

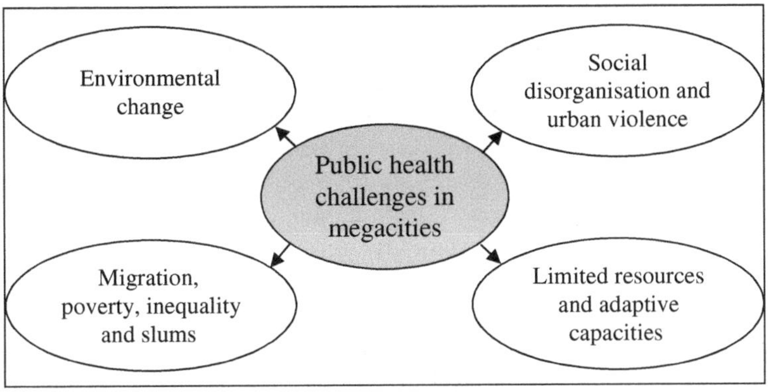

Fig. 1.1 Public health challenges in megacities

will not have any megacity in the next 15 years (UN-HABITAT 2008; United Nations 2008).

The growing number of cities including megacities clearly indicates that these areas are gradually becoming the major settings for human habitation. At present half of the world population lives in cities and by 2030, 60% of the population will reside in urban areas (UN-HABITAT 2008). The social and landscape transformations through urbanisation are literally changing the face of the planet (Cohen 2004). Although cities (1) are focal points of economic development, innovation, and employment; and (2) act as centres of modern living, culture, science, education, health care, politics, and other basic services (Cohen 2006; Leon 2008), the failure to manage the impacts of rapid urbanisation threatens the environment, human health, equity, urban productivity and hence the quality of life (Fadda and Jiron 1999). These areas can be the breeding grounds for poverty, exclusion and environmental degradation (UN-HABITAT 2008). Some visible dimensions of public health challenges in megacities and urban areas which may increase the health risks are: environmental change; uncontrolled rural–urban migration, poverty, inequality and slum development; social transformation, disorganisation and urban violence; lack of resources and adaptive capacities (Fig. 1.1).

1.3 Major Dimensions of Public Health Challenges in Megacities

Global environmental change is a growing and challenging area of multidisciplinary and multisectoral research. It poses a great threat to global public health and human well-being of many populations (Campbell-Lendrum and Corvalan 2007; Costello et al. 2009). Climate change threats the progress in poverty reduction and the achievement of the Millennium Development Goals (Mitchell and Tanner 2006). In fact, climate change will continue to impact on all sectors, from national and economic security to human health, food production, infrastructure, water availability and ecosystems (WWF International 2009). The poorest populations with limited access to health care, located mostly in Asia and Africa, are most vulnerable to the impact of global environmental change (Campbell-Lendrum and Corvalan 2007; Costello et al. 2009; WHO 2003). For instance, about 99% of all extreme climate/weather-related global deaths in 1990 occurred in developing countries (WHO 2003). The lack of necessary institutional, economical and financial capacities, as well as the inability to rebuild the infrastructure damaged by the natural disasters, makes poor nations more vulnerable (Campbell-Lendrum and Corvalan 2007; Costello et al. 2009; Huq et al. 2003).

Environmental change and cities are strongly linked. Cities are the hotspots for climate change (Patz and Kovats 2002). They are key players concerning carbon emissions and other climate change driving activities because most human and economic activities are concentrated in urban areas (UN-HABITAT 2008). Climate change remarkably affects the health of urban populations (Kovats and Akhtar 2008) and the poor environmental quality of cities in developing countries has been recognised as one of the most urgent and severe public health problems (Fadda and Jiron 1999). Although all city inhabitants are affected by global environmental change, inhabitants from the cities of developing countries (e.g. many of the Asian megacities) are more vulnerable to the impact of climate change (UN-HABITAT 2008; WWF International 2009) as compared to the cities in developed countries due to limited resources and adaptive capacity. Cities in developing countries are affected by localised health-threatening environmental issues belonging to the "brown agenda", while cities in developed countries are affected by the "green agenda" (UN-HABITAT 2008). The ecology of cities and megacities is degrading by anthropogenic activities, which is additionally burdened by climate changes (Grimm et al. 2008; Nicholls 1995). Cities cover only 1% of the planet's surface but use 75% of the world's energy and emit 75% of global greenhouse gases (WWF International 2009). Particularly the long-lasting impact of climate change in megacities must be considered as a long-term problem (Nicholls 1995).

Another public health challenge in megacities is attributed to the rapid rural–urban migration. Higher poverty, inadequate basic facilities, and lack of job opportunities in rural areas generally force people to move to cities. The pull factors of migration may include the expectation of higher income and better life. Generally rural migrants come to the cities under the illusion that cities will offer prospects of good employment, better education, a good living standard (Oloruntimehin 1996), and a life with rights and security (Briceno-Leon 2005). Unfortunately, the real situation in the cities, however, mostly does not come up to the migrants' expectations. On the contrary, migrants often find themselves in situations of unemployment, underemployment, hopelessness (Oloruntimehin 1996) and insecurity (Briceno-Leon 2005).

The city's infrastructure and resources are not sufficient to provide facilities according to people's demand. Consequently most of the migrants from low-income families encounter various problems such as insecurity and social discrimination. The majority of the migrants normally settle in slum and squatter settlements (Khan and Krämer 2008). They are often under-served by the municipal authorities, experience social, economic and political exclusion and are exposed to a wide range of health threats (Montgomery 2009). Urban slum settlers are frequently exposed to adverse living conditions including insufficient provision of health care, drinking water, solid waste and waste water management, electricity and fossil fuels for cooking and heating (Khan et al. 2009). They often have no security of tenure and suffer from dense and poor structural housing, overcrowded dwellings and inadequate sanitation facilities (Sclar et al. 2005). Additionally the generally poor-educated slum dwellers often lack knowledge on health threats. The poor environmental conditions of slums exacerbate the risks of waterborne diseases (e.g. diarrhoea, cholera) and airborne diseases (e.g. influenza, pneumonia, and tuberculosis) (Rashid 2009).

Rapid urbanisation and the rising trend of slum populations in urban areas is a public health concern. Already more than one billion people live in slums, mostly in developing countries, and experts project that this figure could rise to 1.7 billion (Sclar et al. 2005) or even double by 2030 (Sclar and Northridge 2003). The proportion of the urban population living in slums (Fig. 1.2) is highest in sub-Saharan Africa, followed by southern Asia and eastern Asia (Fig. 1.2) (UN-HABITAT 2008). Inhabitants of slums often suffer from poor mental and physical health as compared to inhabitants who do not live in slums (Khan and Krämer 2009).

Increasing urban inequality is another challenge in the cities. In many cities, wealth and poverty coexist in close proximity. For instance, rich, well-served neighbourhoods and gated communities are often situated near densely populated slum communities. Slum dwellers of the world's poorest cities often experience multiple deprivations in terms of housing, food, education, health and basic services. The high level of inequality creates social and political fractures within the society, increases political tension, reduces investment and is associated with devastating consequences on societies (UN-HABITAT 2008).

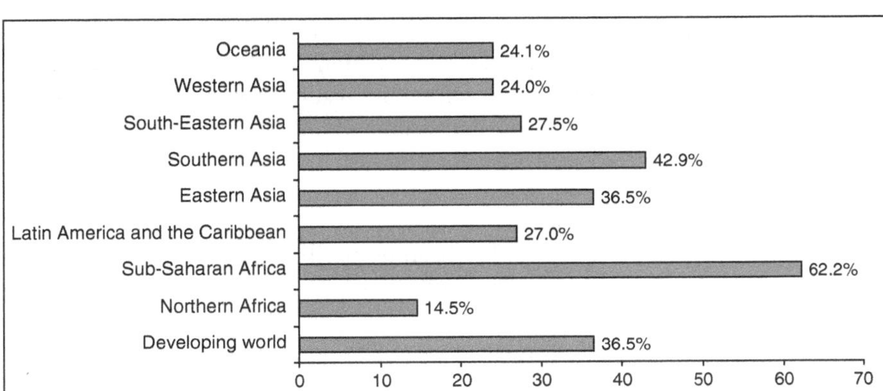

Fig. 1.2 Proportion of urban population living in slums, 2005 (UN-Habitat 2008)

A further public health challenge is related to the urban social environment which is generally quite different from rural areas. Generally megacities are complex communities of heterogeneous groups of people and are often characterized by limited resources, social disintegration, area fragmentation, uncontrolled growth of slums and marginal settlements without enough employment opportunities (Khan et al. 2009). These factors in megacities have led to the continued expansion of slums and marginal settlements and create favourable conditions for various forms of urban violence (Oloruntimehin 1996). Urban violence is an increasing problem in many cities of the world (Urban Violence Subcommittee 2008) and has reached high levels in many nations (Moser 2004). Urban violence is common in every region of the world, and in every culture (Briceno-Leon 2005; Imbusch 2003). This problem has appeared as an important dimension of public health. In cities, different kinds of violence such as political violence, economic violence, social violence and institutional violence are observed (Moser 2004). Like all social developments, urban violence is a multifactorial phenomenon and is influenced by biological, social, cultural, economical and political determinants (de Jesus Mari et al. 2008). Feelings of insecurity, fear of crime and violence are often high in large cities especially among women mainly due to their exposure to sex crimes (Oxfam 2009). The economic and social costs of urban violence has reached alarming proportions (CSPV 1998). Exposure to violence can generate a sense of fear and impair social participation (Moser 2004). Fear of crime affects the quality of life across various demographic and socio-economic social strata (Franklin et al. 2008). Urban violence may produce generalized emotional distress, aggressions and disruptions in interpersonal relationships. It can cause cognitive and psychological impairment and can result in physical symptoms like chronic fatigue (CSPV 1998).

Structural characteristics of urban neighbourhoods have also impact on the degree of urban violence. In many cases persistent poverty, high population turnover, and ethnic heterogeneity – often found in migrant populations living in slums – may reduce social ties, common values and community participation. These conditions can derogate the social and economic viability of local institutions (e.g. churches, schools, and the family) and impede the establishment of social connections and community attachment (Coutts and Kawachi 2006; Sampson 1997). Additionally, social segregation and a high degree of intra-community diversity can lead to distrust within a community resulting in a low level of social capital and social support and isolation (Ryan et al. 2008). Such isolation may in turn promote health-related problems in terms of increasing tolerance for risky lifestyles and detachment from mainstream values and as a result can increase crime, violence and substance abuse.

Although informal social control has been primarily evoked in the context of a community's ability to control deviant behaviour, it can be generalised to health behaviours and health outcomes (e.g. control of smoking, drinking, and drug abuse). Social capital (e.g. trust, civic engagement, social and electoral participation, voluntarism) refers to the resources available to individuals and groups through social connections and may therefore influence human health both positively (mostly) and negatively (Coutts and Kawachi 2006).

Increasing public health challenges in megacities are also related to globalization. Globalisation processes are fuelled by neo-liberal economical deregulation, including less restrictive international trade policies and deregulated financial and labour markets (UN-HABITAT 2003). It appears in the form of increasing mobility of goods, ideas, capital, work force, technologies, services and so forth (Galea et al. 2005). According to neo-liberal market theories, deregulation would lead to the most effective production and distribution of goods as well as to rising gross domestic incomes in involved countries. Between 1970 and 1990, the world trade tripled and the economical growth continuously increased in the 1990s (UN-HABITAT 2003). According to the "neo-liberal economic doctrine", higher productivity and increasing wealth should have led to prosperous developments in countries that were then able to participate in world trade and production (UN-HABITAT 2003). However, not all parts of the world equally benefit from globalisation processes because such development can cause environmental degradation, social inequality, insecurity, conflicts, poverty and insufficient infrastructure (Laaser 2006; Schwefel 2006). These factors are strongly related to the health status of the disadvantaged.

Since megacities are nodal points of globalisation processes (Kraas and Mertins 2008), all health effects of globalisation are concentrated in megacities. Particularly the fast growing megacities in developing countries experience the fastest changes with respect to wealth differences, and other health-determining factors. Globalisation of labour creates new, mostly informal low-income jobs. These job opportunities are pulling rural people into the growing cities. The rural–urban migrants often find their homes in slum settlements after coming to the cities and the adverse living conditions in these settlements cause substantial burden of disease compared to non-slum settlers with higher socio-economic status (Khan et al. 2009). Unhealthy lifestyles (smoking, high caloric nutrition) also become common in populations of developing countries – and at first in the urban centres. Due to globalization processes high caloric food including higher levels of sugar intake and animal products become available first in urban areas. These changes can affect both, city inhabitants of high and low socio-economic status in terms of obesity and increasing non-communicable diseases like diabetes and cardiovascular diseases (Mendez and Popkin 2004).

Also the import and export of infectious diseases is frequently discussed in the context of globalization. It takes place in the world's hotspots of travel, transportation and economical activities. Malaria, tuberculosis, hepatitis, HIV/AIDS are just some examples of typical communicable infections that spread through international mobility including working migration and travel (Gushulak and MacPherson 2000; Gushulak and MacPherson 2004; Harper and Raman 2008).

Public health in megacities in developing countries is also challenged by the lack of resources and limited adaptive capacities. These cities are continuously under demographic, social, environmental and economical change. Resource-poor megacities generally shelter large proportions of poor subpopulations with no or restrictive access to basic needs like education or health care service (Kraas and Mertins 2008). The continuous influx of rural–urban migrants causes further stress to the mega-urban environment. Often authorities fail to keep up with urban growth and to meet the different needs of the diverse subpopulations. On the one hand, these megacities – as complex systems of internal diversity and global interaction – suffer from permanent high levels of internal and external stressors and lack, on the other hand, capacities to activate internal or external resources to cope with these stressors. Such megacities are particularly at risk of increasing stress or shocks like natural disasters.

1.4 Determinants of Public Health in Megacities: Empirical Findings from Dhaka

Dhaka is the ninth largest megacity in the world with more than 13 million inhabitants (United Nations 2008). Out of 11 large vulnerable cities in Asia, Dhaka is the most vulnerable one to the impact of climate change (WWF International 2009). The city is growing fast as compared to other megacities. The total population increased from 0.42 million in 1950 to 3.3 million in 1980, to 10.2 million in 2000, and is expected to increase to 16.8 million in 2015 (Khan and Krämer 2009). Likewise the total population living in slums in Dhaka increased sharply from 20% in 1996 to 37% in 2005 mainly due to rapid rural–urban migration (Centre for Urban Studies 2006; Khan and Krämer 2008). About 300,000–400,000 new migrants stream to Dhaka city every year, and most of them initially settle in slums (World Bank 2007). Crime, violence, and risky lifestyles such as smoking and illicit drug use are commonly reported among the urban poor living in slums and marginal settlements in Dhaka (World Bank 2007). Here we present different health determinants based on our data analyses collected through a cross-sectional study and follow-up surveys conducted in Dhaka and adjacent areas (Fig. 1.3). According to our findings, poor public health for the people living in the megacity of Dhaka is associated with poor socio-economic and environmental conditions, poor lifestyles, migration, informal activities, lack of health facilities, lack of social support and lack of income.

1.5 A Conceptual Framework for Urban Health

Evidence indicates that urban health is the function of various factors ranging from individual to macro (global) determinants (Galea et al. 2006). Because of a varying strength of associations between health determinants and urban health outcomes, these determinants should be placed in the framework in such a way that people can understand their relations with health outcomes. Considering the complex multi-level background of urban health based on our own research as well as on the available literature, we propose a comprehensive urban health framework which includes micro-, meso- and macro-level determinants (Fig. 1.4). According to this

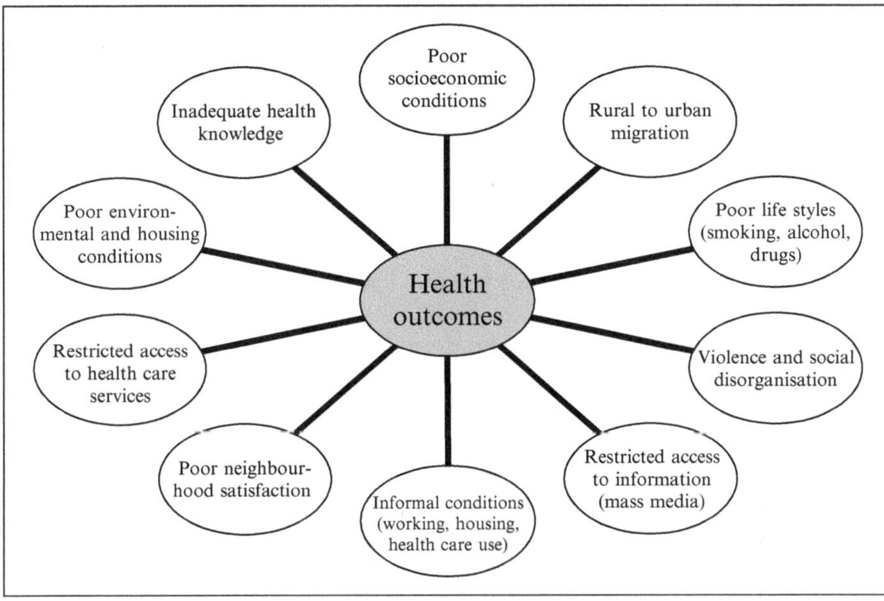

Fig. 1.3 Determinants of health outcomes in megacities

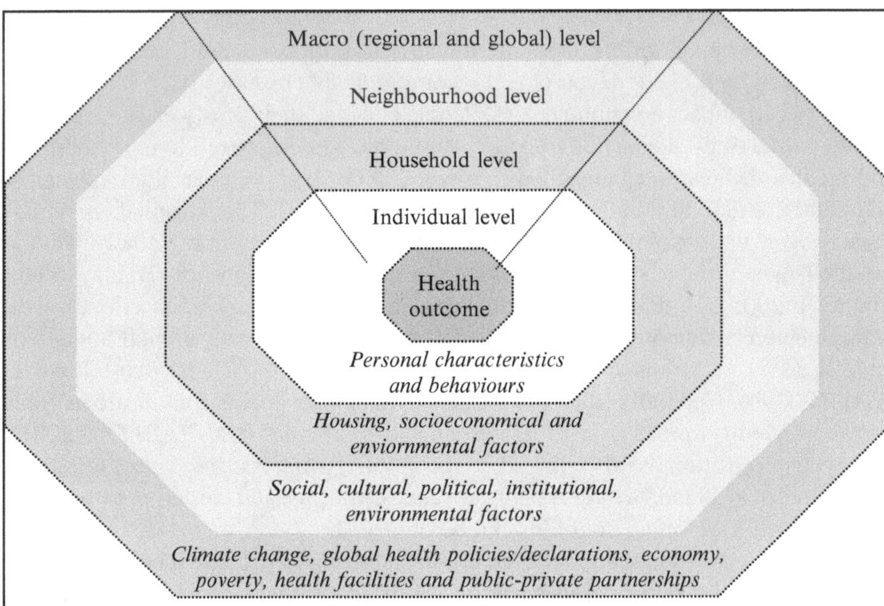

Fig. 1.4 A multilevel conceptual framework for urban health

model, health outcomes are strongly associated with individual determinants, (micro-level) followed by meso-level (e.g. neighbourhood and household) and macro-level determinants.

1.6 Burden of Diseases in Megacities

The burden of disease in megacities has no homogeneous appearance throughout the world. This is true for differences between the cities in different continents or countries and for cities within one country. Even within a city, the distribution of the disease burden can be quite unequal between certain subpopulations. Urban health is influenced by various factors (Fig. 1.3) and is reflected by the proposed multilevel framework of urban health (Fig. 1.4). Therefore a general statement about the disease burden in megacities would not be appropriate. Within the following sections some key issues for megaurban health will be presented and core patterns of disease burden, classified according to the WHO disease grouping, will be introduced.

1.6.1 Group I Diseases: Communicable Diseases, Maternal, Perinatal and Nutritional Conditions

Since the beginning of the age of industrialisation one could observe that – accompanied by societal development – communicable diseases (CD) increasingly had a reduced impact on public health. People changed their lifestyles and were less prone to infectious agents due to better sanitation. The improved living conditions and medical achievements resulted in lower CD burden, in higher life expectancy and higher burden of non-communicable diseases (NCD). The so-called *epidemiologic transition* took place and is, to a certain extent, still ongoing globally. Today, these changes are particularly observable in developing countries (Boutayeb and Boutayeb 2005). In high-income countries, in which this transition already took place, the percentage of years of life lost (YLL) due to CD is only a small fraction of all YLL (8%). In low-income countries, however, 68% of YLL are caused by CD (WHO 2009b). Generally, the epidemiologic transition occurs first in urban areas and subsequently spread over to less urban and rural areas (UN-HABITAT 2001). Since this transition is also linked to economic progress and since wealth is unequally distributed within countries and cities, mixed disease patterns in megacities in developing countries can be observed.

Economic growth in the globalized urban centres mostly led to a better situation in terms of food provision and nutrition, from which these people benefited that were able to participate in the economic upturn. Both life expectancy and prevalence of NCD in this group increased.

On the other hand in megacities in developing countries generally a high percentage of the urban population is not able to take advantage of increasing wealth. A substantial share is forced to live in the urban slums with the addressed health threatening conditions. One of the largest slums in Dhaka, for instance, the Korali Basti slum that is home to more than 12,000 households, does not even have one public toilet or health clinic (World Bank 2007). Such circumstances foster the risk of spreading communicable diseases like acute respiratory infections, tuberculosis, influenza, meningitis (UN-HABITAT 1996), diarrhoeal diseases, measles and cholera (Environmental Health Project 2004). They can lead to even higher disease burden than in rural areas although rural people generally are stronger affected by communicable disease burden than urban populations in developing countries. Particularly young children suffer from communicable disease morbidity and mortality. Depending on the compared areas, the death rates in poor urban children for diarrhocal diseases, tuberculosis and measles can be up to 100 times higher than among children in developed countries (Environmental Health Project 2004). Epidemiologic data from Nairobi shows that the children of the urban poor suffered the highest mortality rates (150 deaths per 1,000 births) compared to urban children, who are better off (approx. 84) and compared to children living in rural Kenya (approx. 113) (Montgomery 2008).

1.6.2 Group II Diseases: Non-communicable Disease

Due to the epidemiologic transition, the burden of non-communicable diseases in cities is rising. This happened in the developed countries and is now a growing problem in the urban areas of developing countries (Leon 2008). As addressed earlier, all groups are not equally at risk. Overall, the affluent people in the cities are at present at higher risk for chronic and non-communicable diseases. For instance, the self-reported risk for non-communicable diseases was significantly higher among affluent people as compared to people living in slums in Dhaka (Khan et al. 2009). The rising urbanisation and continued economic development in developing countries were positively associated with an increasing prevalence of overweight. Other urban characteristics, such as the use of cars and other fuel-based vehicles, limited space for walking and physical activity, the availability, preference and consumption of fast and fatty foods and less preference for vegetables, improved technologies that require less energy, and sedentary and changing lifestyles, all contribute to the rising trend of overweight and obesity in urban areas. Besides, obesity is considered a condition of high socio-economic status in many developing countries (Khan and Krämer 2009).

Living in cities can influence mental health in many ways. Migration was found to be associated with poor mental health in cities. For instance, labour migrants reported an increased risk of psychological disorders associated with reduced social support due to family disruption in Indonesia (Lu 2009). Another group of internal migrant workers in Shanghai, China, reported migration-related stress in terms of financial and employment difficulties and interpersonal tensions and conflicts, which were both associated with mental problems (Wong et al. 2008).

Besides migration-related factors of mental health there are various others like the urban environment, its design and land use patterns as well as socio-economic conditions within the (close) environment. A cohort study from New York City showed that people living in neighbourhoods with low socio-economic status had a more than two times higher chance of developing depression compared to people living in high socio-economic status neighbourhoods (odds ratio: 2.19; 95% confidence interval 1.04–4.59) (Galea et al. 2007). The association between neighbourhood conditions are also supported by another study. It showed that over time the mental health status of people living in one area improved while the living conditions improved in the same area. In contrast, a comparison-community, where no improvement of living conditions took place, showed no improvement in mental health (Dalgard and Tambs 1997).

1.6.3 Group III Diseases: Injuries

Globally injuries are of substantial public health concern because they belong to the leading causes of death and disability in almost all age groups, except among people over 60 years of age (WHO 2002a). The mortality rates caused by injuries are substantially higher in low- and middle-income countries than those in high-income countries (90.3 per 100,000 population vs. 50.7) (Mathers et al. 2006). The most important factors for mortality and morbidity due to injuries are traffic-related accidents (Fig. 1.5). Annually more than 1.2 million traffic-related deaths occur worldwide and between 20 and 50 million people suffer from non-fatal traffic-related injuries.

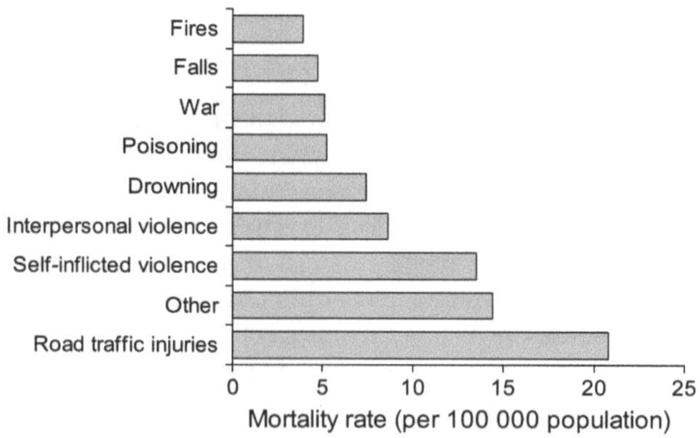

Fig. 1.5 Global injury mortality rates by cause, 2000 (WHO 2002b)

WHO estimated that road traffic injuries ranked 9 in the leading causes of death worldwide with 2.2% of all deaths in 2004 and will increase to 3.6% (rank 5) in 2030 (WHO 2009a). Overall, 90% of all road traffic-related deaths occur in low- and middle-income countries and since megacities are focal points of global production and transportation, they are naturally places in which accidents and injuries frequently take place. The lack of governability and resources in megacities in developing countries led to less regulated and unsafe individual and public transportation causing high levels of traffic-related disease burden (WHO 2004). Regrettably, the literature does neither provide many data with respect to urban/rural differences in traffic-related injuries (WHO 2004) nor about the differences between megacities in developing compared to megacities in developed countries. However, the mentioned conditions in terms of urban population growth, infrastructure, traffic safety and governability in megacities in developing countries suggest a tremendous and increasing burden of traffic-related diseases. Although the overall burden of work-related disease is by far not as high as the traffic-related one, working conditions and the related mortality and morbidity are also of strong public health concern. About 310,000 workers die each year due to work-related injuries. They cause 0.9% of globally occurring Disability Adjusted Life Years (DALYs),[1] an amount of 13.1 million DALYs. The workforce in developing countries is under higher risk than their counterparts in the developed world (WHO 2002c). The highest risks for occupational injuries exist in the agricultural and industrial production sector. Although there are not many reliable data regarding the burden of work-related injuries in developing countries (Concha-Barrientos et al. 2004), the WHO's estimation that the work-related mortality rates in industrializing countries are two to five time higher than in industrialized countries seems to be plausible (WHO 2002d). Particularly the large urban centres of low-cost production suffer from work-related injuries. The under- and unemployment in terms of formal employment opportunities forces workers – in the first place low-skilled ones like rural–urban migrants – to accept any kind of working opportunities, mostly without any social security or work place safety regulations.

In China, for instance, it is estimated that annually 15,000 workers die because of work-related accidents. The annual work-related death rates were estimated to be 11.1 deaths per 100,000 Chinese workers in 2000, which is much higher than the rate in developed countries like in the United States (2.2 deaths) (Concha-Barrientos et al. 2004). Within China, the Pearl River Delta (PRD) belongs to the production centres with about 40 million inhabitants. PRD is a megaurban area with large cities like Dongguan, Foshan, Guangzhou, Hong Kong and Shenzhen (Li et al. 2006) which are home to millions of working migrants. They constitute the majority of cheap labour and suffer from dangerous working conditions. In PRD alone, yearly 30,000 work-related injuries occur (Pareles 2005) and Wen reported that, according to the China Youth Daily, April 27, 2005, yearly 40,000 fingers were cut off due to work-related accidents (Wen 2006). There is an increasing

[1]One lost DALY represents one lost year of healthy life in a given population.

public awareness with respect to migrant workers rights, health care and work safety in China. In many other developing countries, for instance in the Middle-East, South Asia, Africa or South America the situation is worse. Assessment of occupational public health problems is difficult because the data availability about work-related burden of disease is insufficient. Therefore there is a substantial lack of reliable data base for adequate occupational injury reporting.

Furthermore, violence contributes to the burden of injury-related morbidity and mortality. Nearly 1.6 million people lost their lives in the year 2000 due to all kinds of violence including war- and civil conflict-related deaths and deaths from self-inflicted injuries. About 90% of these deaths occurred in low- and middle-income countries (de Jesus Mari et al. 2008). If one disregards war- and civil conflict-related deaths and self-inflicted injuries, globally, about 600,000 deaths were caused by violence in 2004. In low- and middle-income countries alone 489,000 violence-related deaths occurred (WHO 2010). The impacts of violence on people's health, livelihoods and economic prospects are tremendous. Commonly violence is more prevalent in urban areas with cities in developing countries generally more affected than cities in high-income countries (van Dijk et al. 2007). About 60% of the urban population in Europe and North America and about 70% in Latin America and Africa were affected by crime and violence over the last years (UN-HABITAT 2006). Within cities the socio-economically disadvantaged social strata are most affected by urban violence (Moser 2004).

Although the appearance of urban violence is quite diverse in different countries and related to a multidimensional conglomerate of risk factors (UN-HABITAT 2001), megacities in developing countries are likely to particularly suffer from urban violence. They generally accommodate large populations living in urban slums in overcrowded conditions, tenure insecurity and other adverse living conditions as addressed earlier. These circumstances can cause social conflicts, violence, and crime including mental and physical harm (UN-HABITAT 2008).

Urban violence is an increasing phenomenon worldwide (UN-HABITAT 2007). The homicide rate (per 100,000 population) increased from 5.47 in 1975–1979 to 8.86 in 1990–1994 (Briceno-Leon 2005). In the metropolitan region of Sao Paulo, the homicide rate (per 100,000) grew from 14.6 in 1981 to 33.9 in 1993 to 55.8 in 1996 (Cardia 2000). From Cali, a large city in Colombia, a homicide rate of even 90 cases per 100,000 in 1993 was reported (Guerrero and Concha-Eastman 2001). But these figures differ between countries (UN-HABITAT 1996). In some European and Asian countries the violence-related death rates per 100,000 population are below 2, in some countries even below 1 (UN-HABITAT 1996).

1.7 Strategies to Improve Public Health in Megacities

Dealing with public health in megacities is complex and therefore multidisciplinary and multisectoral cooperation between disciplines is necessary. Particularly cooperation between epidemiologists, statisticians, geographers, urban planners, climatologists, hydrologists, environmentalists, ecologists, policy makers, government and non-governmental organisations are clearly important to address the relevant public health issues.

Health sectors should be developed in terms of infrastructure, manpower, resources, accountability and community participation. Training of public health professionals, health education, and community awareness for health and environmental management are necessary. Strategies are needed to address the barriers within and outside of health systems. Strengthening of public health research, strengthening the capacity of the community in terms of decision-making and implementation and most importantly the development of a community-oriented approach are necessary options. Improving administrative services and governability, developing sustainable policies, improving social and environmental justice and ensuring sufficient financial support are needed. Public health services should be provided according to needs and not be influenced by the ability to pay and profit. All these services should be of high quality irrespective of socio-economic groups (Farrell et al. 2008). Relevant stakeholders contributing to the development of the health system should further facilitate and create more flexible legal procedures to allow greater access to low-cost medication and treatment (United Nations 2008).

To reduce the health gap between different subpopulations (e.g. slum and non-slum dwellers), three broad approaches built on the principles of equity and quality may be useful. These are (1) focusing on the most disadvantaged groups through specific measures; (2) setting targets to improve the health of the poorest groups and (3) tackling social determinants of health inequalities (Farrell et al. 2008). Policy makers and health managers should become aware of the magnitude and trend of inequalities including the most affected subgroups (Countdown 2008 Equity Analysis Group).

Changes in lifestyle are also important because the burden of non-communicable diseases (mostly lifestyles related) like cardiovascular and metabolic diseases are increasing particularly in urban areas. Focused and coordinated action and interventions designed at local, regional and global levels, national commitment to implement global policy and developing better infrastructure at the country side to reduce rural to urban migration are necessary. As a public health prerequisite, surveillance and assessment of the disease burden among (sub)populations and of important health determinants are necessary in order to inform health care stakeholders and health policy decision makers.

References

Boutayeb A, Boutayeb S (2005) The burden of non communicable diseases in developing countries. Int J Equity Health 4:2

Briceno-Leon R (2005) Urban violence and public health in Latin America: a sociological explanatory framework. Cad Saude Publica 21:1629–1648; discussion 1649–1664

Campbell-Lendrum D, Corvalan C (2007) Climate change and developing-country cities: implications for environmental health and equity. J Urban Health 84:i109-117

Cardia N (2000) Urban violence in Sao Paulo. Washington DC: Woodrow Wilson International Center for Scholars

Centre for Urban Studies, National Institute of Population Research and Training (NIPORT), MEASURE Evaluation (2006) Slums of Urban Bangladesh. Mapping and Census, 2005. Dhaka, Bangladesh and Chapel Hill, USA.

Cohen B (2004) Urban growth in developing countries: a review of current trends and a caution regarding existing forecasts. World Dev 32:23–51

Cohen B (2006) Urbanization in developing countries: Current trends, future projections, and key challenges for sustainability. Technol Soc 28:63–80

Concha-Barrientos M, Nelson DI, Driscoll T, Steenland NK, Punnett L, Fingerhut MA, Prüss-Üstün A, Leigh J, Tak S, Corvalan C (2004) Comparative quantification of health risks. Global and regional burden of disease attributable to selected major risk factors. In: Ezzati M, Lopez AD, Rodgers A, Murray CJL, eds. Comparative Quantification of Health Risks Global and Regional Burden of Disease Attributable to Selected Major Risk Factors. Geneva: WHO:1651–1801

Costello A, Abbas M, Allen A, Ball S, Bell S, Bellamy R, Friel S, Groce N, Johnson A, Kett M, Lee M, Levy C, Maslin M, McCoy D, McGuire B, Montgomery H, Napier D, Pagel C, Patel J, de Oliveira JA, Redclift N, Rees H, Rogger D, Scott J, Stephenson J, Twigg J, Wolff J, Patterson C (2009) Managing the health effects of climate change: Lancet and University College London Institute for Global Health Commission. Lancet 373:1693–1733

Countdown 2008 Equity Analysis Group. Mind the gap: equity and trends in coverage of maternal, newborn, and child health services in 54 Countdown countries. *Lancet* 2008; 371: 1259–1267

Coutts A, Kawachi L (2006) The urban social environment and its effects on health. In: Freudenberg N, Galea S, Vlahov D, eds. Cities and the health of the public. Nashville: Vanderbilt University Press:49–60.

CSPV (1998) Exposure to urban violence. boulder: University of Colorado at Boulder.

Dalgard OS, Tambs K (1997) Urban environment and mental health. A longitudinal study. Br J Psychiatry 171:530–536

de Jesus Mari J, de Mello MF, Figueira I (2008) The impact of urban violence on mental health. Rev Bras Psiquiatr 30:183–184

Environmental Health Project (2004) Improving the health of the urban poor. Learning from USAID experience environmental health project.

Fadda G, Jiron P (1999) Quality of life and gender: A methodology for urban research. Environ Urban 11:261–270

Farrell C, McAvoy H, Wilde J, Combat Poverty Agency. Tackling health inequalities- an all-ireland approach to social determinants. Dublin: Combat Poverty Agency. http://www.publichealth.ie/files/file/Tackling%20health%20inequalities.pdf, 2008.

Franklin TW, Franklin CA, Fearn NE (2008) A Multilevel analysis of the vulnerability, disorder, and social integration models of fear of crime. Soc Just Res 204–227

Galea S, Freudenberg N, Vlahov D (2005) Cities and population health. Soc Sci Med 60:1017–1033

Galea S, Freudenberg N, Vlahov D (2006) A framework for the study of urban health. In: Freudenberg N, Galea S, Vlahov D, eds. Cities and the health of the public: Vanderbilt University Press:277–293.

Galea S, Ahern J, Nandi A, Tracy M, Beard J, Vlahov D (2007) Urban neighborhood poverty and the incidence of depression in a population-based cohort study. Ann Epidemiol 17:171–179

Grimm NB, Faeth SH, Golubiewski NE, Redman CL, Wu J, Bai X, Briggs JM (2008) Global change and the ecology of cities. Science 319:756–760

Guerrero R, Concha-Eastman A (2001) An epidemiological approach for the prevention of urban violence: the case of Cali, Colombia. World Health Popul 4:1–11

Gushulak BD, MacPherson DW (2000) Population mobility and infectious diseases: the diminishing impact of classical infectious diseases and new approaches for the 21st century. Clin Infect Dis 31:776–780

Gushulak BD, MacPherson DW (2004) Globalization of infectious diseases: the impact of migration. Clin Infect Dis 38:1742–1748

Harper I, Raman P (2008) Less than human? diaspora, disease and the question of citizenship. Int Migration 46:3–26

Huq S, Rahman A, Konate M, Sokona Y, Reid H (2003) Mainstreaming adaptation to climate change in least developed countries (LDCs). Nottingham, UK: Russell Press

Imbusch P (2003) The Concept of Violence. In: Heitmeyer W, J. H, eds. International handbook on violence research: Kluwer Academic Publishers:13–39.

Khan MMH, Krämer A (2008) Socio-economic factors explain differences in public health-related variables among women in Bangladesh: a cross-sectional study. BMC Public Health 8:254

Khan MMH, Krämer A (2009) Factors associated with being underweight, overweight and obese among ever-married non-pregnant urban women in Bangladesh. Singapore Med J 50:804–813

Khan MMH, Krämer A, Grübner O (2009) Comparisons of health-related outcomes between urban slums, urban affluent and rural areas in and around Dhaka megacity, Bangladesh. DIE ERDE 140:69–87

Kovats RS, Akhtar R (2008) Climate, climate change and human health in Asian cities. Environ Urban 20:165–175

Kraas F, Mertins G (2008) Megastädte in entwicklungsländern. vulnerabilität, informalität, regier- und steuerbarkeit. Geographische Rundschau 60:4–10

Laaser U (2006) Schlaglicht 2: Verstädterung und Gesundheit. In: Razum O, Zeeb H, Laaser U, eds. Globalisierung - gerechtigkeit - gesundheit einführung in international public health. first ed. Bern: Verlag Hans Huber:91–94.

Leon DA (2008) Cities, urbanization and health. Int J Epidemiol 37:4–8

Li J, Zhang G, Li X-d, Qi SH, Liu GQ, Peng X (2006) Source seasonality of polycyclic aromatic hydrocarbons (PAHs) in a subtropical city, Guangzhou, South China. Sci Total Environ 355:145–155

Lu Y (2009) Rural-urban migration and health: Evidence from longitudinal data in Indonesia. Soc Sci Med published online: 412–419

Mathers CD, Lopez AD, Murray CJL (2006) The burden of disease and mortality by condition: data, methods and results for 2001. In: Lopez AD, Mathers CD, Ezzati M, Jamison DT, Murray CJL, eds. Global Burden of Disease and Risk Factors. New York: Oxford University Press:45–240.

Mendez MA, Popkin BM (2004) Globalization, urbanization and nutritional change in the developing world. electronic Journal of Agricultural and Development Economics 1:220–241

Mitchell T, Tanner T (2006) Adapting to climate change. Challenges and opportunities for the development community. Brighton, UK: Institute of Development Studies.

Montgomery MR (2008) The health of the urban populations in developing countries. New York: United Nations Populations Division.

Montgomery MR (2009) Urban poverty and health in developing countries. Popul Bull 64

Moser CON (2004) Urban violence and insecurity: an introductory roadmap. Environ Urban 16:3–16

Nicholls RJ (1995) Coastal megacities and climate change. GeoJournal 37:369–379

Oloruntimehin O (1996) Urban violence: health consequences and costs. Soc Sci Med 43:887–888

Oxfam (2009) Urban poverty and vulnerability in Kenya. Background analysis for the preparation of an Oxfam GB Urban Programme focused on Nairobi. Ottawa: Oxfam.

Pareles M (2005) Crushed. A Survey of work injuries and treatment in the Pearl River Delta. Report New York: China Labor Watch September

Patz JA, Kovats RS (2002) Hotspots in climate change and human health. BMJ 325:1094–1098

Rashid SF (2009) Strategies to reduce exclusion among populations living in urban slum settlements in Bangladesh. J Health Popul Nutr 27:574–586

Ryan L, Sales R, Tilki M, Siara B (2008) Social networks, social support and social capital: the experiences of recent Polish migrants in London. Sociology 4:672–690

Sampson RJ (1997) Collective regulation of adolescent misbehavior: validation results from eighty chicago neighborhoods. J of Adolesc Res 12:227–244
Schwefel D (2006) Gesundheit in globalisierter entwicklung. In: Razum O, Zeeb H, Laaser U, eds. Globalisierung-Gerechtigkeit-Gesundheit Einführung in International Public Health. first ed. Bern: Hans Huber:47–64.
Sclar ED, Northridge ME (2003) Slums, slum dwellers, and health. Am J Public Health 93:1381
Sclar ED, Garau P, Carolini G (2005) The 21st century health challenge of slums and cities. Lancet 365:901–903
UN-HABITAT (1996) An urbanizing world: global report on human settlements 1996. Oxford.
UN-HABITAT (2001) Cities in a globalizing world. Global report on human settlements 2001. London.
UN-HABITAT (2003) The challenge of slums. Global report on human settlements 2003. London.
UN-HABITAT (2006) State of the world's cities report 2006/2007. The millennium development goals and urban sustainability: 30 years of shaping the habitat agenda. London: Earthscan Publications
UN-HABITAT (2007) Enhancing urban safety and security. Global report on human settlements 2007. London: Earthscan Publications
UN-HABITAT (2008) State of the world's cities report 2008/2009. Harmonious cities. London: Earthscan Publications
United Nations (2008) World urbanization prospects. The 2007 revision: executive summary. New York.: United Nations.
Urban Violence Subcommitee (2008) Urban Violence in the commonwealth: prevention, intervention & rehabilitation: Commonwealth of Massachusetts.
van Dijk J, van Kesteren J, Smit P (2007) Criminal Victimisation in International Perspective. Key findings from the 2004–2005 ICVS and EU ICS. Den Haag: Tilburg University, UNICRI, UN ODC.
Wen D (2006) Wie China die Globalisierung bewältigt. Ein kritischer blick auf die reformpolitik seit 1978. Essen: Asienstiftung.
WHO (2002a) Injury. A leading cause of the global burden of disease 2000. Geneva: WHO.
WHO (2002b) The injury chart book. A graphical overview of the global burden of injuries. Geneva: WHO.
WHO (2002c) World health report. Reducing risks, promoting healthy life. Geneva: World Health Organization.
WHO (2002d) The world health report 2002. Reducing risks, promoting healthy life. In: WHO.
WHO (2003) Climate change and human health. Risk and responses. Geneva: WHO.
WHO (2004) World report on road traffic injury prevention. Geneva: WHO.
WHO (2009a) Global status report on road safety. Time for action. Geneva: WHO.
WHO (2009b) World health statistics 2009. Geneva: WHO.
WHO (2010) Summary: Deaths by cause, and income group, estimates for 2004. Geneva: WHO.
Wong DFK, He X, Leung G, Lau Y, Chang Y (2008) Mental health of migrant workers in China: prevalence and correlates. Soc Psychiatry Psychiatr Epidemiol 43:483–489
World Bank (2007) Dhaka: improving living conditions for the urban poor. Dhaka: World Bank Office, Dhaka.
WWF International (2009) Mega-Stress for mega-cities. A climate vulnerability ranking of major coastal cities in asia. Gland: WWF International.

Chapter 2
The Burden of Disease Approach for Measuring Population Health

Paulo Pinheiro, Dietrich Plaß, and Alexander Krämer

2.1 Introduction

Quantitative assessments of the health status of a population are undisputedly an important source of information to support decision-making and priority-setting processes in the field of Public Health. A common practice to (a) indicate the average level and the distribution of health in a population and (b) identify the impact of diseases on population health has been the use of findings on the epidemiology of diseases and injuries, their causes and risk factors. One major part of such efforts has targeted the determination of mortality patterns based on death and causes of death statistics. In addition, findings on mortality and its derivative life expectancy have widely been used as surrogates to inform about the overall health status as well as to identify the most important health problems in a population.

The remarkable changes in demographic and epidemiological factors and risk patterns in virtually all populations across the world over the last decades (Rowland 2003; Omran 1971; Smith 1988) have a significant impact on the health status of a population. Scientific as well as public discussions about the health effects associated with the transition models are also ongoing. The observation of decreasing death and birth rates, increasing life expectancies at birth and disease patterns shifting from infectious towards chronic conditions in nearly all populations over the world has e.g. raised the issue whether increases in the quantity of life have been accompanied by benefits in the quality of life. Several hypotheses on health in ageing populations have since then been postulated and scenarios ranging from a compression to an expansion of the lifetime burden due to morbidity have been presented (for more details see Nusselder 2003). Also, because of growing importance of non-communicable diseases and their often non-fatal impact on health, it has been concluded that death and causes of death statistics have increasingly

P. Pinheiro • D. Plaß • A. Krämer (✉)
Department of Public Health Medicine, School of Public Health, Bielefeld University, Bielefeld, Germany
e-mail: alexander.kraemer@uni-bielefeld.de

A. Krämer et al. (eds.), *Health in Megacities and Urban Areas*,
Contributions to Statistics, DOI 10.1007/978-3-7908-2733-0_2,

become inaccurate measures when exclusively used as surrogates to describe the overall health status of a population (for an updated discussion on health statistics see e.g. Murray 2007). Assessing the impact of non-fatal health outcomes on health has thus become an issue of major concern. One approach to meet the need for new methods when assessing population health has been the use of burden of disease studies and development of measures that combine information on mortality and non-fatal health outcomes to a single number (Field and Gold 1998). Such measures are usually referred to as Summary Measures of Population Health (SMPH) and have become key measures in many of the current burden of disease (BoD) assessments.

This chapter aims at providing basic information on the BoD approach and health measures from the SMPH group. A focus will be set on the measure Disability Adjusted Life Year (DALY) to exemplify the level of complexity inherent in a SMPH. To outline the informative value of DALY estimates, a selection of findings from the Global Burden of Disease (GBD) study will then be presented. Finally, potentials and limitations of the burden of disease approach will be discussed and conclusions about the value of BoD data that require linking health with spatial information will be drawn.

2.2 The BoD: A Definitional Approach

Obviously, there is no unambiguous understanding of the burden of disease idea in the literature. In a broader sense, BoD or sometimes burden of ill-health (e.g. Smith et al. 1999; Allender and Rayner 2007; Balakrishnan et al. 2009), is frequently used to include a wide range of different approaches that aim at assessing the impact of disease events on various dimensions of human life including health. Among the large number of attempts to define BoD, a definition given by the Connecticut Department of Public Health in 1999 appears to be useful to determine some key characteristics of a BoD approach. They defined BoD as

> a general term used in public health and epidemiological literature to identify the cumulative effect of a broad range of harmful disease consequences on a community, including the health, social, and economic costs to the individual and to society (Connecticut Department of Public Health 1999).

This definition plausibly illustrates that, in general, a BoD framework (a) targets the identification of consequences resulting from disease events, (b) might not be restricted to the impact on health but also relates to effects on social and economic realities, and (c) is related to communities, or populations rather than to individuals. This rather unspecific understanding of burden of disease allows for assessing the impact of diseases on a population with a wide range of outcomes from virtually all areas of life and enables many different disciplines such as epidemiology, social sciences, or economic sciences to develop their particular burden of disease approach by use of their routine methodologies and indicators.

The understanding of BoD has in the recent past increasingly been associated with a particular approach jointly developed by the World Bank, the World Health Organization (WHO) and the Harvard School of Public Health in the late 1980s: The Global Burden of Disease (GBD) Project. A main objective of this groundbreaking project was to generate a comprehensive and internally consistent comparable set of estimates of mortality and morbidity by age, sex, and regions of the world (Murray and Lopez 1996). First estimates were made for the year 1990. Also, the GBD Project provided the public health community with a new conceptual und methodological framework that was developed for integrating, validating, analyzing, and disseminating partial and fragmented information on the health of populations (e.g. Murray 1994). As a result of the fast dissemination and general acceptance of this particular burden of disease technique, though its results and its relevance for public health have critically been discussed (e.g. Arnesen and Nord 1999; Anand and Hanson 1997), the BoD understanding has since then become narrowed and is now predominantly associated with the WHO GBD approach. According to Colin Mathers

> BoD analysis provides a standardized framework for integrating all available information on mortality, causes of death, individual health status, and condition-specific epidemiology to provide an overview of the levels of population health and the causes of loss of health (Mathers 2006).

Using this definition, BoD can be considered as a conceptual and methodological approach that aims at (a) a consistent and comprehensive assessment of disease and injury consequences, (b) an assessment of population health in terms of health losses by using common metric for mortality and morbidity outcomes. To meet these objectives, the WHO GBD framework included the development of methods to assess the quality of available data and to estimate non-available data, the integration of information on non-fatal health outcomes with information on premature death into SMPH, and the development of a new metric, the DALY, to summarize the BoD (Murray and Lopez 1996, 1997). The GBD Project is an ongoing effort and since the original 1990 GBD Study there have been some major revisions of the methodology resulting in improved updates of the global BoD (e.g. Mathers et al. 2003; Lopez et al. 2006a; WHO 2008).

BoD estimates have in recent time increasingly been accepted and used in public health as an additional source to inform about the level of health in a given population. The number of publications that include "burden of disease" in the title or abstract and are listed in PubMed (the most popular database for accessing articles on life sciences and biomedical topics) has continuously increased over the last years starting from the time when the results from the first GBD were initially published in 1996 (Murray and Lopez 1996) (see Fig. 2.1). A major part of the studies were based on the WHO GBD approach that mainly made use of DALYs as BoD indicator. Such estimates have been presented for many populations and with different spatial resolutions, from local (e.g. Andra Pradesh) (Mahapatra 2001), over national (e.g. US, the Netherlands, South Africa) (Michaud et al. 2006; Melse et al. 2000; Bradshaw et al. 2003), to international levels (e.g. WHO 2002).

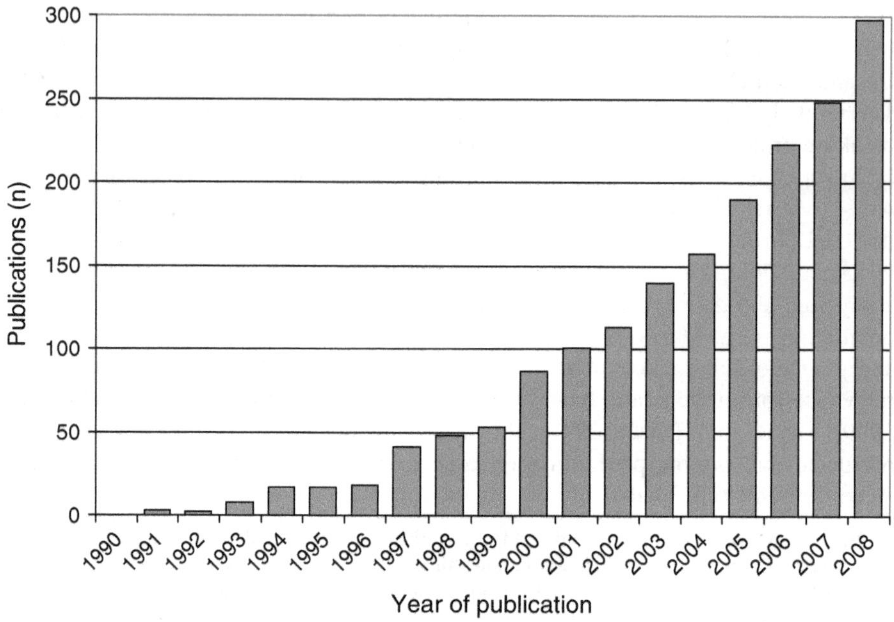

Fig. 2.1 Number of PubMed listed publications with "burden of disease" in title or abstract (date of query: 29.12.2009)

Additionally, estimates are available for some selected diseases and risk factors (chikungunya, dengue, food borne pathogens) (Krishnamoorthy et al. 2009; Luz et al. 2009; van Lier and Havelaar 2007).

2.3 The GBD Project

The first GBD study was designed to meet various objectives. A major objective was the quantification of health losses caused by diseases and injuries in a comprehensive and comparable way. Comprehensiveness and comparability referred to the inclusion of the whole spectrum of diseases and injuries as well as to the inclusion of populations up to a global level. Also, the study aimed at assessing the impact of non-fatal health outcomes on population health, thus, adding the morbidity to the mortality perspective. Further, it was demanded to develop and use a metric that together allowed for the assessment of the disease burden and for an economic appraisal of intervention options. The implementation of the GBD study can roughly be characterised by a four step procedure. The initial step focuses on the assessment of the current BoD using a SMPH. For the GBD study, the DALY was developed to assess estimates of disease burdens. SMPH and the DALY measure will be described in detail at a later stage of this chapter. In a second step, it is intended to attribute the identified amount of burden to various known risk factors by applying the Comparative Risk Assessment (CRA) methodology. Having current and past burden of disease estimates available, it is then intended to make projections of the future BoD in a further step. Here, it is also aimed to identify BoD trends when the current exposures to a risk factor are changed to a specified counterfactual exposure in order to assess the amount of burden that is potentially avoidable. In a last step, burden of disease estimates are linked to cost-effectiveness analyses to allow for an economic appraisal of the impact of different intervention options on the burden reduction (Shih et al. 2009). The GBD study has quantified the burden of premature mortality and disability by age, sex, and region for more than 100 disease and injury causes. The disease and injury causes are closely related to the diagnostic categories of the International Classification of Diseases (ICD) and are classified using a tree structure with four levels of disaggregation. In the GBD classification system, the first level of disaggregation defines three broad cause groups: Group I causes include communicable, maternal, perinatal, and nutritional conditions; Group II and Group III causes comprise non-communicable diseases and injuries, respectively (Mathers et al. 2006). For more detailed information about the GBD concept see (Murray and Lopez 1996).

The GBD study is an ongoing effort and various milestones have been reached after the presentation of the first estimates for the year 1990 (Murray and Lopez 1996). Since then, annual assessments were published in the World Health reports between 1999 and 2004 (e.g. WHO 2000, 2002). Findings from the comparative risk assessment were presented for 26 global risk factors (WHO report 2002; Ezzati et al. 2004). A comprehensive overview and discussion of the measures from the SMPH group was edited in 2002 (Murray et al. 2002). Country tools for national as well as environmental BoD assessment were developed and made freely available for the Public Health community (see www.who.org). Also, first projections of the future BoD and injuries from 2002 to 2030 were published (Mathers and Loncar 2006). Currently, the efforts are focused on the new GBD 2010 Study, which commenced in Spring 2007, to produce estimates of the BoD, injuries and risk factors for two time periods, 1990 and 2005. The study is expected to produce a first set of estimates by November 2010 (Global Burden of Disease Study 2010).

2.4 The SMPH Measures

These are

> measures that combine information on mortality and non-fatal health outcomes to represent the health of a particular population as a single numerical index (Field and Gold 1998).

According to this definition, the SMPH assess the health status of a population by integrating information on mortality and morbidity into a single number and thus are qualified to meet the demands of many BoD assessments on a health measure. Also, SMPH are considered to be a health indicator of use as they include non-fatal health outcomes in their estimates and thus reasonably extend the traditionally available set of population health indicators. Since the idea of a population health indicator that brings together data on mortality and morbidity was first presented in the mid 1960s (Sanders 1964), much efforts have been put in the conceptualisation and implementation of composite health measures (Robine et al. 2003; Murray et al. 2002) resulting in a marked increase of the availability of SMPH.

The SMPH family can broadly be divided into two groups: health expectancy (HE) and health gap (HG). Summary measures from the HE group basically aim at estimating years of life that can be expected to live in full health (Mathers 2002). The HE concept can be considered as an extended notion of the life expectancy concept that adds some information on the health status of a population (e.g. prevalence of disability) to information on the mortality. Widely accepted HE measures in use are e.g. the Healthy Life Years (HLY), the Disability Free Life Expectancy (DFLE) or the Disability Adjusted Life Expectancy (DALE). A core methodology for HE estimates is the so-called Sullivan-Method. In brief, this method requires to build up a period life table based on age- and sex-specific death numbers in a population and to include information on the age- and sex-specific prevalence of people living in a state less than full health such as disability (Sullivan 1966). The HLY indicator is currently in use as part of the European Union's structural core indicators to represent the health of the European population (Jagger et al. 2008). The DFLE and DALE measures differ in the way that the DALE measure includes a graduated valuation of the severity of disability, e.g. indicated by disability weights, while the DFLE uses a dichotomous graduation of disability versus non-disability. DALEs were presented as a part of the findings from the GBD study to represent the life expectancy of a population taking current prevalence rates of disability into account (Murray and Lopez 1997; Mathers et al. 2001).

The HG measures on the other hand provide information on years of healthy life lost and thus, focus on the quantification of health losses in a population. The most common member from the HG family is the DALY measure. The DALY indicator was developed to meet the objectives of the first GBD study in 1990 and has since then largely diffused into the field of Public Health and been used for many global, national, regional and local burden of disease assessments (Michaud et al. 2006; Melse et al. 2000; Bradshaw et al. 2003; Chapman 2006; Kominski et al. 2002; Dodhia and Philips 2008; Mahapatra 2001).

The HG measures are normative measures because the calculation of health losses calls for the definition of a health goal in order to allow for estimates of losses of health. Figure 2.2 illustrates the basic idea behind the HG approach and shows the survivorship curve of a hypothetical initial birth cohort with the x-axis showing the age in years and y-axis the percentage of survivors over a lifespan of 100 years. The upper curve in the figure indicates for each age along the x-axis the proportion of the hypothetical cohort that will remain alive at that age and includes people living in an ideal health state as well as people living in a state worse than perfect health. To distinguish people living in ideal health from people living in a health state worse than perfect, a second curve (in this example indicated by the lower curve) needs to be identified in order to allow for estimates of the burden due to non-fatal health outcomes. While areas A and B under the survivorship curve can be

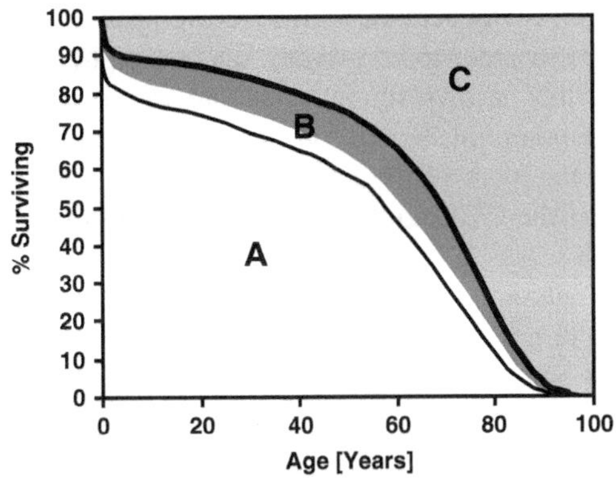

The philosophy of a health gap measure is illustrated on the basis of a survivorship curve for a hypothetical cohort. Upper horizontal line: health goal; upper curve: survivorship curve; lower curve: proportion of people living in ideal health; area A: years of life lived in ideal health; Area B: years of life lived in a health state worse than ideal, including a proportion shaded in gray indicating years of life lost due to living in a health state worse than ideal; area C: years of life lost due to premature death.

Fig. 2.2 The basic idea behind the concept of a health gap measure

used to represent life expectancy at birth, health expectancies can be derived from these areas by taking into account some lower weights for area B, i.e. the years lived in health states worse than perfect. For HG estimates, additional information on the health goal is needed in order to assess the difference between the current health of the population and the goal for population health. In Fig. 2.2, the health goal is indicated by the upper horizontal line enclosing area C and assuming that everyone in the hypothetic cohort lives in ideal health until the maximum age indicated. Only the definition of a health goal enables to assess the life lost due to premature mortality and to identify the mortality gap in a population. In the example of Fig. 2.2, the mortality gap is represented as the area C. To finalize the HG assessment, there is the need to additionally account for the health losses due to living in health states worse than perfect and to add losses identified in area B to the losses in area C due premature mortality. Health losses due to living in health states worse than perfect can be assessed by weighting health states less than ideal health and using a scale between 0 and 1 where a weight of one implies that the time lived in a particular health state is equivalent to the time lost due to premature mortality.

2.5 The DALY Measure

Among the composite HG measures, the DALY is undisputedly the one that has attracted most attention over the last years. Though, the DALY seems readily understandable at a first glance, its construction is characterised by a high degree of complexity. The following section will therefore provide the basic information on the DALY concept in order to contribute to a comprehensive understanding of the DALY measure that allows for an adequate interpretation of findings and enables to outline the potential as well as limitations when using the DALY.

The conceptual framework of the DALY measure was developed to explicitly meet the objectives of the GBD study. As the DALY was claimed to comprehensively quantify health losses, a concept was required to incorporate both mortality and non-fatal health outcomes into a single measurement unit (Murray and Lopez 1996). Another main target defined for the DALY was to assess burden of disease amounts and patterns up to a global level. Meeting this objective, a basic assumption was made to treat like events equally to ensure comparability between different populations. So e.g. a loss of a finger in Zimbabwe should contribute to the same burden as a loss of a finger in Turkey (Murray 1994). Further, DALY uses time as unit of measure to represent the disease burden in a given population. Chosen time as the unit of measure, the DALYs can then be based on both, incidence or prevalence data. In the past, there has been much debate about the choice of the adequate epidemiologic input measure for the DALY. For fatal health outcomes, it is obvious that there is no other way than using the incidence approach for calculating the burden due to premature death. For non-fatal health outcomes, the use of an incidence as well as a prevalence perspective is basically feasible (Murray 1994). It was argued that estimates of the non-fatal health outcomes can lead to different amounts of DALYs when the structure and dynamics of a population or a disease are not constant over time. For this reason, it was decided for the GBD study to calculate DALYs based on an incidence perspective in order to achieve a higher sensitivity towards burden of disease trends (Murray 1994). More technically, the DALY is calculated as the sum of the Years of Life Lost (YLL) representing mortality as years of healthy life lost due to premature death and the Years of Life Lost due to Disability (YLD) representing years of healthy life lost due to non-fatal health outcomes. Thus, YLLs represent the impact of fatal outcomes on population health whereas YLDs account for the impact of non-fatal health outcomes based on the concept of disability. YLLs and YLDs as calculated for the first GBD study are then based on further specifications. YLLs are estimated as standard expected years of life lost reflecting the reference that is used as the ideal population health goal. Technically, the calculation of years of healthy life lost due to premature death refers to a standard life table for a hypothetical cohort with a life expectancy at birth of 82.5 years for women and 80 years for men. These values were chosen based on the observation that approx 82.5 years were the highest observed life expectancy at birth at that time (Japanese women) and based on the assumption that the sex-specific gap of about 2.5 years explains the differences attributable to the human biology when leaving out gender-specific causes due the different social roles of men and women. Thus e.g. a death of a woman at age 40 would contribute to 42.5 healthy years of life lost. The idea of using a hypothetical cohort with standard life expectancies is basically similar to the technique of standardised mortality rates. Using an ideal standard also allows for treating events equally even if they occur in different social and physical environments all over the world and thus enables to draw cross-national comparisons of the BoD and injuries which is a major objective of the GBD study.

To comprehensively assess the disease burden in a population, DALYs include the YLDs to estimate the years of healthy life lost due to non-fatal health outcomes. An essential demand for the YLD implementation decision is the clarification of how non-fatal health outcomes are understood. For the YLDs in the GBD study, the concept of disability according to the International Classification of Impairments, Disabilities and Handicaps (ICIDH) of the WHO was chosen because it was regarded to be most suitable for the objectives of the project. Besides, the reason of data availability, using disability as definition of non-fatal health outcomes also allows for cross-national comparisons, leaving out the social and environmental background. Beyond the conceptualisation of non-fatal health outcomes, the quantification and comparability of disease and injury specific severity of a disability is a further issue of relevance. Here, a common approach is to define disability weights for the different diseases and injuries. There are many approaches to derive disability weights (e.g. visual analogue scale, standard gambling method, person trade off, time trade off) (for an overview see Gold et al. 2002; Murray and Lopez 1996; Torrance 1976, 1986) in the first GBD study the Person Trade-Off (PTO) method was used to derive disability weights for the different disease and injury events from the GBD classification system (Murray and Lopez 1996). In the PTO exercises, a group of health professionals were asked to trade off the life extension of people living in different health stages. These exercises resulted in disease and injury specific disability weights ranging from 0 reflecting a health state equivalent to perfect health and 1 reflecting a health state equivalent to death. A complete list of disability weights for all diseases included in the GBD classification system was provided by Lopez and colleagues (Lopez et al. 2006b). To finalize the calculations of the YLD component, information on disease and injury specific incidence and duration is needed.

To complete the outline of the DALY framework, other specifications that apply to YLL as well as YLD have to be considered. The first GBD study incorporated two social value choices into the DALY measure, namely time discounting and age weighting (Murray and Lopez 1996). Time discounting describes preferences of time as they are commonly used in the field of economics. These preferences are based on observations that people prefer benefits today rather than in the future and, thus, discount future benefits. The existence of time preferences was also assumed in the context of health and for the assessment of the burden on health. People prefer to have a healthier life now rather than in the future. Time preferences were integrated into the DALY framework and implemented with an annual 3% time discounting for future health losses. Additionally, the initial GBD study also included an age-weighting function in the DALY measure. This concept is based on the theory of human capital (Drummond 1997). According to this rationale, people give higher weights to an individual in productive age, and lower weights to very young and older people. This refers to the understanding, that younger and older people are often dependent on the social and financial support of people in productive age. Thus, for the first GBD study, higher weights for people in

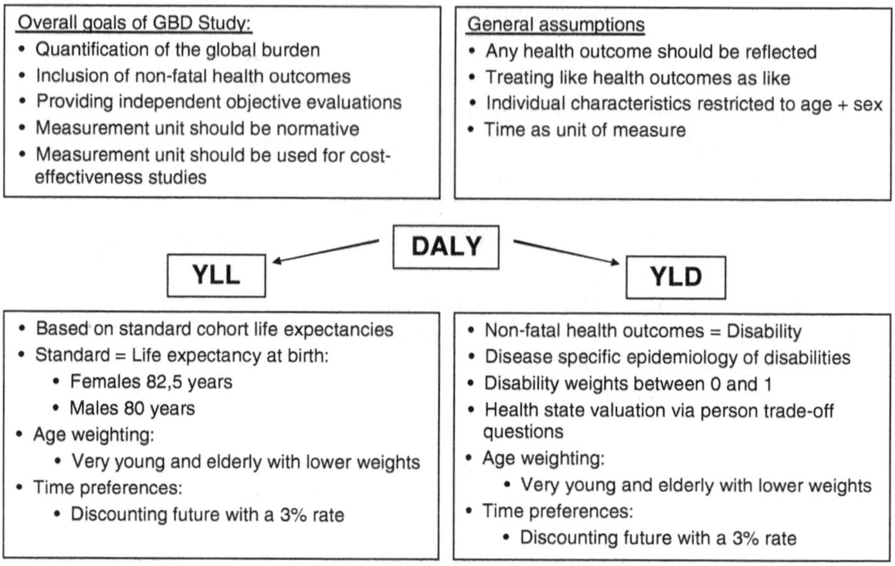

Fig. 2.3 The DALY (Disability Adjusted Life Year) concept

productive age were used. Figure 2.3 gives a comprehensive summary and overview of the DALY concept.

Although the original GBD DALY measure, its components and methodology have been debated in the literature and various international forums since its first publication in 1996 (Arnesen and Kapiriri 2004; Anand and Hanson 1997, 1998), the DALY measure has increasingly been used in various national and sub-national burden of disease studies (e.g. national studies: USA, the Netherlands, South Africa, Zimbabwe; e.g. regional studies Los Angeles, London, Andra Pradesh) (Michaud et al. 2006; Melse et al. 2000; Bradshaw et al. 2003; Chapman 2006; Kominski et al. 2002; Dodhia and Philips 2008; Mahapatra 2001).

2.6 Core Findings from the GBD Study

The GBD study has provided the public health community with numerous findings over the last decades (see Murray et al. 1994; Lopez et al. 2006a; WHO 2008). The GBD project is an ongoing effort resulting in refined concepts, methods and updated results. Regional findings are usually presented in low-, middle- and high-income categories as defined by the World Bank. Here, countries are not only grouped geographically but also based on their gross national income. This section provides a selection of some main global and regional findings on the BoD as measured in DALYs.

In 2001 the global average BoD across all regions of the world was 250 DALYs per 1,000 population, of which about two-thirds were due to premature death. YLL

Table 2.1 The 20 leading causes of global burden of disease, 2001

	Cause	DALYs (million of years)	% of total DALYs
1	Perinatal conditions	90.48	5.9
2	Lower respiratory infections	85.92	5.6
3	Ischemic heart disease	84.27	5.5
4	Cerebrovascular disease	72.02	4.7
5	HIV/AIDS	71.46	4.7
6	Diarrheal diseases	59.14	3.9
7	Unipolar depressive disorders	51.84	3.4
8	Malaria	39.97	2.6
9	Chronic obstructive pulmonary disease	38.74	2.5
10	Tuberculosis	36.09	2.3
11	Road traffic accidents	35.06	2.3
12	Hearing loss, adult onset	29.99	2.0
13	Cataracts	28.64	1.9
14	Congenital anomalies	24.95	1.6
15	Measles	23.11	1.5
16	Self-inflicted injuries	20.26	1.3
17	Diabetes mellitus	20.00	1.3
18	Violence	18.90	1.2
19	Osteoarthritis	17.45	1.1
20	Alzheimer's and other dementias	17.11	1.1

Source: Lopez et al. (2006b)

varied substantially across regions, with e.g. YLL rates nearly five times higher in Sub-Saharan Africa than in high-income countries. YLD rates varied less, with Sub-Saharan Africa having again higher rates than high-income countries.

The 20 leading causes of global BoD in 2001 are shown in Table 2.1. There are four usually non-fatal conditions among the top 20 causes of burden of which unipolar depressive disorders are identified to be the most relevant non-fatal contributor to the global burden. This finding illustrates not only the relevance of non-fatal conditions for population health but also the importance to include non-fatal health outcomes into burden assessments.

In low- and middle-income countries, the leading causes of the BoD included five communicable and four non-communicable causes among the top ten, whereas the top ten causes in high-income countries exclusively consisted of non-communicable conditions. The burden of non-communicable diseases is becoming increasingly important, not only because of a global increase of absolute DALY levels but also because of an increase in the proportion of the non-communicable burden on the total burden in low- and middle-income countries. While the proportion of the burden from non-communicable disease in high-income countries has remained fairly stable over the last decades, the proportion in low- and middle-income countries has increased with now almost 50% of the adult disease burden being attributable to non-communicable conditions with the conclusion that the populations living in many developing countries are suffering from a double BoD (Fig. 2.4 and Table 2.2).

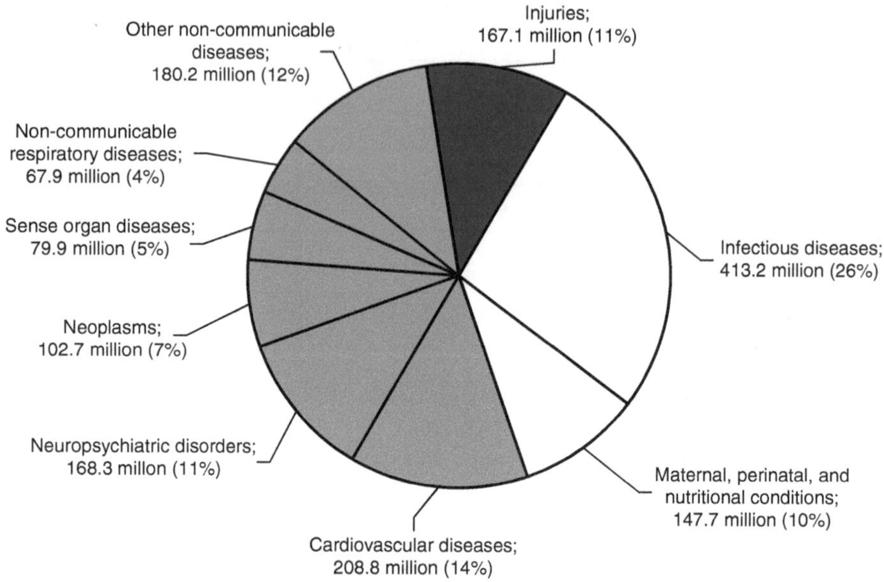

Fig. 2.4 The global burden of disease estimated by DALYs, 2001 (GBD group I conditions: *white*; group II conditions: *gray*; group III conditions: *black*) (*Source*: Lopez et al. 2006b)

Table 2.2 The ten leading causes of global burden of disease, by broad income group, 2001

Low- and middle-income countries				High-income countries			
	Cause	DALYs (millions of years)	% of total DALYs		Cause	DALYs (millions of years)	% of total DALYs
1	Perinatal conditions	89.07	6.4	1	Ischemic heart disease	12.39	8.3
2	Lower respiratory infections	83.61	6.0	2	Cerebrovascular disease	9.35	6.3
3	Ischemic heart disease	71.88	5.2	3	Unipolar depressive disorders	8.41	5.6
4	HIV/AIDS	70.80	5.1	4	Alzheimer's and other dementias	7.47	5.0
5	Cerebrovascular disease	62.67	4.5	5	Trachea, bronchus, and lung cancers	5.40	3.6
6	Diarrheal diseases	58.70	4.2	6	Hearing loss, adult onset	5.39	3.6
7	Unipolar depressive disorders	43.43	3.1	7	Chronic obstructive pulmonary disease	5.28	3.5
8	Malaria	39.96	2.9	8	Diabetes mellitus	4.19	2.8
9	Tuberculosis	35.87	2.6	9	Alcohol use disorders	4.17	2.8
10	Chronic obstructive pulmonary disease	33.45	2.4	10	Osteoarthritis	3.79	2.5

Source: Lopez et al. (2006b)

Injuries, both unintentional and intentional, accounted for about 11% of the global BoD and have been identified as the "hidden" epidemic (see Fig. 2.3). A proportion of the burden due to injuries on the total burden of even up to 30% has been reported for male adults aged 15–44 years in various parts of the world (e.g. Europe and Central Asia, Latin America and the Caribbean). In this age group, road traffic accidents, violence, and self-inflicted injuries were usually among the top ten leading causes of the BoD. Furthermore, the burden of road traffic accidents is increasing and especially affects the health of the young male population in developing countries of Sub-Saharan Africa and South and Southeast Asia.

The GBD study provides information not only on the burden at a global or regional but also at a national level. Country-specific data on the burden are readily accessible (see http://www.who.int/healthinfo/global_burden_disease/estimates_country/en/index.html) and represent the highest spatial resolution that is available from the global BoD assessments. An example that illustrates the opportunity for cross-country comparisons is given for Bangladesh, China and Germany. Table 2.3 shows the age-adjusted DALY rates per 100,000 population in 2002 for these countries. DALY rates are presented for the total burden as well as for the burden due to group I, II, and III conditions. Figure 2.5 additionally informs about the proportion of

Table 2.3 Age-standardized DALY rates per 100,000 population in Bangladesh, China, and Germany, 2002 (group I: communicable, maternal, perinatal, and nutritional conditions; group II: non-communicable conditions; group III: injuries)

	DALYs per 100,000 population			
Country	All causes	Group I	Group II	Group III
Bangladesh	25,292	9,877	12,455	2,960
China	15,149	3,162	9,710	2,276
Germany	10,114	581	8,671	862

Source: http://www.who.int/entity/healthinfo/statistics/bodgbddeathdalyestimates.xls (date of query: 29.08.2009)

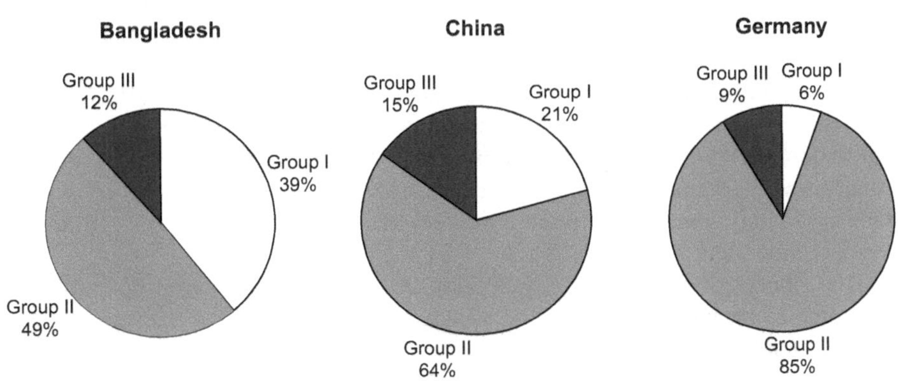

Fig. 2.5 The burden of disease in Bangladesh, China, and Germany estimated by DALYs, 2002 (group I: communicable, maternal, perinatal, and nutritional conditions, group II: non-communicable conditions; group III: injuries)

the group I, II, and III conditions on the total BoD and injuries. In brief, the Bangladesh population suffers not only the highest overall burden but also the highest burden when stratified by each of the three groups. This finding confirms that non-communicable diseases affect not only high-income countries such as Germany but also low- and middle income countries such as Bangladesh or China. Also, Fig. 2.5 points out that Bangladesh – alike many other developing countries – suffers a double BoD by communicable and non-communicable diseases.

2.7 Linking Health with Spatial Information: Potentials and Limitations of the BoD Approach

There is increasing demand for coherent and comprehensive information on the vulnerability and adaptation of populations to changes in the natural and physical environment because issues such as climate change and urbanisation or mega-polisation have become top of the agenda of many policy-making and research institutions. The creation of a harmonised data set that allows for e.g. conclusions on the impact of climate change or urbanisation on the overall health of populations requires the combination of data sets from different disciplines such as geography, climatology, public health, or epidemiology. Using population health as outcome of interest and as a proxy for a population's vulnerability to environmental changes is undisputedly of high value but is also limited due to several characteristics in the collection and processing of health data. Although the quantity and quality of health data have markedly increased in the past, there are still many difficulties in the handling of these quantitative datasets, especially when policy-maker and researcher in public health aim at comprehensive assessments of the overall health of populations. One frequent limitation of health information is the comparability of data, e.g. with regard to different health status, diseases, health determinants, or populations. Also, the global coverage of health data is still unequally distributed especially in low-income countries which still lack information on mortality and on a wide range of important diseases (Boerma and Stansfield 2007). Health data that are routinely collected within surveillance systems usually show a level of spatial and temporal resolution that is of limited use. The spatial resolution if available usually covers administrative boundaries often at a coarse level and is not consistent with the spatial domains preferred by others like climatologists who use climatic zones or modellers who use grids.

The concept of the GBD study as outlined above offers several potentials to overcome some basic problems when merging health data with data from other sources. With the objectives to assess overall levels of population health and to produce comprehensive and comparable estimates, the GBD study basically complies with some requirements on the structure of health information to allow for a spatial arrangement of findings other than administrative boundaries. Also, focusing the measurements on health losses rather than health expectancies and selecting an approach stratified by sex, ages, diseases, injuries and risk factors facilitate the assessment of the impact of various environmental determinants on population health. The disease-specific approach and the attribution of the prevalent burden to known risk factors can further be considered useful because of greater availability of and access to health data. Moreover, the GBD concept offers solutions for the handling of missing data and low data quality to ensure the comprehensiveness of the burden findings. Another non-negligible aspect of the GBD approach is the fact that it is an ongoing effort with updated results that has obtained increasing acceptance in Public Health over the last years.

However, the GBD estimates as currently presented have their limitations when used for the purpose of spatial analyses. A major limitation is the fact that a stratification of results is restricted to age and sex. Other important determinants of health such as socioeconomic status, or living and occupational conditions are not assessed by the GBD study. Further, the spatial resolution of the findings from the GBD project is fairly coarse and limited to national levels representing the highest level of resolution available. Thus, when looking at an urban level, data on burden of disease as presented by the GBD is not available. Identifying the burden of disease patterns in urban areas poses in turn the need for gathering data. Using GBD methods, data on both mortality and morbidity as described in the previous sections is needed and requires the collection of various epidemiologic variables. Traditional surveillance methods (e.g. death registries) as implemented in developed societies are of limited use in highly informal settlements such as urban slums. High informal movement from rural areas to urban settlements hamper tracking both acute and chronic disease events. Since many studies aim at gathering data about the epidemiology of different diseases in urban areas, the combination of data from different studies and possible modelling and validation of data with methods provided by the WHO (e.g. DisMod Software) may help to shed more light on disease burden patterns and to approach a comprehensive view of population health in megacities. Combining burden of disease with spatial information could then also help to investigate hot spots of disease burden in areas prone to different risk factors.

Also, there are in general difficulties in the understanding of the DALY measure and in the interpretation of DALY estimates, especially when contrasting the DALY with other health proxies such as death rates or life expectancies. Finally, focussing on a disease-specific approach might be considered a limitation because it does not allow for investigating health domains other than the absence of disease.

In conclusion, the BoD approach offers several potentials when health information are sought to be included in spatial analyses. A major advantage of the WHO GBD approach over other approaches used in public health is the possibility to generate comprehensive and comparable estimates of a population's health status and thus to represent overall health in spatial arrangements. The use of currently available BoD assessments is however limited by the level of stratification and resolution of the available data. This in turn implies that the arrangement and harmonisation of BoD data with spatial data from other disciplines needs to be clarified in advance when considering the WHO BoD approach for small scale analyses.

References

Allender S, Rayner M (2007) The burden of overweight and obesity-related ill health in the UK. Obesity Reviews. Vol. 8, pp 467 – 473

Anand S, Hanson K (1997) Disability-adjusted life years: a critical review. Journal of Health Economics; Vol. 16; pp 685–702

Anand S, Hanson K (1998) DALYs: efficiency versus equity. World Development Report 26:307–310

Arnesen T, Kapiriri L (2004). Can the value choices in DALYs influence global priority-setting? Health policy Amsterdam, Netherlands 70:137–149

Arnesen T, Nord E (1999) The value of DALY life: problems with ethics and validity of disability adjusted life years. British Medical Journal 319:1423–1425

Balakrishnan R, Allender S, Scarborough P, Webster P, Rayner M (2009) The burden of alcohol-related ill health in the United Kingdom. Oxford Journal of Public Health Vol. 31, pp 366–73

Boerma JT, Stansfield SK (2007). Health statistics now: are we making the right investments? The Lancet Vol. 369, pp 779–786

Bradshaw D, Groenewald P, Laubscher R, Nannan N, Nojilana B, Norman R, Pieterse D, Schneider M, Bourne DE, Timaeus IM, Dorrington R, Johnson L (2003) Initial burden of disease estimates for South Africa, 2000. South African Medical Journal. Vol. 93; pp 682–688

Chapman G (2006) The burden of disease in Zimbabwe in 1997 as measured by disability-adjusted life years lost. Tropical Medicine and International Health. vol. 11, pp 660–671

Connecticut Department of Public Health (1999) Looking Toward 2000 – An Assessment of Health Status and Health Services; Hartford, Connecticut, page 368

Dodhia H, Philips K (2008) Measuring burden of disease in two inner London boroughs using Disability Adjusted Life Years. Journal of Public Health. pp 1–9

Drummond MF (1997) Methods for the Economic Evaluation of Health Care Programmes. Oxford: Oxford University Press

Ezzati M, Lopez AD, Rodgers AA, Murray CJL (2004). Comparative quantification of health risks: global and regional burden of disease attributable to selected major risk factors. Geneva, World Health Organization

Field MJ, Gold MR (1998) Summarizing Population Health – Directions for the Development and Application of Population Metrics. Washington DC; National Academy Press

Global Burden of Disease Study (2010) www.globalburden.org [last visit 14.07.2010]

Gold MR, Stevenson D, Fryback DG (2002). HALYS AND QALYS AND DALYS, OH MY: Similarities and Differences in Summary Measures of Population Health. Annual Review of Public Health 23:pp 115–134

Jagger C, Gillies C, Moscone F, Cambois E, Van Oyen H, Nusselder W, Robine JM; EHLEIS team (2008) Inequalities in healthy life years in the 25 countries of the European Union in 2005: a cross-national meta-regression analysis. Lancet 372:2124–2131

Kominski GF, Simon PA, Ho A, Luck J, Lim YW, Fielding JE (2002) Assessing the burden of disease and injury in Los Angeles County using disability-adjusted life years. Public Health Report 117: pp 185–191

Krishnamoorthy K, Harichandrakumar KT, Krishna KA, Das LK (2009) Burden of chikungunya in India: estimates of disability adjusted life years (DALY) lost in 2006 epidemic. Indian Journal of Vector Bourne Diseases 46:pp 26–35

Lopez AD, Mathers CD, Ezzati M, Jamison DT, Murray CJL (2006a) Global and regional burden of disease and risk factors, 2001: systematic analysis of population health data. The Lancet 367: pp 1747–1757

Lopez AD, Mathers CD, Ezzati M, Jamison DT, Murray CJL (2006b) Global burden of disease and risk factors. Oxford University Press and World Bank

Luz PM, Grinsztejn B, Galvani AP (2009) Disability adjusted life years lost to dengue in Brazil. Journal of Tropical Medicine and International Health 14: pp 237–246

Mahapatra P (2001) Estimating national burden of disease: sensitivity to local data; the burden of disease in Andhra Pradesh, 1999. Institute of Health Systems. Hyderabad, India

Mathers C (2006) Introduction to Burden of Disease. BoD Workshop Bielefeld, January 2006

Mathers CD (2002) Health expectancies: an overview and critical appraisal. In: Murray CJL, Salomon JA, Mathers CD, Lopez AD (2002) Summary Measures of Population Health; pp 177–204

Mathers CD, Bernard C, Moesgaard Iburg K, Inoue M, Ma Fat D, Shibuya K, Stein C, Tomijima N, Xu H (2003) Global Burden of Disease in 2002: data sources, methods and results. World Health Organisation, Geneva

Mathers CD, Loncar D (2006). Projections of Global Mortality and Burden of Disease from 2002 to 2030. PLoS Medicine 3:pp e442

Mathers CD, Lopez AD, Murray CJL (2006) The burden of disease and mortality by condition: data, methods and results for 2001. In: Lopez AD, Mathers CD, Ezzati M, Murray CJL, Jamison DT, eds. Global burden of disease and risk factors. New York, Oxford University Press: pp 45–240

Mathers CD, Sadana R, Salomon JA, Murray CJL, Lopez AD (2001) Healthy life expectancy in 191 countries, 1999.The Lancet 357:1685–1691

Melse JM, Essink-Bot ML, Kramers PG, Hoeymans NA (2000) National burden of disease calculation: Dutch disability-adjusted life-years. Dutch Burden of Disease Group; American Journal of Public Health. Vol. 90; pp 1241–1247

Michaud C, McKenna M, Begg S, Tomijima N, Majmudar M, Bulzacchelli M, Ebrahim S, Ezzati M, Salomon J, Gaber Kreiser J, Hogan M, Murray CJL (2006): The burden of disease and injury in the United States 1996; Population Health Metrics, Vol. 4:11

Murray CJL & Lopez AD (1996) The Global Burden of Disease. Cambridge; Harvard School of Public Health

Murray CJL (1994) Quantifying the burden of disease: the technical basis for disability- adjusted life years. Bulletin of the World Health Organization 72: pp 429–445

Murray CJL (2007) Towards good practice for health statistics: lessons from the Millennium Development Goal health indicators. Lancet 2007 369; pp 862–73

Murray CJL, Lopez AD (1997) Regional patterns of disability-free life expectancy and disability-adjusted life expectancy: Global Burden of Disease Study. The Lancet 349; pp 1347–1352

Murray CJL, Lopez AD, Jamison DT (1994). The global burden of disease in 1990 : summary results, sensitivity analysis and future directions. Bulletin of the World Health Organisation, 1994 72; pp 495–509

Murray, CJL, Salomon JA, Mathers CD, Lopez AD (2002). Summary measures of population health: concepts, ethics, measurement and applications. Geneva, World Health Organization

Nusselder WJ (2003) Compression of Morbidity. In: Robine JM, Jagger C, Mathers CD, Crimmins EM, Suzman RM (2003) Determining Health Expectancies. John Wiley and Sons Ltd. The Atrium, Southern Gate, Chichester, West Sussex, England: pp 35–58

Omran AR (1971) The Epidemiologic Transition: A Theory of the Epidemiology of Population Change. The Milbank Memorial Fund Quarterly 49; pp 509–538

Robine JM, Jagger C, Mathers CD, Crimmins EM, Suzman RM, (2003) Determining Health Expectancies Chichester UK: John Wiley

Rowland DT. The demographic transition. In: Demographic Concepts and Methods. 2003

Sanders, B. S. (1964). "Measuring community health levels." American Journal of Public Health 54; pp 1063–1070

Shih STF, Cater R, Sinclair C. Mihalopoulos C, Vos T (2009) Economic evaluation of skin cancer prevention in Australia; Journal of Preventive Medicine 49; pp 449–453

Smith KR (1988) The Risk Transition. East-West Center

Smith KR, Corvalán CF, Kjellström T (1999) How Much Global Ill Health Is Attributable to Environmental Factors? Epidemiology. Vol. 10; pp 573–584

Sullivan DF (1966) Conceptual problems in developing an index of health. Vital and Health Statistics. vol. 2:1–18

Torrance GW (1976) Social preferences for health states: an empirical evaluation of three measurement techniques. Socio-Economic Planning and Science 10: pp 129–136

Torrance GW (1986) Measurement of health state utilities for economic appraisal; Journal of Health Economics. Vol. 5; pp 1–30

van Lier EA, Havelaar, AH (2007) Disease burden of infectious disease in Europe: a pilot study

WHO (2008) The global burden of disease: 2004 update. World Health Organisation, Geneva

World Health Organisation (2000) The World Health Report 2000. Health Systems: Improving Performance. Geneva, World Health Organisation

World Health Organisation (2002) The World Health Report 2002. Reducing Risk, Promoting Healthy Life. Geneva, World Health Organisation

Chapter 3
Megaurbanisation and Public Health Research: Theoretical Dimensions

Heiko J. Jahn, Md. Mobarak Hossain Khan, and Alexander Krämer

3.1 Introduction

Human health is a complex phenomenon influenced by socioeconomic, demographic, psychological, genetic, social, behavioural and environmental factors. Human health in megacities and urban areas is even more complex. Megacities in developing and transitional countries (e.g. in China) experience fast urbanisation processes due to continuing rural-to-urban migration (Ping and Pieke 2003; Tunon 2006; Wong et al. 2007). For instance, an estimated number of 150 million Chinese working migrants moved to cities from rural areas to find new opportunities (Tunon 2006). The migrant population is particularly affected by difficult living conditions because they mostly suffer from low socioeconomic status and experience restricted access to health care and education (Li et al. 2006). They often pay higher health costs as compared to non-migrants (Zheng and Lian 2006) and are more frequently exposed to low-standard living and working conditions (Ping and Pieke 2003; Zheng and Lian 2006). These living conditions often coined by poor hygiene and crowded living space increasing the risk for infectious diseases (Zheng and Lian 2006). Besides the somatic health risks, migrants are also threatened by psychological diseases and symptoms. For instance, Wong et al. found in their study on mental health among Chinese migrant workers that about 25% of their male participants could be classified as mentally unhealthy (Wong et al. 2008:486).

Many megacities in developing countries also suffer from deficient governability partly due to fast population growth and insufficient resources to overcome the challenges of rapid mega-urbanisation. The loss of governability affects urban planning and control (Kraas 2003). Rapid urbanisation leads to numerous changes causing risks for human health in megacities. Environmental pollution due to e.g. increasing traffic and industrial activities is one of the most

H.J. Jahn (✉) • M.M.H. Khan • A. Krämer
Department of Public Health Medicine, School of Public Health, Bielefeld University, Bielefeld, Germany
e-mail: heiko.jahn@uni-bielefeld.de

A. Krämer et al. (eds.), *Health in Megacities and Urban Areas*,
Contributions to Statistics, DOI 10.1007/978-3-7908-2733-0_3,

important risk factors (WHO and UNDP 2001). Such urbanisation processes may result in a lack of formal public health care and educational services and in unhealthy living and working conditions. Water supply, waste water- and solid waste management are also often deficient in megacities threatening the health of the inhabitants (Lu and Liu 2006). However, the development of megacities also offers several positive effects like employment and the opportunity to improve people's income (e.g. Ping and Pieke 2003).

These complex conditions in megacities are also reflected by the core topics proposed by the Priority Programme SPP 1233 *Megacities – Megachallenge: Informal Dynamics of Global Change* funded by the German Research Foundation (DFG).[1] These topics comprehend the major aspects of urbanisation and related informal dynamics in megaurban regions. These dynamic processes can have influence on various lifeworld dimensions of megacities' inhabitants, which may affect their health status as well. In addition to the influence of the core topics on various dimensions of urbanisation, these dimensions also influence each other. Health is therefore somehow influenced by all these aspects and is therefore to perceive as a cross-sectional topic (Fig. 3.1).

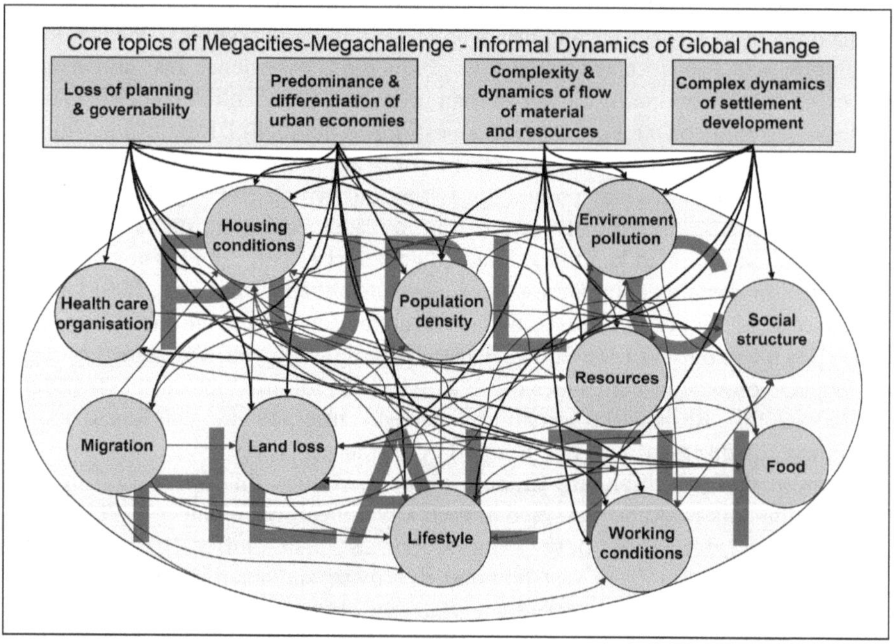

Fig. 3.1 Core topics of SPP 1233 and their influences on lifeworld dimensions of inhabitants in megaurbanised areas

[1]This paper presents our theoretical background of public health research in megaurban environments, which is also basis of our activities in Guangzhou, South China. They are part of the SPP 1233 and are jointly conducted with colleagues from the School of Public Health, Sun Yat-sen University in Guangzhou.

Against this multidimensional complex background of megaurban health, epidemiological public health research requires a broad approach and a complex theoretical framework. It needs to be based on scientific health-related theories in order to study the health determinants in certain populations in megacities. Hereby not only local phenomena are of importance but also global developments influencing public health. Therefore fundamental health-related policies, goals and declarations, introduced by international institutions carrying high authority like the World Health Organization (WHO) and the United Nations (UN), should be considered. They have an impact on public health on a global, regional and local scale like the Millennium Development Goals (MDG) declared by the UN. The MDG influence global health policy strategies and thus are embedded in global change. On the other hand they have impact on a local level as they may influence the national decision making or can be the motive for international support of certain parts of the world.

3.2 Public Health-Related Theoretical Orientation

In order to obtain health-related information about different subpopulations in megacities, a dynamic research process in cooperation with collaborators from different scientific disciplines is required. It needs a theoretical orientation based on health-related theories and concepts but it is also influenced by theoretical issues stemming from other disciplines like geography.

Within this chapter we explain which theoretical concepts and theories are applied within our urban health research activities. We first refer to our understanding of health as a multidimensional concept based different definitions and health-related theories taking in account Aaron Antonovsky's health concept. In the subsequent paragraphs our understanding of the concepts of vulnerability, resistance and resilience will be introduced because they play an important role in public health research. Additionally, important public health-related transition theories are introduced.

3.2.1 The Theoretical Concept of Health

Since there are long lasting discussions about the definition of health, we provide a brief insight in frequently cited definitions and describe our multidimensional concept of health.

Our research is based on a specific understanding of the theoretical concept of health. The basic definition is the one proposed by the WHO in 1946: "Health is a state of complete physical, mental and social well-being and not merely the absence of disease or infirmity" (Preamble of the constitution of the WHO 1948).

This definition established a comprehensive and positive view on health taking into account also psychological aspects, contrary to a narrowly focussing bio-medical view on human health (Young 2005). In 1984, the WHO Regional Office for Europe suggested another definition comprising a population dimension and emphasising that health is an ability or resource or capacity to realise health-related necessities and to overcome health-threatening external influences: "[Health] is the extent to which an individual or group is able on the one hand to realize aspirations and satisfy needs, and, on the other hand, to change and cope with the environment. Health is therefore seen as a resource for everyday life, not the objective of living; it is a positive concept emphasizing social and personal resources as well as physical capacities" (Young 2005:1).

Keeping in mind that public health research in urban areas requires a population based and multidimensional orientated framework, our understanding of health relies on an interdisciplinary approach incorporating findings of various health-related theories:

1. According to *learning and personality theories*, an individual's personality traits determine the extent and profile of the ability to cope with physical and mental demands, influencing the individual's health.
2. *Stress and coping theories* emphasise the individual's competencies, which are required to cope with internal and external demands. They accent the reciprocal relation between the individual and his or her environment. These stress and coping theories consider health as an instable state, which necessitates the individual's effort to sustain an equilibrium of health.
3. *Theories of socialisation* further broaden this perspective by taking into account the lifelong process of handling reality and coping with it. They also consider personal and social resources as requirements to sustain a dynamic balance between risk and protective factors and point out that there are in-between stages between absolute health and absolute disease.
4. *Theories of interaction and social structure* refer to institutional and social factors which are related to human health and disease. Health and disease are hypothesised to be related to the society's social and power structure or as a reaction to these structures.
5. *Public health theory* concentrates on the analysis of interlinkages between social traits and the states of health and disease of a population. Based on this approach or perspective, public health experts determine, which activities are needed to improve the population's health in societies (Hurrelmann 2003).

In addition, Aaron Antonovsky's interpretation of health is of particular importance for our understanding of the concept of health: Generally, people are either considered as ill or as healthy, a dichotomous classification. If people are classified as healthy, they may be left unnoticed by the public health care system (Bengel et al. 1999). For example, primary prevention measures, which take place before any adverse health consequence is identified (Kickbusch 2003), could take a back seat under this perspective. Antonovsky, an American-Israeli medical sociologist, "juxtaposed this dichotomy with a continuum he called the 'health ease/disease continuum' on which people can be rated as more or less ill or healthy" (Bengel et al. 1999:24). Besides biological and environmental risk factors, Antonovsky underlines the significance of mental risk factors, which can have substantial influence on health. Antonovsky proposes three attitudes which support individuals coping with mental burden.

The *sense of comprehensibility* describes the expectation or the ability of the person to "process both familiar and unfamiliar stimuli as ordered, consistent, structured information and not to be confronted with stimuli that are chaotic, random, accidental and inexplicable" (Bengel et al. 1999:26)

The *sense of manageability* describes a person's belief that there are resources to cope "with the inherent stressors of human existence" (Antonovsky 1996:15) – "[. . .] a person's conviction that difficulties are soluble" (Bengel et al. 1999:27).

The *sense of meaningfulness* means the extent to which a person feels that life makes sense, that at least some of the problems and demands in life are worth investing energy in and are worthy of commitment and engagement (Bengel et al. 1999). It emphasises the importance of a person's wish/motivation to cope with inherent stressors of human existence (Antonovsky 1996).

While elaborating this theory, Antonovsky introduced the term *Sense of Coherence* (SOC), which integrates these three attitudes. He postulated that the more a person experiences a SOC – the level of comprehensibility, manageability and meaningfulness – the healthier he or she might be and the more quickly this person will regain health and remain healthy (Antonovsky 1996; Bengel et al. 2001).

Antonovsky's view on health and his theory of SOC has important implications on public health interventions. According to Antonovsky's health continuum, also people who are not diagnosed as ill should be considered with respect to public health promotion and preventive measures. Following Antonovsky, every person has certain health potentials which are worth to support in order to move the health status to the more healthy side of the health continuum.

Additionally, supposing that Antonovsky's SOC influences an individual's coping capacities, public health experts can utilise this assumption to design SOC supporting measures. Such measures would be appropriate to develop a higher extent of resistance/resilience against internal and external stressors. Surtees et al. (2006) show in their study on the influence of SOC on resilience and mortality that those subjects with a weak SOC report significantly slower adaptation to the adverse effects/stressors than those with a strong SOC. The authors suggest that SOC is a potential marker of an individual's social stress adaptive capacity, which is predictive of mortality (Surtees et al. 2006).

Considering these different theoretical approaches, the following definition of health was chosen within our research activities. It was proposed by Hurrelmann and reflects the holistic approach proposed by the WHO, the integration of the above mentioned health-related theories and Antonovsky's theoretical understanding of the health continuum: *Relative health and relative disease, respectively, is the state of a partially disturbed equilibrium of risk factors and protective factors, which takes place if an individual is only partially or merely for a certain period able to cope with both internal (physical and mental) and external (social and material) demands. Relative health and relative disease are a state that provides only limited well-being and vitality* (Hurrelmann 2006).[2]

3.2.2 Vulnerability, Resistance and Resilience

Vulnerability, resistance and resilience can play an important role in public health research because public health measures aim e.g. to reduce vulnerability and to increase the level of resistance and resilience in individuals and communities. Since these terms are strongly related it is necessary to primarily put some light on these concepts' theoretical background.

3.2.2.1 Vulnerability

The term vulnerability is broadly used in various contexts, such as disaster research (Parker 1995), global change (Leichenko and O'Brien 2002; Schröter 2005), environmental studies (Cross 2001) and development studies (Dercon 2005). Vulnerability plays also a vital role in geographical research (e.g. Uitto 1998) and has been in use since almost 30 years. Vulnerability is also of high significance regarding health-related sciences. A cursory literature search within the medical data base Medline® identified 3,437 articles with the term vulnerability in the titles. Correspondingly, the concept of vulnerability is used in a variety of related different meanings (Dercon 2005), the use is vague (Chambers 2006) and there is no non-ambiguous, widely accepted definition (Kremer 2004).

However, within the different definitions in the literature, there are some frequently shared aspects. According to Chambers (2006), e.g., vulnerability has two sides: "[...] an external side of risks, shocks, and stress to which an individual or household is subject; and an internal side which is defencelessness, meaning a lack of means to cope without damaging loss" (Chambers 2006:1). Also Cutter broadly defines vulnerability as the "potential for loss" (Cutter 1996:529; Cutter et al. 2003:242). Bohle et al. (1994) refer to Chambers definition (first published 1989) and emphasise three aspects related to vulnerability: First the "risk of exposure to crises, stress and shocks". Second the "risk of inadequate capacities to cope with stress, crises and shocks" and third the "risk of severe consequences of, and the attendant risk of slow or limited recovery from, crises, risk and shocks" (Bohle et al. 1994:38). Bogard (1989) also stresses the meaning of capacities to react against possible stressors: "Vulnerability is operationally defined as the inability to take effective measures to insure against losses" (Bogard (1989) in Cutter 1996:531). According to these considerations *we understand vulnerability as the lack of capacities to activate internal or external resources to cope with stressors.*

3.2.2.2 Resistance

Resistance is to perceive as the opposite of vulnerability. Whereas vulnerability is the lack of resources and capacities to cope with stressors, resistance means the stability against these stressors. If a system is resistant, a stressor was not able to alter the system function. There had been sufficient resources to buffer or block the stressor. From this point of view, absolute resistance is the ideal outcome after a stressor affects a system. There are at least two aspects that have strong influence on resistance: First, if a stressor is of strong force and long lasting, meaning a great deal of exposure, it is likely that the stressor will have a considerable impact. Second, a system will not be prepared against stressors if they are unlikely to occur. Therefore resistance of a population affected by strong and unexpected disasters is very unlikely (Norris et al. 2008:132). On the basis of Norris et al.'s (2008) explanations, we understand *resistance as the capacity to activate internal or external resources to react immediately to buffer or block appearing stressors and their effects to preferably avoid dysfunctions.* Nonetheless, our understanding of resistance deviate from Norris et al.'s (2008) point of view (a systems ability to keep the system stable without any occurring dysfunctions) (Norris et al. 2008:130). We view resistance also as present if a system reacts immediately to prevent dysfunction against an appearing stressor, even if the counteracting resources are just partly able to prevent the system against negative impacts (limited resistance). Our basic assumption is that systems commonly contain both vulnerable and resistant characteristics depending on the stressor, its effects and the system's resources and capacities. This view is supported by Rutter (1993), who pointed out that no individual has absolute resistance. He rather proposes "to consider susceptibility to stress as a graded phenomenon" (Rutter 1993:626). In other words, a system can appear to be resistant against stressors on a continuum between maximum survivable vulnerability and maximum attainable resistance with a corresponding level of dysfunction. The concept of resistance has important implications in the health domain. Besides the earlier addressed psychological aspects, resistance also play an important role concerning physical health. Norris et al., e.g., exemplarily refers to the human immune system which is able to fight against a causal agent entering the body. The immune system can block the pathogen's effect (Norris et al. 2008). In public health it is important to identify capacities and resources of individuals and populations to cope with health threats, such as unhealthy living and working conditions and lacking access to health care service. Thus adequate public health measures can be designed to support resistance against health risks.

3.2.2.3 Resilience

Resilience, similar to resistance, is to perceive as a system's capacity to response to internal and external stressors. In contrast to resistance, when a system immediately tries to buffer or block a stressor and its effects, resilience can be viewed as the capacity to response to the dysfunctions that have already taken place (in case the system's resistance was not strong enough to keep absolute stability). Similar to the terms vulnerability and resistance, resilience is differently defined in the scientific literature (Luthar et al. 2000). According to Yehuda and Flory, resilience has been defined in the psychosocial literature "as the process of adapting well in the face of adversity, trauma, tragedy, threats of harm, or even significant sources of stress" (Yehuda and Flory 2007:438). Fagg et al. (2008) describe resilience responses as dynamic phenomena and state that it can be conceptualised as processes of adaptive functions (Fagg et al. 2008). Norris et al. characterise resilience as an ability to adapt in response to adverse stressors, what strongly corresponds with our understanding of health. They define resilience as "a process linking a set of adaptive capacities to a positive trajectory of functioning and adaptation after a disturbance" (Norris et al. 2008:130).

The concept of resilience plays a role in several scientific disciplines, e.g., in disaster research (Norris et al. 2008), psychology (Bonanno 2004, 2005; Bonanno et al. 2002a; Bonanno et al. 2002b; Bonanno et al. 2004; Rutter 1987, 1993), environmental research (Adger 2000; Gunderson 2000) and geography (Martin et al. 1993). Resilience is of vital interest concerning health as well. In particular within psychology research, resilience has a crucial meaning in the sense of individual and community resilience. Resilience is also an important concept with respect to physical health. Using the earlier mentioned example, the immune response to a causal agent can also be seen as a resilience response. After, e.g. an influenza virus enters the human body the affected person's immune system will not be able to provide absolute resistance against all symptoms. He or she will experience symptoms like fever, headache, myalgia, malaise, etc. – a systemic dysfunction takes place. Nevertheless, even without any medication, the human immune system is commonly able to combat the influenza infection after one to more weeks (Treanor 2005), thanks to the resilience of the human immune system. According to the aforementioned definitions and explanations, *we conceive the term resilience as the capacity to activate internal or external resources to counteract appearing stressors and its effects, so that already occurred dysfunctions can be reversed.*

3.2.2.4 Our Conceptualisation of Vulnerability, Resistance and Resilience

Carthey et al. (2001), who examined the strategies of health care organisations to cope with health care service problems, propose that the "ideas of resistance and vulnerability can be represented as the extremes of a notional space [. . .]" with one axis from an extreme of maximum attainable resistance on one side to a maximum of survivable vulnerability (Carthey et al. 2001:29). The concept vulnerability was earlier defined as a kind of "potential for loss" (Cutter 1996:529; Cutter et al. 2003:242) or the defencelessness against loss (Chambers 2006). On the other hand Bonanno et al. (2002a, 2004) opposes loss with resilience (Bonanno 2004; Bonanno et al. 2002a). These viewpoints suggest that resilience is a kind of antipole to vulnerability. Moreover, Norris et al. (2008) pointed out that vulnerability occurs "when resources were not sufficiently robust, redundant, or rapid to create resistance or resilience, resulting in persistent dysfunction" (Norris et al. 2008:130).

According to these explanations, we conceptualise vulnerability, resistance and resilience as follows:

1. We understand vulnerability as the opposite of both resistance and resilience. Whereas vulnerability is considered as the lack of capacity to activate internal or external resources to cope with stressors, both resistance and resilience are similar concepts of being able to activate internal or external resources to cope with stressors and their effects.
2. In line with Rutter, ("no one has absolute resistance" [Rutter 1993:626]) and bearing Antonovsky's SOC and his proposed health continuum (Antonovsky 1996) in mind, we considers the attributes of vulnerability and resistance on the one hand and vulnerability and resilience on the other hand as not mutually exclusive. Rather these poles are the end poles of two continuums (I. *vulnerability-resistance-continuum* and II. *vulnerability-resilience-continuum*).
3. The extent of resistance and vulnerability (vulnerability-resistance-continuum) determines the influence of an occurring stressor on the system function. Maximum survivable vulnerability leads to a high degree of temporary dysfunction. Conversely, a maximum attainable resistance would result in unhindered function.
4. The degree of temporary dysfunction determines the need for resilience.
5. After the demand for the needed resilience is manifest, resilience processes take action in order to reverse appeared temporary dysfunctions. The extent of resilience (vulnerability-resilience-continuum) ultimately decides on the degree of the persistent system dysfunction. The range here goes from a maximum survivable persistent dysfunction or even total system breakdown to unhindered function. The latter case means the total recovering from the stressors' impacts on the system.
6. The resilience process means not merely a process of system protection but also a learning process. A system, which experienced the need to reverse dysfunctions, will aim to prepare itself against possible reoccurrence of similar stressors. This learning process will increase future resistance.
7. In most cases, a system is neither totally vulnerable nor is it absolutely resistant or resilient. In fact, it is more likely that systems inhere both, vulnerable and resistant and/or resilient attributes (Fig. 3.2).

Exemplarily, a community that is threatened or was struck by a flood has still some resources to counteract against the effects of destruction. Before or at the beginning of the flood people would organise groups of helper and transportation to protect themselves and their property/dwelling by removing their possessions to secure places (→ resistance). After the flood has reached the peoples' dwellings causing a certain degree of destruction, the normal daily life will be impaired within this community (→ temporary dysfunction). The affected people would develop emergency plans to guarantee the provision of food, drinking water, and medication to prepare against and reduce health threats due to hunger and infectious disease.

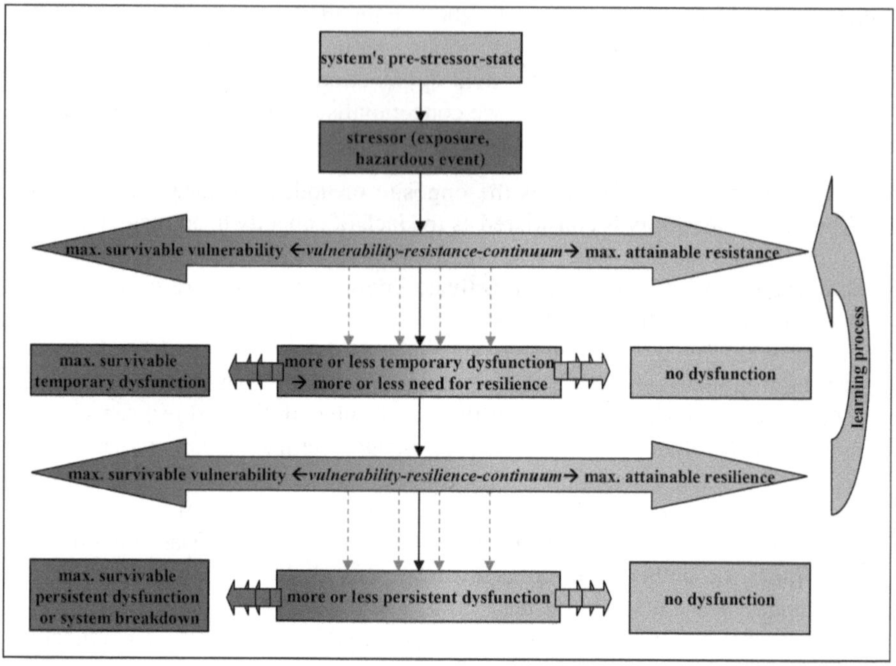

Fig. 3.2 PRD 4's conceptualisation of the relations between vulnerability, resistance and resilience

They would also build water drainages and would restore their houses after the water level got back to normal (→ resilience process). The best case scenario would be that the community is able to reverse all adverse flood effects and can return to normal daily life as before the flood (→ no persistent dysfunction). During the resilience process, the affected community will identify the flood vulnerability characteristics (→ learning process) and aim to decrease vulnerability to increase flood resistance.

In public health research it can be useful to identify certain threats, the level of the respective resistance and the threshold level, which would – after exceedance – lead to temporary dysfunction. The same applies to the knowledge about the kind and degree of dysfunction, the expectable resilience processes and its potential degree. Knowing these characteristics offer public health experts valuable prospects to develop effective interventions in order to strengthen the protective capabilities of populations.

3.3 Public Health-Related Transition Theories

The theory of *demographic transition* describes the shift in societies during development from a situation of young populations with high fertility and high mortality rates and stable population size to slower growth and aging societies. Over a long period of human development, humankind was able to reduce mortality through social and economic changes. Fertility still remained high leading to a strong population growth – the first step of the demographic transition (Smith and Ezzati 2005). In some European countries at the ending nineteenth century, in Latin America and Asia in the 1970s, and in Africa during the 1990s, fertility rates started to decline resulting in slower population growth (Ulrich 2006). With further societal development and decline of fertility, similarly low levels of fertility and mortality were observable in various countries. That resulted in a stable population size and aging societies (Smith and Ezzati 2005). In the year 1971, Abdel R. Omran developed the theory of *epidemiological transition* and pointed out:

> Conceptually, the theory of epidemiological transition focuses on the complex change in patterns of health and disease and on the interactions between these patterns and their demographic, economic and sociologic determinants and consequences. An epidemiologic transition has paralleled the demographic and technologic transitions in the now developed countries of the world and is still underway in less-developed societies (Omran 1971:161).

The theory of epidemiologic transition describes the changes in health characteristics in developing societies preceding and during the demographic transition. It says that there is a shift in the disease patterns and causes of death from infectious diseases, such as malaria, bronchitis, influenza, pneumonia or diarrhoea, to non-infectious or chronic diseases, such as cardiovascular diseases, cancer, and diabetes (Grundy 2004; Lucas 2004; Smith and Ezzati 2005).

Another societal change is described by the theory of *risk transition* proposed in the 1990s. It identified a shift in the character of environmental risk during the period of societal development. This theory was based on the idea that before a shift in mortality and disease patterns (epidemiologic transition), a shift in risk factors responsible for disease and death occurs (Smith and Ezzati 2005).

Demographic, epidemiologic and risk transition are interlinked. These theories have implications on populations' health and therefore need to be taken into account when assessing populations' health status.

Particularly the health statuses of people living in developing and transitional countries are affected by these societal changes. These countries are challenged by the so-called *double burden of disease*. On the one hand they still suffer from disease patterns related to food insecurity and poverty (e.g. high rates of communicable infections/diseases, perinatal conditions, traffic-related injuries). On the other hand they also experience increasing morbidity and mortality due to chronic and non-communicable diseases (e.g. cardiovascular diseases, cancer, diabetes), which cause the main burden of disease in developed countries (Amuna and Zotor 2008; Boutayeb 2006). In this respect, theories of transitions can set a framework for designing public health research activities and measures. The actual risk characteristics, the demographic stage and the epidemiologic health and disease patterns of the target population should be considered.

3.4 Conclusions

Epidemiological research in complex urban lifeworlds in megacities require a broad interdisciplinary approach and needs to be rooted in health-related theories and concepts as well as in globally recognised health-related policies. Furthermore megaurban public health research should consider the effects of global change and urbanisation on the cities' inhabitants. An interlinked global, regional and local perspective is necessary since particularly megacities are involved in globalisation processes. They are strongly affected by globalisation and urbanisation but they are drivers of global change as well. With respect to public health research design and practical field work on a more local level, a specific understanding of health needs to be applied and the target population's internal and external bio-psycho-social health determinants should be considered. The knowledge, e.g. about certain risk or protective factors or the level of vulnerability/resistance/resilience has profound influence on designing a questionnaire or on the planning of adequate public health interventions. Also the stage of demographic or the epidemiological transition of a society should be considered during the preparation of research activities and interventions. The described theoretical orientation serves as a basis to properly perform our scientific activities and to obtain a deeper interdisciplinary understanding of the influence of megaurban lifeworlds on human health.

Acknowledgements We thank the German Research Foundation for funding this research. We are grateful to our colleagues, Mrs. Prof. Dr. Li LING, Mrs. Lu HAN and Mrs. Yinghua XIA, School of Public Health, Sun Yat-sen University, Guangzhou, which performed the interviews and supported us in designing the questionnaire and during the project coordination. Special thanks are given to our friend Mr. Fei FANG, PhD candidate at the School of Biomedical Sciences, Faculty of Medicine, Chinese University of Hong Kong, who continuously supported us by social and practical support.

References

Adger NW (2000) Social and ecological resilience: are they related? Progress in human geography 24(3): 347–364

Amuna P, Zotor FB (2008) Epidemiological and nutrition transition in developing countries: impact on human health and development. Proc Nutr Soc 67(1): 82–90

Antonovsky A (1996) The salutogenic model as a theory to guide health promotion. Health promotion international 11: 11–18

Bengel J, Strittmatter R, Willmann H (1999) What keeps people healthy? The current state of discussion and the relevance of Antonovsky's salutogenetic model of health. Federal Centre for Health Education, Cologne

Bengel J, Strittmatter R, Willmann H (2001) Was erhält Menschen Gesund? Antonovskys Modell der Salutogenese - Diskussionsstand und Stellenwert. erweiterte ed. Bundsezentrale für gesundheitliche Aufklärung, Köln

Bohle HG, Downing T, E., Watts MJ (1994) Climate change and social vulnerability. Global Environmental Change 4: 37–48

Bonanno GA (2004) Loss, Trauma, and Human Resilience. American Psychologist 59(1): 20–28

Bonanno GA (2005) Resilience in the face of potential trauma. American Psychological Society 14(3): 135–138

Bonanno GA, Papa A, Kathleen ON (2002a) Loss and human resilience. Applied & Preventive Psychology 10: 193–206

Bonanno GA, Wortman CB, Lehman DR, Tweed RG, Haring M, Sonnega J, Carr D, Nesse RM (2002b) Resilience to Loss and Chronic Grief: A Prospective Study From Preloss to 18-Months Postloss. Journal of Personality and Social Psychology 83(5): 1150–1164

Bonanno GA, Wortman CB, Nesse RM (2004) Prospective Patterns of Resilience and Maladjustment During Widowhood. Psychology and Aging 19(2): 260–271

Bogard W C (1989) Bringing social theory to hazards research: conditions and consequences of the mitigation of environmental hazards. Sociological Perspectives 31: 147–168

Boutayeb A (2006) The double burden of communicable and non-communicable diseases in developing countries. Trans R Soc Trop Med Hyg 100(3): 191–199

Carthey J, de Leval MR, Reason JT (2001) Institutional resilience in healthcare systems. Quality in Health Care 10: 29–32

Chambers R (2006) Vulnerability, Coping and Policy. Institute of Development Studies Bulletin 37(4): 33–40

Cross JA (2001) Megacities and small towns: different perspectives in hazard vulnerability. Environmental Hazards 3: 63–80

Cutter SL (1996) Vulnerability to environmental hazards. Progress in human geography 20: 529-539

Cutter SL, Boruff BJ, Shirley WL (2003) Social vulnerability to environmental hazards. Social Science Quarterly 84(2): 242–261

Dercon S. "Vulnerability: a micro perspective". Oral presentation at the *Annual bank conference on development economics*. Amsterdam. May 23–24, 2005

Fagg J, Curtis S, Stansfeld SA, Cattell V, Tupuola A-M, Arephin M (2008) Area social fragmentation, social support for individuals and psychosocial health in young adults: Evidence from a national survey in England. Social Science & Medicine 66: 242–254

Grundy E. (2004) Demography and public health. In R. Detels, J. McEwen, R. Beaglehole & H. Tanaka (eds.), Oxford Textbook of Public Health. Oxford University Press, New York, pp. 807–828

Gunderson LH (2000) Ecological resilience - in theory and application. Annual Review of Ecology and Systematics 31: 425–439

Hurrelmann K (2003) Gesundheitssoziologie. Eine Einführung in sozialwissenschaftliche Theorien von Krankheitsprävention und Gesundheitsförderung. 5. ed. Juventa, Weinheim und München

Hurrelmann K (2006) Gesundheitssoziologie. Eine Einführung in sozialwissenschaftliche Theorien von Krankheitsprävention und Gesundheitsförderung. 6., revised ed. Juventa, Weinheim und München

Kickbusch I. (2003) Gesundheitsförderung und Prävention. In F. W. Schwartz, B. Badura, R. Busse, R. Leidl, H. Raspe, J. Siegrist & U. Walter (eds.), Das Public Health Buch. Gesundheit und Gesundheitswesen. Urban & Fischer, München, Jena, pp. 181–225

Kraas F (2003) Megacities as Global Risk Areas. Petermanns Geographische Mitteilungen 147: 6–15

Kremer A. (2004) Urbane Umwelt und Gesundheit: Exposition und Risikowahrnehmung vulnerabler Risikogruppen in Pondicherry, Indien. Mathematisch-naturwissenschaftliche Fakultät (p. 271). Bonn: Rheinische Friedrich-Wilhelm-Universität

Leichenko RM, O'Brien KL (2002) The Dynamics of rural vulnerability to global change: The case of southern Africa. Mitigation and Adaptation Strategies for Global Change 7: 1–18

Li X, Stanton B, Fang X, Lin D (2006) Social Stigma and Mental Health among Rural-to-Urban Migrants in China: A Conceptual Framework and Future Research Needs. World Health Popul 8(3): 14–31

Lu D, Liu H. (2006) Urbanization and environmental issues in China. In W. Wuyi, T. Krafft & F. Kraas (eds.), Global change, urbanization and Health. China Meteorological Press, Beijing, pp. 3–10)

Lucas AO. (2004) Health policies in developing countries. In R. Detels, J. McEwen, R. Beaglehole & H. Tanaka (eds.), Oxford Textbook of Public Health. Oxford University Press, New York, pp. 281–295)

Luthar SS, Cicchetti D, Becker B (2000) The Construct of Resilience: A Critical Evaluation and Guidelines for Future Work. Child Development 71(3): 543–562

Martin R, Sunley P, Willst J (1993) The geography of trade union decline: spatial dispersal or regional resilience? Transactions of the Institute of British Geographers, New Series, 18(1): 36–62

Norris FH, Stevens SP, Pfefferbaum B, Wyche KF, Pfefferbaum RL (2008) Community Resilience as a Metaphor, Theory, Set of Capacities, and Strategy for Disaster Readiness. American Journal of Community Psychology 41: 127–150

Omran AR (1971) The epidemiologic transition. A theory of the epidemiology of population change. The Milbank Memorial Fund Quarterly 49: 509–538

Parker D (1995) Disaster vulnerability of megacities: An Expanding problem that requires rethinking and innovative responses. GeoJournal 37: 295–301

Ping H, Pieke FN. China Migration Country Study. Oral presentation at the *Regional Conference on Migration, Development and Pro-Poor Policy Choices in Asia* Dhaka, Bangladesh. June 22–24, 2003

Preamble to the Constitution of the World Health Organization as adopted by the International Health Conference, New York, 19–22 June, 1946; signed on 22 July 1946 by the representatives of 61 States (Official Records of the World Health Organization, no. 2, p. 100) and entered into force on 7 April 1948. WHO, 1948. (Accessed July 16, 2008, at http://www.who.int/about/definition/en/print.html.)

Rutter M (1987) Psychosocial resilience and protective mechanisms. American Journal of Orthopsychiatry 57(3): 316–331

Rutter M (1993) Resilience: some conceptual considerations. Journal of Adolescent Health 14(8): 626–631

Schröter D (2005) Ecosystem service and vulnerability to global change in Europe. Science 310: 1333–1337

Smith KR, Ezzati M (2005) How environmental health risks change with development: The epidemiologic and environmental risk transitions revisited. Annual Review of Environment and Resources 30: 291–333

Surtees PG, Wainwright NW, Khaw KT (2006) Resilience, misfortune, and mortality: evidence that sense of coherence is a marker of social stress adaptive capacity. Journal of psychosomatic research 61(2): 221–227

Treanor JJ. (2005) Influenza virus. In G. L. Mandell, J. E. Bennett & R. Dolin (eds.), Mandell, Douglas and Bennett's principles and practice of infectious diseases. Elsevier, Philadelphia, pp. 2060–2085

Tunon M. (2006) Internal Labour Migration in China: Features and Responses. Beijing: International Labour Organization

Uitto JI (1998) The geography of disaster vulnerability in megacities. Applied geography 18(1): 7–16

Ulrich RE. (2006) Weltweite Bevölkerungsentwicklung. In O. Razum, H. Zeeb & U. Laaser (eds.), Globalisierung - Gerechtigkeit - Gesundheit. Einführung in International Public Health. Verlag Hans Huber, Bern, pp. 79–89

WHO, UNDP. (2001) Environment and people's health in China. Geneva, New York: WHO, UNDP

Wong DFK, Chang Y, He X (2007) Rural migrant workers in urban China: living a marginalised life. International Journal of Social Welfare 16: 32–40

Wong DFK, He X, Leung G, Lau Y, Chang Y (2008) Mental health of migrant workers in China: prevalence and correlates. Social Psychiatry and Psychiatric Epidemiology 43(6): 483–489

Yehuda R, Flory JD (2007) Differentiating Biological Correlates of Risk, PTSD, and Resilience Following Trauma Exposure. Journal of Traumatic Stress 20(4): 435–447

Young TK (2005) Population health. Concepts and Methods. 2nd ed. Oxford University Press, New York

Zheng Z, Lian P. (2006) Health vulnerability among temporary migrants in urban China. In W. Wuyi, T. Krafft & F. Kraas (eds.), Global change, urbanization and Health. China Meteorological Press, Beijing, pp. 197–207

Chapter 4
Urban Health Research: Study Designs and Potential Challenges

Md. Mobarak Hossain Khan and Arina Zanuzdana

4.1 Introduction

According to World Health Organization (1948), health is defined as a state of complete physical, mental and social well-being and not merely the absence of disease or infirmity. In terms of this definition, urban health is referred to as the health of population living in the city or town (Galea and Vlahov 2005). More than half of the world population currently live in urban areas (approximately one-third of them are estimated to live in marginal settlements or slums (UN-Habitat 2003)) and virtually most of the world population growth from now on will be in cities (Leon 2008). For example, the urban population is projected to increase by 1.6 billion by 2030 while the rural population shrinks by 28 million. Although people migrate to cities for a better life and income (Cohen 2004), urbanisation is also considered as a health hazard for certain vulnerable populations. The demographic shift due to rapid and uncontrolled urbanisation also creates a humanitarian disaster (Patel and Burke 2009).

Urban health is of recent vintage and offers a perspective on health and disease. The health of urban dwellers represents a convergence of powerful biologic, social and contextual forces. A comprehensive approach to study urban health integrates clinical and public health communities and draws on the social and political sciences to seek understanding of the impact of cities on the health of the populations and individuals (Fleischman and Barondess 2004).

An urban health study is highly complex and the success of urban health research depends on many factors. Application of appropriate study designs and overcoming the challenges specific for urban health research are some of the major pillars for a successful urban health study (Table 4.1). In this chapter first some of the common epidemiologic study designs applicable for studying complex urban health

M.M.H. Khan (✉) • A. Zanuzdana
Department of Public Health Medicine, School of Public Health, Bielefeld University, Bielefeld, Germany
e-mail: mobarak.khan@uni-bielefeld.de

A. Krämer et al. (eds.), *Health in Megacities and Urban Areas*,
Contributions to Statistics, DOI 10.1007/978-3-7908-2733-0_4,

Table 4.1 Potential challenges of an urban health study

Major challenges	What is meant
Definitional	• Inconsistencies of urban definition • Inconsistencies of urbanization processes
Disciplinary	• Multidisciplinary • Transdisciplinary • Interdisciplinary • Lack of co-ordination
Methodological	• Triangulation • Sampling for hard-to-reach population • Use of adequate sample size
Informational	• Lack of secondary data • Lack of surveillance • Lack of data quality
Interpretational	• Causation • Bias • Confounding
Others	• Lack of resources • Frequent movement of vulnerable populations • Settlement changes • Generalisability of results

problems are overviewed. Next we discuss some potential challenges specific for urban health research, partly based on available literature and partly on the basis of our own megacity research experience which we have gathered over the last few years in frames of the multidisciplinary program "Megacities - Mega challenges: Informal Dynamics of Global Change", funded by the German Research Foundation.

4.2 Common Study Designs in Urban Health Research

A range of common epidemiological study designs can be used to study urban health. Some of the study designs are used only to generate hypotheses and some of them are employed to test hypotheses. Some study designs assess relationship between exposures and outcomes based on past histories and on prospective data. However, an appropriate research design always aims to establish the truth by reducing bias, confounding and chance (Clancy 2002).

Broadly study designs can be classified into different types:

Experimental and observational
Descriptive and analytical
Qualitative and quantitative.

The framework for classification of types of epidemiological studies is presented on Fig. 4.1, modified from Grimes and Schulz 2002a.

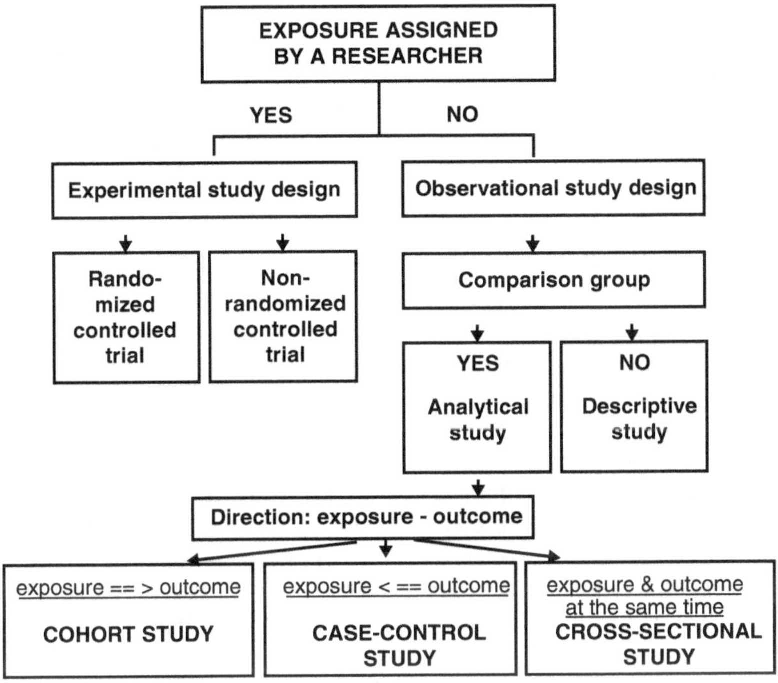

Fig. 4.1 Framework for classification of study designs (Modified from Grimes and Schulz 2002a)

4.2.1 Experimental and Observational Studies

In experimental studies, investigators have freedom to control research setting, manipulate the study factors and randomly assign subjects to the exposed and non-exposed groups. Clinical trial is an example of an experimental study design. This type of study design is commonly used when researchers want to test the effectiveness of a new drug or therapy over existing drug or therapy. Experimental studies (e.g., randomised controlled trial, RCTs) are usually conducted after observational studies provide strong evidence of associations. Experimental studies can be expensive, ethically unacceptable and may lack generalisability because of exclusion criteria. However, experimental studies are preferred if they are ethical, practical and appropriate (Clancy 2002). They can provide much stronger evidence than observational studies, because randomisation of the study participants to treatment and control groups prevents many biases typical for observational studies (Barnett and Hyman 2006).

In contrast, investigators have no or only a little control over study setting, subjects, and exposures in observational studies. Researchers attempt to make valid comparisons between people with or without diseases or between those naturally exposed or unexposed to a factor of interest (Clancy 2002). Cohort, cross sectional, and case–control studies are collectively referred to as observational studies. Often these studies are the only practicable method of studying various problems, especially when a randomised controlled trial is unethical, or when the condition to be studied is rare.

Cohort studies are prospective studies in which groups of subjects (cohorts) are selected on the basis of exposure and followed prospectively in order to see how many members of each group develop the target disease (Barnett and Hyman 2006). Cohort studies are used to study incidence, causes, and prognosis; they measure events in chronological order, which allow us to distinguish between cause and effect (Mann 2003). Only one risk factor can be assessed for each cohort study but multiple outcomes can be measured (Clancy 2002). Typically these studies require large samples, if the outcome disease is rare and needs long time period, which often makes cohort studies expensive (Clancy 2002). Data from a cohort study are more accurate than the data of a case–control study, as cohort study can eliminate recall and minimise selection biases (Barnett and Hyman 2006) (see Fig. 4.2 for relative risk (RR) calculation using cohort data).

Case–control studies compare groups retrospectively. Case means a person with the target disease, whereas control means a person without the target disease regardless of other diseases. Normally this design is applied for rare diseases as well as for diseases which are new or unusual and can measure multiple exposures. They are often used to generate hypotheses that can then be studied via prospective cohort or other studies (Mann 2003). As compared to cohort studies, case–control studies are relatively short with respect to duration and less expensive as they involve smaller number of cases. However, selection of control groups is difficult and often introduces selection bias. Recall bias by patients and measurement bias by investigators may also distort the exposure-outcome relationships (Clancy 2002; Barnett and Hyman 2006) (see Fig. 4.2 for odds ratio calculation using case–control data).

Cross-sectional studies are like a snapshot and measure exposure and outcome at one point in time (Grimes and Schulz 2002a). These studies are primarily used to determine prevalence but are not efficient if the conditions or diseases are rare (Mann 2003). The term prevalence simply means the number of cases in a population at a given time point. Subjects are recruited without considering the outcome of interest. These studies are also useful at indentifying associations that can be more rigorously studied using cohort or randomised controlled study. Multiple outcomes can be studied at the same time. Cross-sectional surveys are relatively quick and easy but do not permit distinction between cause and effect. For example, if people living in marginal settlements are interviewed about their income and their health, then in the cross-sectional study it might be difficult to identify whether their income is so low at this point of time because they are sick and cannot work or they are sick because they do not have any income and cannot afford healthy food and treatment (see Fig. 4.2 for odds ratio calculation using cross-sectional data).

Ecological studies allow us to study exposure and outcome at the population level. They are suitable, for example, for quantification of the associations between exposure and response for some climate-sensitive diseases. Such studies take into account spatial and temporal aspect of exposure and outcome, as well as they utilize

Example 1: Calculation of RR for cohort data

Relative risk is the ratio of the risk of a disease/an outcome among the exposed persons to the risk among the unexposed persons (Last 2001).

	Disease	No disease	Total
Risk group	a	b	a+b
Non-risk group	c	d	c+d
Total	a+c	b+d	n

Relative Risk (RR) = [a / (a + b)] / [c / (c + d)]
95% Confidence Interval = RR * $e^{(\pm 1.96 * [SE\ RR])}$
where Standard Error of the RR (SERR)=
Square root(*[b / {a*(a+b) }] + [d / { c*(c+d) }]*)

Let us assume that in an urban male cohort study, some male participants reported hard physical works and some of them not at the baseline survey. During three months of follow-up period, some developed back pain and some of them not. The distribution was as follows:

	Back pain	No back pain	Total
Hard physical work	334	121	455
No hard physical work	45	212	257
Total	379	333	712

Therefore, RR = [334 / 455] / [45 /257)] = 0.734/0.175=4.19

Interpretation: Male participants who had been working hard had approximately 4 times higher chance to develop a back pain as compared to the study participants who had not been working physically hard.

Example 2: Calculation of OR for case-control data/cross-sectional data

For a case-control data the exposure-odds ratio is the ratio of odds in favour of exposure among the cases to the odds in favour of exposure among controls. For a cross-sectional study the disease-odds ratio or the prevalence-odds ratio is the ratio of the odds in favour of disease among the exposed to the odds in favour of disease among the unexposed (Last 2001).

Exposure Disease	Cases	Controls	Total
Yes	a	b	a+b
No	c	d	c+d
Total	a+c	b+d	n

Odds Ratio (OR) = (a/b)/(c/d) = ad / bc
95% Confidence Interval = $e^{\ln OR \pm 1..96 * SE}$ $_{\ln OR}$

Let us assume that in poor urban settlements some neighbourhoods have got a permanent access to hand-washing items and modern toilet and other neighbourhoods still had no or very few hand-washing items and no modern toilet. The following table presents information about exposure (hygiene) and outcome (diarrhoea).

Hygiene / Diarrhoea	Cases	Controls
No (exposure Yes)	90	80
Yes (exposure No)	10	120
Percentage exposed	90%	40%

OR= 90*120/10*80 = 10800 / 800 = 13.5

Here the OR of 13.5 indicates a strong association between hygiene (e.g. hand-washing) and diarrhoea. A lack of hygiene is a risk factor for having diarrhoea. Similarly, an OR for a cross-sectional data can be calculated.

Fig. 4.2 Examples for relative risk (RR) and odds ratio (OR) calculation

large aggregated databases of routinely reported health outcomes (fatal cases, hospital admissions) (Kovats et al. 2003). Confounders, ecological fallacies and other biases should be carefully controlled or addressed.

4.2.2 *Descriptive and Analytical*

Descriptive studies such as case study or case report and case-series report (more than one case) are mainly used to introduce into a new area of research, to collect basic information and to generate hypothesis. Such studies can be used to describe the natural history of certain disease. Its frequency and other determinants are important for the further research (Kelsey et al. 1996; Grimes and Schulz 2002a). A result of a descriptive study in urban area can be, for example, a description of development of a disease (e.g., dengue fever) among children in an urban area, the characteristics of this disease and of the group of affected children. On the basis of this information further hypotheses about possible source and cause of the disease can be developed and tested with the help of analytical studies. Descriptive studies do not provide any comparison and thus cannot assess any associations or dose–response relationship (Grimes and Schulz 2002a). In analytical studies a temporal component should be identified, in other words the direction of an exposure and an outcome. As seen from Fig. 4.1, in different types of observational analytical studies (cohort study, case–control study and cross-sectional study) temporal relation of exposure and outcome are determined.

4.2.3 *Qualitative and Quantitative Studies*

As no study design is completely suitable for studying urban health, combination of both qualitative and quantitative study designs (see above) is frequently used by urban health researchers. Qualitative studies are not used to test the hypothesis and there is no strict role about sample size. In contrast, quantitative studies are used to test the hypothesis and sample size is important. Qualitative studies can provide high quality information but all such studies can be influenced by known and unknown confounding variables (Mann 2003).

Qualitative research, in contrast to quantitative research, does not necessarily start with formulating a hypothesis. It rather aims to develop a concept in the process of research in order to understand some social phenomena in its natural settings. The focus of research lies on experiences, meanings, images, perceptions and views. Researcher makes an attempt to understand personal reasons or motivations, beliefs or decisions of participants. An example of qualitative research in urban settings can be a study of practices, perceptions and decision-making processes regarding condom use among young women in low-income urban settlements (Fadda and Jirón 1999). In qualitative research sample size is usually fairly small and personality of the researcher plays an important role, which is often a point of criticism due to subjectivity and limited reproducibility (Clancy 2002). The most common qualitative methods include in-depth interviews, key informant method, focus group discussions, phenomenological interpretation, action research, simulated client methods, and documents reviews. Although application of qualitative research methods alone for studying urban health is questionable, there are some circumstances (mentioned below), when they can precede or complement other quantitative methods (Clancy 2002; Curry et al. 2009):

Initiating research into new areas to collect pilot information and describe a subject of interest
Supplementing of quantitative methods
Explaining unexpected or not logical findings from quantitative research.

Research within quantitative study designs requires a correct formula for proper sample size calculation and deals with different types of biases (Bartlett et al. 2001), which are described in further text.

4.3 Complexities of Modern Urban Health Research

Urban health emerged as a distinct field of inquiry in international public health in the mid 1980s, highlighting issues of poverty, urban morbidity and mortality, and burden of communicable and non-communicable diseases in low-income urban populations (Harpham and Molyneux 2001). Generally, population health depends on many factors ranging from micro to macro level factors. Therefore assessment of urban health means study of several multilevel urban factors which may influence the health of urban populations. Planning and performing research in urban areas is complicated because of several issues, which may include specification of research question and choice of appropriate study design, complexity of causation in urban context, and application of a common language for urban health (Galea and Vlahov 2005).

The complexity of causation in urban health research is another important challenge mainly attributed to the nature of complex societies. Compared to rural societies, urban societies are more heterogeneous, for instance in terms of races, ethnicities, and cultures. All these factors play an important role in shaping the health of the urban populations. In this context studying urban health requires more sophisticated designs and methods rather than simple analytical and descriptive approaches.

4.3.1 Definitional Challenges

Research of urban health poses many challenges, some of which concern definitions of most common terms like "urban" and "urbanization". These terms vary from country to country, as uniform definitions of these words are not found. To define urban areas, for example, some countries use administrative boundaries or size and density of population, or some functional characteristics like economic activity. Urbanisation can, for instance, be described in terms of "pushing out" factors (people are compelled to leave less attractive rural areas) and "pulling in" factors (people move in to more attractive urban areas). While the former can comprise such factors as limited employment opportunities, low-quality or absent social services, lack of educational and health facilities, the latter often include better and diverse employment opportunities, freedom of choice of religion and education, and better chances for finding a life partner (de Leeuw 2009).

Problematic definitional issues of "urban" and "urbanization" across different countries are only "the top of iceberg" in the research on urban health (Leon 2008; Cohen 2004). Further challenges which may arise are related, for example, to the identification and comparability of cities in different countries. By defining a city and estimating its size one has to take into account different aspects, like an estimate of the central city and the greater metropolitan area (compare: Mexico City and Greater Mexico City), as well as a wider region and suburbs. If the administrative boundaries of a city are too broad and include agricultural or other nonurban areas, then some areas are misclassified as urban areas (e.g., Shanghai). In contrast, if the boundaries are drawn too tight, then some populations residing in peri-urban areas can be missed (e.g., Bangkok, Manila, Taipei) (Cohen 2004; Bayoumi and Hwang 2002). Thus international differences in city definitions pose additional challenges when comparing study results from different countries.

4.3.2 Disciplinary Challenges

One of the most important steps for any etiologic research is to specify clearly the research question at the beginning. According to Galea and Vlahov (2005), specification of a research question in urban areas is difficult due to several reasons. One of them is interdisciplinarity nature of urban health research and application of different theoretical frameworks and terminologies typical for certain fields (e.g., epidemiology, geography and molecular biology). The need for inter- and trans-disciplinarity research is apparent for researching the urban phenomena because social and environmental changes are multi-causal and require combinations from multiple disciplines. The problems of urban society are increasingly complex and interdependent. They are not isolated to any particular discipline. Also traditional disciplinary approaches that focus on one aspect of the problem are inadequate to elicit the necessary information and to provide theoretical framework that reflect the realities we observe in the urban areas (Goebel et al. 2009). Multidisciplinary techniques, knowledge and interpretations are clearly required to study interdependent research questions in urban health, which are often interlinked and do not meaningfully exist in isolation (Galea and Vlahov, 2005; Goebel et al. 2009). For instance, environmental researchers are challenged by complex and urgent environmental problems which require insights from both natural and social sciences, and the participation of ordinary people and other stakeholders to find some solutions to those problems (Goebel et al. 2009). It is important to find common terms, languages and interpretations which are equally meaningful and understandable for different disciplines. However, according to Ramadier (2004), transdisciplinarity raises the problem of methodology, because it encourages researchers to unify their methodology to identify more easily the theoretical points that do not pertain to the same level of reality. Conflicts may appear because researchers are often systematically sceptical about the methods and results applied in their fields (Ramadier 2004). According to Goebel et al. (2009), one challenge of the transdisciplinarity approach is the difficulty in transforming a real life problem into a research problem that can be addressed with available academic tools, and within a theoretical framework.

4.3.3 Methodological Challenges

4.3.3.1 Hard-to-Reach Populations

So called hard-to-reach or hidden populations which may include homeless, street dwellers, floating population, sex workers and their clients, undocumented migrants, injecting drug users, single parents, people with disabilities, elderly, high rise apartment dwellers, gamblers, culturally and linguistically diverse communities (Nomura et al. 2007; Brackertz 2007) are relatively common in cities and urban areas. Homeless people can be seen not only in the cities of developing countries (Koehlmoos et al. 2009) but also in the cities of developed countries (Hwang 2001). For instance, in 9 largest metropolitan areas of Canada, about 5/1,000 population are homeless (Hwang 2001). In the city of Dhaka, the estimated number of homeless people who sleep on streets, railway terminals and platforms, bus stations, parks and open spaces, religious centres, construction sites, around graveyards, and in other public spaces without roof were about 15,000 in 1997 (Koehlmoos et al. 2009). Homeless people are extremely vulnerable in terms of personal security and high-risk behaviours (Koehlmoos et al. 2009). They suffer from a variety of medical problems with higher severity and therefore have higher risk of death compared to the general population. For example, mortality rates among street youth in Montreal are 9 times higher for males and 31 times higher for females (Hwang 2001). Providing health care facilities for them also might be challenging at least for developing countries because of higher health care costs. Hidden populations are hard to reach because of the difficult access due to stigmatisation or illegal status in the societies. Lack of reliable sampling frames or difficulties in applying systematic sampling methods also limit researchers to study those populations. Community-based studies based on random sampling are rarely used to study hard-to-reach populations. The most frequent methods are facility-based (e.g., medical facilities) and use convenience sampling. For instance, out of 285 studies that focused on hard-to-reach populations in Japan, 284 studies used convenience sampling and only one study used random sampling (Nomura et al. 2007). If the proportion of hard-to-reach population is relatively small, it is difficult to find a sufficient sample using a usual probability sample design. In such a case, the study will be very time consuming and expensive. If questions/variables are sensitive and threatening for the person, a usual probability sample design is not adequate either because of unreliable answers or because of an expected high non-response. These populations are generally reluctant to co-operate researchers. For detailed information about link-tracking sampling designs (e.g., network sampling, snowball sampling and the random walk approach), which are mainly applied because of the impracticability of standard survey methods, consult the article of Spreen (1992).

4.3.3.2 Triangulation

Generally any particular study design to study the urban health problems is not sufficient to represent the scenario which is very close to the reality. It is partly because of the complex nature of health problems in the cities and of the inherent limitations of any particular study design. Therefore, it is strongly recommended to apply and combine both qualitative and quantitative methods within the same project. Triangulation actually refers to this concept and is defined as the use of multiple methods or sources for the collection and interpretation of data about a given phenomenon (Foss and Ellefsen 2002; Jones and Bugge 2006; Fadda and Jirón 1999; Begley 1996). Triangulation has been proposed as a technique for studying complexity (Jones and Bugge 2006). It is being used increasingly to have an accurate impression about the reality. The two general purposes of triangulation are confirmation and/or completeness of the results. It provides a better understanding of the given problem (i.e., completeness) as well as it validates the methods and instruments (i.e., confirmation).

Through triangulation, bias originated from a single-method or single observer can be reduced and the confidence about the findings can be increased. Different methods may inform each other and can act as partial correctives to each other. Considering the advantages of triangulation, in our public health study in the megacity of Dhaka under the German Research Foundation (DFG) priority programme "Megacities – Megachallenges: Informal Dynamics of Global Change", we applied multiple study designs namely cross-sectional, cohort, focus group discussions, and key informant method. We have also validated our study findings with other sources of information. However, research designs that combine different methodologies within the same study is a challenging issue because it is associated with a high degree of complexity. One particular reason is that these methods belong to traditionally different paradigms with fundamentally different epistemological frameworks (Foss and Ellefsen 2002). It is expensive, which might be another big challenge especially in the developing countries. Like our study in Dhaka, both qualitative and quantitative methods were applied to complete formative research at the patient, provider, and system levels at the urban community health centres in USA. They also identified several system-level challenges (Lemon et al. 2006).

4.3.3.3 Sample Size

In any of the chosen study designs the calculation of sample size is an essential element, which helps to prevent either unnecessary expenditures of time and resources or limitation in statistical power and thus limited scientific conclusions of the study. Sample size and power are important measures which define the number of cases needed for a study (Jones et al. 2003). These estimations are a crucial step preceding any research and necessary not only to rationally calculate costs and resources needed, but also to obtain meaningful results. In studies of urban health which often take place in poor-resource settings or in difficult access areas, sample size estimations should be an inevitable component of the research process. In almost all types of quantitative studies, whether it is a clinical trial or a comparative study, sample size calculation serves the precision of final results (Jones et al. 2003) It should be noted that determination of sample size does not set a goal of obtaining the biggest sample possible, but the most adequate-sized one. Cost-effectiveness, clinically important difference and ethics of research are further important issues of sample size estimation (Naing et al. 2006).

To avoid ambiguity, it is necessary to distinct between sample size and power. Generally, these two terms can be used interchangeably. However, power refers to all sample size estimations in a study, or to the number of subjects needed to avoid a type II error in comparative studies; sample size estimation is more universal and broad term, applicable to all other study types (Jones et al. 2003). Sample is a selected group of a population, which can be random or non-random, representative or non-representative (Last 2001). Determination of sample size depends of several factors, such as incidence or prevalence of the studied outcome, the relationship between variables in the study, the desired power and the allowable magnitude of type I error (Last 2001) (for advanced reading on the type I and II errors please see, for example, Gordis 2009; Bartlett et al. 2001).

Estimation of a sample size in descriptive studies, which do not have any hypotheses, can be done based on the concepts of confidence intervals. In observational studies, in which two or more groups are compared with each other (regarding exposure and outcome), the calculation of the sample size is different from descriptive studies. In cohort studies the estimation depends on (a) the proportion of the cases in the unexposed group which are expected to exhibit the outcome of interest and (b) the proportion of cases in the exposed group who are expected to exhibit the outcome of interest. In case–control studies the sample size is calculated based on the proportion of individuals among the exposed cases, and the proportion of individuals among the exposed controls.

4.3.3.4 Modeling

Not only epidemiological methods can be applied to study urban health and diseases. There are situations when dynamic mathematical models can also be used to predict outbreaks of diseases, e.g., climate-sensitive diseases (Patz and Balbus 1996). Furthermore, such outbreak prediction models can be integrated into broader systems approach, which enclose more complex relationships between climate and its changes, ecosystem changes, human health and human adaptive capacity (Patz and Balbus 1996). Socioeconomic factors are essential part of research on urban health, however, human diseases are determined by many other factors (adequate food and water provision, secure housing), which in turn are related to sectors of agriculture and water resources. Integrated mathematical modelling is a method which represents in this regard an incorporation of all relevant factors and systems into human health assessment, making it possible to accurately predict changes in health and susceptibility to disease, including climate change (Patz and Balbus 1996).

4.3.4 Informational Challenges

Cities often suffer from a lack of reliable and up-to-date socio-demographic data. Collection of census data in cities usually takes place once in a decade (in some countries irregularly) and provides information with significant temporal delays. Use of the United Nations (UN) data on urban health is also limited. The officially published UN data is based on countries individual reports thus on countries internal definitions and standards, e.g., definitions of "urban" and "rural", which makes international comparisons of populations living in urban and rural areas difficult. Other issues concern data availability and quality of the calculation of summary measures of health, such as disability-adjusted life-years (DALYs) and quality-adjusted life-years (QALYs).

High-income countries have a better system of census and routine data collection than low-income countries. In this respect it is worth mentioning the Demographic and Health Surveys as a reliable and highly-standardized source of representative socio-demographic information from more than 80 developing countries, available for free for all researchers (http://www.measuredhs.com). For example, models of relation of climate change and health have been developed for certain health outcomes; however, modeling, or scenario-based modeling strongly depends on the availability and quality of data and has limited generalisability potential. To sources of uncertainty in data count furthermore missing components and errors in data, biased and incomplete observations, and issues of limited representativeness of a sample.

4.3.5 Interpretational Challenges

4.3.5.1 Causal Associations

Epidemiologic studies have inherent limitations that preclude establishing causal associations between exposures and outcomes (Barnett and Hyman 2006). Each study design has limitations that can distort the findings. In epidemiological studies we always emphasize on significant associations between two or more diseases or factors, however, statistical significance does not necessarily means causal relation. There are many possibilities for which significant associations can occur:

True causal association between exposure and outcome
Statistical significant association between exposure and outcome due to confounding and/or bias
Statistical significant association by chance

Causality cannot definitely be established by epidemiologic studies. Hill proposed several features to assume causal associations. He called these features his "viewpoints" and did not claim that the fulfillment of these viewpoints proof causality (Hill 1965). However, they are still helpful in order to derive some probability of causation between e.g., exposure and outcome. The most common features (in the literature often called Hill criteria of causation) are given in Table 4.2.

One further Hill causation feature, *specificity*, is considered as a weak criterion for causation (Grimes and Schulz 2002b). Specificity means, exposure leads to only one outcome. In reality, only few exposures can be characterized in this way (e.g., polluted water leads to numerous outcomes), so non-fulfilment of this criteria does not reject the causation. Although Hills viewpoints on causation are useful guidelines, there are many instances of exposures which failed to meet the criteria but showed causal association. Similarly there are some examples of exposures which met the criteria but proved no causal associations (Barnett and Hyman 2006).

Table 4.2 Major criteria for causation by Hill (1965)

Strength of association	Expressed in odds ratio or relative risk. Some authors suggest that OR/RR >3 is a strong support for causation (Sackett et al. 1991)
Consistency of association	Effect has been also seen in other studies, with different designs and time scales
Temporality	Exposure precedes the outcome
Dose–response relationship	Increased exposure leads to more of the outcome
Biological plausibility	Findings support known biological and disease mechanisms and findings from other fields
Experimental evidence	Evidence from clinical trials (not always possible out of ethical reason); indirect evidence

Adapted from Grimes and Schulz (2002b) and Mann (2003)

4.3.5.2 Biases

Bias is a systematic error which can occur in the design, conduct or analysis of a study (Barnett and Hyman 2006). There are more than 30 known and well-studied biases described in literature (Sackett 1979). However, the most important and frequently occurring biases are those produced in the definition and selection of study population, data collection and the association between different determinants of an effect in the population (Delgado-Rodriguez and Llorca 2004). They are briefly described below and are reinforced through examples.

Selection bias is an error that occurs in the method of participant selection. It is introduced when the study population does not represent the target population and may emerge due to poor definition of the eligible population and sampling frame (Delgado-Rodriguez and Llorca 2004). For instance, subjects who attend a remotely situated antenatal care clinic may not be representative of all other women with an outcome of interest, which may affect the generalisability of study results obtained from this sample.

Information bias occurs during data collection and may lead to misclassification. Recall bias, more common in case–control studies, is an information bias which occurs if cases recall past exposure better than controls. Non-respondent bias occurs when participants of the study differ from those who refuse to participate.

Self-selection bias is a case when there are differences between people who volunteer to participate and who do not. Ecological fallacy is a bias which can occur when the analysis is done at the group level but inferences are made at the individual level (Delgado-Rodriguez and Llorca 2004; Barnett and Hyman 2006).

Intervention bias might occur if some cases are highly compliant and motivated to follow the intervention procedures and other cases are less engaged and show low motivation to complete an intervention. These extremities may lead to over- or underestimation of potential benefit of interventions, respectively (Clancy 2002).

Confounding and effect modification are also common biases. These issues are not discussed in this chapter as they are elaborated elsewhere (Rothman et al. 2008; Barnett and Hyman, 2006). Other possible biases, which are not discussed in this chapter, include: disease spectrum bias, referral bias, participation bias, image-based selection bias, verification bias, clustering bias, and context bias (Sica 2006).

Bias in observational studies can be prevented through a good and thorough planning, effective sampling strategy and choice of objective outcome indicators (e.g., standardised instruments validated in previous studies). Also such strategies like matching, stratified analysis or use of two or more control groups can be useful to overcome a sampling bias (Mann 2003; Jepsen et al. 2004). Information biases can be avoided, when the expected outcome is objectively assessed by researchers without knowledge of the real exposure status of a participant (Adamson 2004). Unfortunately, if a bias is discovered after the study is completed, there are no ways of improving or correcting obtained results.

A hypothetical example based on a case–control study is given below to explain all these biases including confounding and effect modification. Using a case–control study, investigators assessed the relationship between drinking coffee at dinner and car accidents at night. They recorded that coffee drinkers had two times more accidents than those who did not drink coffee. The association between coffee drinking and car accidents was statistically significant at 5% level of significance (i.e., $p < 0.05$). On the basis of this finding, investigators interpreted that drinking coffee could be the cause of increased car accidents. This interpretation may be correct. Some other interpretations can also be made on the basis of this finding. This result could appear only due to chance, perhaps there is no real association between them. Perhaps people who drank coffee were more likely to be tired (i.e., fatigue) and hence fatigue is a risk factor for significantly higher number of accidents (confounding). Perhaps a higher percentage of the coffee drinkers were male compared with non-coffee drinkers. Normally car accidents are higher among males than females (confounding, selection bias). Caffeine might have a higher effect on people when they drive if they also smoke (effect modification). Some people might not have correctly remembered whether they drank coffee that night (information bias). The memory of those who made accidents was worse because of high stress (recall bias). Some participants might give that information according to desire of investigators (information bias). Coffee drinkers might have been less likely to participate in the study if they had an accident (non-response bias). Some errors can occur during data management and recording (misclassification bias). This example clearly demonstrates problems inherent in the study design that could seriously distort results. It also points to the need for care in the design, conduct and analysis of observational studies (Barnett and Hyman 2006).

4.3.6 Other Challenges Due to High Mobility of Vulnerable Population, Poor Resources and Settlement Changes

According to different sources, slum settlements are increasing in numbers in urban areas of developing countries (Khan et al. 2009; UN-Habitat 2003). In the previous chapter, Krämer et al. mentioned that rural–urban migration is one of the driving forces of rapid urbanisation. According to their report, migrant people initially settle in slums because of cheap accommodation and no need of special residence permission. Data based on our recent public health cohort study in Dhaka indicates that within a short period of time, these migrants are compelled to change their place of residence due to various adverse factors, among which the most common are: insecure housing due to land authority and eviction, lack of basic amenities and health services, low level of social cohesion, pollutions, high seasonal migrations and natural disasters. In our 1 year cohort study, 662 families from three different slums and 600 families from three rural villages were recruited at the baseline survey. All of them gave consent that they will stay at the same residence until the next 1 year. However, 160 slum families changed their place of residence within year, of which 64 families changed within the first 3 months preceding the baseline survey. In contrast, no rural family left their areas within 1 year period. This data clearly indicates remarkable differences between urban slums and rural villages. The rate of changing place of residence is higher among slum dwellers as compared to people living in rural areas. Some potential challenges should be discussed based on this particular phenomenon. First, slum results may suffer from potential biases and uncertainties due to higher rate of lost-to-follow-up families than rural results especially when lost-to-follow-up families bear some significantly distinct characteristics than available families. Second, cohort study for a longer period of time might not be suitable in the urban slums in context of high mobility rate. Third, high mobility of people within urban areas makes it difficult for researchers to keep records prospectively. Because of these potential limitations, conducting a cohort study is somehow challenging in the cities particularly in the slum settlements as compared to rural areas.

Further factors which may influence the success of urban health research count lack of resources, high level of advancement of urban areas and high level of conflicts over limited resources. For instance, in a situation when the urban health researchers are restricted by limited resources, it is difficult to apply several methods (called triangulation, see above) in the same study which provides better results about the focused problem. Cities and megacities particularly in developing countries which contain many slums and informal settlements are experiencing rapid changes in terms of infrastructure and development. Very often we observe that slums of the inner city areas (e.g., in Dhaka) are replaced by the improved settlements or high rise modern buildings. Generalisation of public health results from one city to another city is another challenge because cities generally differ by multiple factors such as geographical location, population density, ethnicity, environment, governance and infrastructure, and pace of urbanisation (Galea and Vlahov 2005). Even within the same city, results are different by different sub-groups and geographical locations. For instance, slum people suffer more from communicable diseases whereas affluent people suffer more from non-communicable diseases in Dhaka (Khan et al. 2009).

4.4 Outlook

One of the major priorities for urban health study is to reduce the burden of disease among the population living in cities and highly urbanised areas and to reduce the health disparities. In this section we try to outline the major areas of urban health studies which prevail in modern and will most likely dominate future research.

Areas of modern urban health research investigate increasing urban-rural disparities in developing countries, in respect to increasing urbanisation and migration and the rise of non-communicable diseases, like diabetes, obesity, cardiovascular diseases, cancer. Disease spectrum of urban research also includes injuries, violence and conflict, traffic accidents (e.g., in countries of South-East Asia and countries of the former Soviet Union), mental ill-health in developed and developing countries and other "lifestyle" preventable diseases.

Climate change and human health constitute a huge area of urban health research. Researchers investigate changes in disease patterns in the context of urbanisation, globalisation and climate change consequences (e.g., devastating Earthquake on Haiti in January 2010 caused numerous disease outbreaks among survived population). An important tool in urban health study is mapping. For example, mapping deaths attributed to flood may be useful for predicting future populations at risk in coastal areas (Kovats et al. 2003) or mapping data on vector-borne disease distribution may help to predict the patterns of disease distribution in relation to climate and temperature variations.

Apart from the methods and tools described in this chapter, such new methods as multilevel analysis are obtaining more and more attention in urban health research, as it provides an opportunity to examine how features of urban environment and living affect health and how these influences differ between various urban units, like families and communities. Finally, one of the dimensions in the modern research on urban health is a gender perspective, which becomes increasingly important in developing countries. Although researchers often operate in terms of "communities", "families" or "households", heterogeneity of these groups and diversity of gender relations is gaining weight in urban health studies (Fadda and Jirón 1999). Independent from the area of research, urban health studies should provide an evidence base for policy and action, base for strategies of poverty reduction and elimination of extreme intra-urban health inequities (Harpham and Molyneux 2001).

4.5 Concluding Remarks

Urban health research is a research of urban diversity, a research of multiple factors which shape each city and health of its inhabitants. Combination of different disciplines that apply both quantitative and qualitative methods and that use proper sampling strategies provide better answers to questions about both how and why urban characteristics affect health. Just as in any other research area, studying urban health requires application of study designs relevant to the objectives of this research. Observational studies are still the most common and suitable to study urban health research questions, and such challenges as confounding, bias and chance, as well as interpretation of results should be taken into consideration. In any case, a transparency in methodological approaches is required from the investigators before drawing strong conclusions. New and improved methods for collecting precise and accurate data on the health of urban populations are needed. Besides, urban health research has to work for its main objective, which is improving health of urban populations, establishing an effective dialogue and involvement of urban communities, communicating effectively research results to policy-makers and end-users.

References

Adamson J (2004) Evaluation of studies of causation (aetiology). Evidence-Based Nursing, 7: 36–40

Barnett ML, Hyman JJ (2006) Challenges in interpreting study results: the conflict between appearance and reality. JADA, 137(suppl 10):32 S-36 S.

Bartlett JE, Kotrlik JW, Higgins CC (2001) Organizational Research: Determining Appropriate Sample Size in Survey Research. Informat Technol, Learning, Performance J, 19:43–50

Bayoumi AM, Hwang SW (2002) Methodological, practical, and ethical challenges to inner-city health research. J Urban Health, 79(4):S35–S42

Begley CM (1996) Using triangulation in nursing research. J Adv Nurs, 24:122–128

Brackertz N (2007) Community consultation and the hard-to-reach. Victoria

Clancy MJ (2002) Overview of research designs. Emerg Med J, 19:546–549

Cohen B (2004) Urban growth in developing countries: a review of current trends and a caution regarding existing forecasts. World Development, Vol. 32, No. 1:23–51

Curry LA, Nembhard IM, Bradley EH (2009) Qualitative and mixed methods provide unique contributions to outcomes research. Circulation, 119:1442–1452

De Leeuw E (2009) Mixing urban health research methods for best fit. Journal of Urban Health: Bulletin of the New York Academy of Medicine

Delgado-Rodriguez M, Llorca J (2004) Bias. J Epidemiol Community Health, 58:635–641

Fadda G, Jirón P (1999) Quality of life and gender: a methodology for urban research. Environment and Urbanization, Vol. 11, No. 2, 261–270

Fleischman AR, Barondess JA (2004) Urban health: a look out our windows. Acad Med, 79(12): 1130–1132

Foss C, Ellefsen B (2002) The value of combining qualitative and quantitative approaches in nursing research by means of method triangulation. J Adv Nurs, 40(2):242–248

Galea S, Vlahov D eds. (2005) Handbook of urban health: populations, methods, and practice. Springer, New York, USA

Goebel A, Hill T, Fincham R, Lawhon M (2010) Transdisciplinarity in urban South Africa Futures, 42(5):475–483

Gordis L (2009) Epidemiology, 4th edn. Saunders Elsevier, USA

Grimes DA, Schulz KF (2002a) An overview of clinical research: the lay of the land. Lancet, Vol. 359:57–61

Grimes DA, Schulz KF (2002b) Bias and causal associations in observational research. Lancet, 359:248–52

Harpham T, Molyneux C (2001) Urban health in developing countries: a review. Prog Dev Stud, 1(2):113–137

Hill AB (1965) The environment and disease association or causation. Proc R Soc Med, 58: 295–300

Hwang SW. Homelessness and health. CMAJ 2001; 164:229–233

Jepsen P, Johnsen SP, Gillman MW, Sørensen HT (2004) Interpretation of observational studies. Heart, 90:956–960

Jones A, Bugge C (2006) Improving understanding and rigour through triangulation: an exemplar based on patient participation in interaction. J Adv Nurs, 55(5):612–621

Jones SR, Carley S, Harrison M (2003) An introduction to power and sample size calculation. Emerg Med J, 20:453–458

Kelsey JL, Whittemore AS, Evans AS and Thompson WD (1996) Methods in observational epidemiology, 2nd edn., Oxford University Press, USA

Khan MMH, Kraemer A, Gruebner O (2009) Comparison of health-related outcomes between urban slums, urban affluent and rural areas in and around Dhaka megacity, Bangladesh. Die Erde, 140:69–87.

Koehlmoos TP, Uddin MJ, Ashraf A, Rashid M. Homeless in Dhaka: violence, sexual harassment, and drug-abuse (2009) J Health Popul Nutr.2009, 27:452–461

Kovats S, Ebi K, Menne B (2003) Methods of assessing human health vulnerability and public health adaptation to climate change. World Health Organization, Copenhagen, Denmark

Last JM, ed. (2001) A dictionary of epidemiology, 4th edn. Oxford University Press, New York, USA

Lemon SC, Zapka JG, Estabrook B, Benjamin E (2006) Challenges to research in urban community health centers. Am J Public Health, 96:626–628

Leon DA (2008) Cities, urbanization and health. Int J Epidemiol, 37:4–8

Mann CJ. Observational research methods. Research design II: cohort, cross sectional, and case-control studies (2003). Emerg Med J, 20:54–60

Nomura Y, Poudel KC, Jimba M Hard-to-reach populations in Japan (2007). Southeast Asian J Trop Med Public Health, 38(2):325–327

Naing L, Winn T, Rusli BN (2006) Practical issues in calculating the sample size for prevalence studies. Arch Orofac Sci, 1:9–14

Patz JA, Balbus JM (1996) Methods for assessing public health vulnerability to global climate change. Clim Res, 6:113–125

Patel RB, Burke TF. Urbanization- an emerging humanitarian disaster (2009) N Engl L Med, 361:741–743

Ramadier T. Transdisciplinarity and its challenges: the case of urban studies. Futures 2004, 36: 423–439

Rothman KJ, Greenland S, Lash TL (2008) Modern epidemiology, 3rd edn. Lippincott Williams & Wilkins, Philadelphia, USA

Sackett DL (1979) Bias in analytic research. J Chron Dir, 32:51–63

Sackett DL, Haynes RB, Guyatt GH, Tugwelll P (1991) Clinical epidemiology: a basic science for clinical medicine, 2nd edn. Boston: Little, Brown and Company

Sica GT (2006) Bias in research studies. Radiology, 238:780–789

Spreen M (1992) Rare populations, hidden populations, and link-tracing designs: what and why? Bulletin de Methodologie Sociologique, 36:34–58

UN-Habitat (2003) The challenge of slums. Global report on human settlements 2003. UN-Habitat, Kenya

WHO. Preamble to the constitution of the World Health Organization as adopted by the International Health Conference, New York, 19 June - 22 July 1946; signed on 22 July 1946 by the representatives of 61 States (Official Records of the World Health Organization, no. 2, p. 100) and entered into force on 7 April 1948

Part II
Cases Studies and Examples

Chapter 5
Intervention Programme for Promoting Physical Activities in the Citizens of Sapporo City, Japan

Mitsuru Mori, Asae Oura, Erhua Shang, Fumio Sakauchi, Hirofumi Ohnishi, Aklimunnesa Khan, Md. Mobarak Hossain Khan, and Alexander Krämer

5.1 Introduction

In accordance with nutritional improvement, the environmental sanitation, and advance in medical technology after World War II, Japan has achieved almost the longest lifespan in the world. This change has two implications. Firstly, this has led to a drastic increase in elderly population in Japan. Secondly, a part of Japanese has faced the problem of over-nutrition due to e.g. an inappropriately increased intake of total energy and total fat, a more sedentary lifestyle or insufficient physical activity in association with use of various mechanic devices or transportation including a car in their life. Their lifestyles are not only associated with an increased risk of lifestyle-related morbidity, but also with increased medical expenditure. Some studies reported higher prevalences of obesity, glucose intolerance (Kawamori 2002), hypertension (Ueshima et al. 2000), and/or hyperlipidemia (Koba and Sasaki 2006) in recent years as compared to past. These lifestyles factors and morbidity are suggested as possible risk factors for a higher mortality related to cardiovascular diseases and certain types of cancer.

To reduce lifestyle related morbidity for elderly population, a number of community or clinical trials focusing on the usefulness of exercise programmes have been reported worldwide (Anderssen et al. 2007; Blumenthal et al. 2000; Copper et al. 2000; Corpeleijn et al. 2006; Elmer et al. 2006; Green et al. 2002; Higashi et al. 1999; Hinderliter et al. 2002; Irwin et al. 2003; Jakicic et al. 2003; Jancey et al. 2008a; Knowler et al. 2002; Kraus et al. 2002; Lindström et al. 2003;

M. Mori (✉) • A. Oura • E. Shang • F. Sakauchi • H. Ohnishi • A. Khan
Department of Public Health, Sapporo Medical University School of Medicine, Sapporo, Japan
e-mail: mitsurum@sapmed.ac.jp

M.M.H. Khan • A. Krämer
Department of Public Health Medicine, School of Public Health, Bielefeld University, Bielefeld, Germany

A. Krämer et al. (eds.), *Health in Megacities and Urban Areas*,
Contributions to Statistics, DOI 10.1007/978-3-7908-2733-0_5,

Mattila et al. 2003; Miller et al. 2002; Ohkubo et al. 2001; Orchard et al. 2005; Poston et al. 2001; Ross et al. 2000; Simons-Morton et al. 2001; Slentz et al. 2004; Steptoe et al. 2001; Stevens et al. 2001) including Japan (Higashi et al. 1999; Ohkubo et al. 2001). A part of these studies have intervened not only in exercise, but also in dietary habits (Anderssen et al. 2007; Copper et al. 2000; Corpeleijn et al. 2006; Hinderliter et al. 2002; Knowler et al. 2002; Lindström et al. 2003; Mattila et al. 2003; Miller et al. 2002; Ross et al. 2000; Steptoe et al. 2001; Stevens et al. 2001).

Despite the documented benefits of physical activity, it is still difficult to motivate older adults to start and maintain regular physical activity (Jancey et al. 2008b). Therefore, the Japanese Ministry of Health, Labor and Welfare encouraged local governments to develop sustainable programmes to increase the physical activity among the elderly in collaboration with public health specialists. In response, The Sapporo City Bureau and Department of Public Health of Sapporo Medical University jointly performed an intervention study to assess whether home-based or gym-based increased physical activities reduce the risk of lifestyle-related morbidity (Oura et al. 2008; Sakauchi et al. 2008). Here we reported the effect of these exerise-based interventions among elderly population living in Sapporo city. Sapporo is the capital city of Hokkaido Prefecture located in the northernmost island of Japan. This city, with a population of over 1.8 million, is the sixth largest city in Japan following Tokyo, Osaka, Nagoya, Yokohama, and Kyoto.

5.2 Subjects and Methods

We performed a series of intervention programmes in 2003 and 2004. First we selected our study subjects from those citizens who were participants in health check-ups financially supported by the Sapporo City Bureau. Detailed information about subjects and methods of the study have been published elsewhere in Japanese text (Oura et al. 2008; Sakauchi et al. 2008). However, we briefly explained them by year in the following section.

5.2.1 Study Subjects and Intervention Programme in 2003

In 2003, a total of 4,930 subjects were randomly selected from about 130,000 participants in health check-ups financially supported by the Sapporo City Bureau. Inclusion criteria in 2003 were set up using the data at the health check-up as having just 2 of the following 4 criteria satisfied; (1) 24.2 $\leq$ body mass index (BMI) $<$ 35.0, (2) 130 mmHg $\leq$ systolic blood pressure (SBP $<$ 180 mmHg, (3) 120 mg/dL $\leq$LDL cholesterol (LDL-cho)$<$220 mg/dL, (4) 110 mg/dL $\leq$ fasting blood glucose (FBG) $<$ 140 mg/dL or 140 mg/dL $\leq$ postprandial blood glucose (PBG) $<$ 200 mg/dL and HbA1C $>$ 5.8%.

A letter asking for participation in the intervention programme was sent to them, and 361 subjects agreed to participate in the programme via his or her written informed consent. As a control group, 585 persons were selected from the participants of a health check-up in 2002 with the same inclusion criteria as the intervention group. Frequency matching for sex and 5-year age strata with the intervention group was used at the time of selecting the control group.

For the intervention group, a 6-month programme was started in 2003 with self-assertions regarding the type, duration, and frequency of home-based exercise (Type 1 Intervention) such as walking, jogging, and light gymnastics. These subjects attended several educational seminars concerning health. They also reported their physical activity status on a monthly basis. They were periodically encouraged to maintain their own exercise via letter or fax from well-trained staff of the intervention programme.

Among the 361 participants, 296 persons actually started their own exercise and 260 persons completed the programme activities. After completion, they were monitored from 2004 to 2006 with regard to their health check-up results such as body weight (BW), BMI, SBP, diastolic blood pressure (DBP), total-cholesterol (T-cho), LDL cholesterol (LDL-cho), HDL cholesterol (HDL-cho), triglyceride (TG), FBP, PBP, and HbA1c. Average ages ($\pm$standard deviation, SD) of the intervention and the control groups were 67.7 $\pm$ 6.5 and 67.6 $\pm$ 6.4, respectively, and male to female ratios in both groups were 0.51 and 0.51, respectively.

5.2.2 *Study Subjects and Intervention Programme in 2004*

In 2004, 21,990 study subjects were randomly selected, under the following inclusion and exclusion criteria, from about 110,000 participants in a health check-up financially supported by the Sapporo City Bureau. Inclusion criteria in 2004 were set up using the data of the health check up as having 2 of the following 4 criteria satisfied inevitably including the first one; (1) 24.2 $\leq$ BMI $<$ 35.0, (2) 130 mmHg $\leq$ SBP $<$ 180 mmHg, (3) 120 mg/dL $\leq$ LDL-cho $<$ 220 mg/dL, (4) 110 mg/dL $\leq$ FBG $<$ 140 mg/dL, or 140 mg/dL $\leq$ PBG $<$ 200 mg/dL and HbA1C $>$ 5.5%. In addition, the following exclusion criteria were set up: (5) DBP $\geq$ 110 mmHg, (6)TG $\geq$ 400 mg/dL.

A letter asking for participation in the intervention programme was sent to them, and 547 people agreed to participate in the programme via his or her written informed consent. As a control group, 1,142 persons were selected from the participants of the health check-up in 2003 with the same criteria as the intervention group. Like previous year, frequency matching for sex and 5-year age strata with the intervention group was used at the time of selecting the control group in 2004.

Three types of intervention programmes were developed in 2004. Namely, Type 1 Intervention was the same home-based training exercise as in 2003, Type 2 Intervention was gym-based training exercise with a frequency of once a week, and Type 3 Intervention was gym-based training exercise with a frequency of twice a week. These subjects attended several educational seminars concerning health and monthly reported their physical activity status. They were periodically encouraged to maintain their own exercise via letter or fax from well-trained staff.

Among the 547 subjects agreed to participate, 114, 268, and 165 subjects belonged to the Type 1, Type 2, and Type 3 Interventions, respectively. However, a total of 495 subjects completed the programme activities, of which 108, 240, and 147 subjects belonged to the Type 1, Type 2, and Type 3 Interventions, respectively. After completion, they were monitored from 2005 to 2007 with regard to their health check-up results such as BW, BMI, SBP, DBP, T-cho, LDL-cho, HDL-cho, TG, FBP, and HbA1c. Average ages (±SD) of total intervention groups and the control groups were 67.4 ± 6.8, and 67.3 ± 7.0, respectively, and the male/female ratios were 0.80 and 0.83, respectively.

5.2.3 *Statistical Analysis*

We compared the data of the primary outcomes such as BMI, SBP, and FBG between the intervention group and the control group. A linear mixed model was used to examine interaction between the groups and the years at measurement of the primary outcomes. This model was applied to the data of both the intervention programmes in 2003 and 2004. SPSS statistical software was used for analysis. Statistical significance was denoted at $P < 0.05$.

5.3 Results

5.3.1 *Results of the Intervention Programme in 2003*

The analysis of the linear mixed model revealed that FBG was significantly lower among the subjects of Type 1 Intervention group as compared to the subjects of control group (Table 5.1). The differences are schematically shown in Fig. 5.1. However, variables such as body weight, BMI, SBP, DBP, T-cho, LDL-cho, HDL-cho, TG, PBG, and HbA1c were not significantly different in the analysis of the mixed linear model.

Table 5.1 Results of Type 1 Intervention Programme in 2003: comparison between the intervention group and the control group by analysis of the linear mixed model

			Year of 2003		Year of 2004		Year of 2005		
Items	Unit	Group	Number	Mean ± SD	Number	Mean ± SD	Number	Mean ± SD	*P* value
Body weight (BW)	kg	Intervention	293	59.0 ± 9.4	220	58.5 ± 9.4	205	58.0 ± 9.2	0.32
		Control	428	57.7 ± 9.5	403	57.2 ± 9.1	370	57.1 ± 9.0	
Body mass index (BMI)	kg/m^2	Intervention	293	24.2 ± 3.0	220	24.0 ± 2.9	205	23.9 ± 2.9	0.68
		Control	428	24.0 ± 3.0	403	23.9 ± 2.9	370	23.9 ± 2.9	
Systolic blood pressure (SBP)	mmHg	Intervention	293	133.7 ± 16.1	220	134.1 ± 15.1	205	133.7 ± 13.5	0.46
		Control	428	136.7 ± 14.8	403	136.3 ± 14.5	371	134.9 ± 14.1	
Diastolic blood pressure (DBP)	mmHg	Intervention	293	79.3 ± 9.5	220	78.5 ± 8.8	205	77.8 ± 8.5	0.96
		Control	428	79.6 ± 8.8	403	78.8 ± 9.1	371	78.1 ± 9.5	
Total cholesterol (TG)	mg/dL	Intervention	293	220.9 ± 31.1	220	215.3 ± 30.0	205	214.6 ± 30.0	0.66
		Control	428	217.2 ± 34.0	403	214.7 ± 34.9	371	213.6 ± 34.0	
LDL cholesterol (LDL-cho)	mg/dL	Intervention	235	138.9 ± 27.4	146	132.5 ± 27.2	138	132.3 ± 27.2	0.83
		Control	288	136.3 ± 29.8	250	132.6 ± 30.3	217	132.6 ± 28.7	
HDL cholesterol (HDL-cho)	mg/dL	Intervention	293	59.5 ± 13.5	220	59.2 ± 13.2	205	59.8 ± 14.5	0.59
		Control	428	59.5 ± 15.6	403	58.7 ± 15.5	371	59.0 ± 15.0	
Triglyceride (TG)	mg/dL	Intervention	235	116.0 ± 62.3	146	120.3 ± 57.2	138	114.9 ± 64.0	0.87
		Control	288	110.2 ± 52.4	250	112.5 ± 54.8	217	112.2 ± 62.9	
Fasting blood glucose (FBG)	mg/dL	Intervention	235	100.2 ± 19.5	146	96.8 ± 16.7	138	95.1 ± 17.1	0.001
		Control	288	96.7 ± 15.4	250	100.3 ± 20.9	217	96.1 ± 13.6	
Post-prandial blood glucose (PBG)	mg/dL	Intervention	58	104.2 ± 24.1	74	104.0 ± 34.9	67	102.3 ± 18.2	0.78
		Control	141	107.1 ± 33.8	153	107.2 ± 24.1	154	108.3 ± 31.9	
HbA_1c	%	Intervention	279	5.2 ± 0.60	204	5.2 ± 0.62	197	5.3 ± 0.60	0.19
		Control	404	5.2 ± 0.59	381	5.2 ± 0.63	356	5.3 ± 0.59	

Adapted from the article by Sakauchi et al. (2008)
SD standard deviation

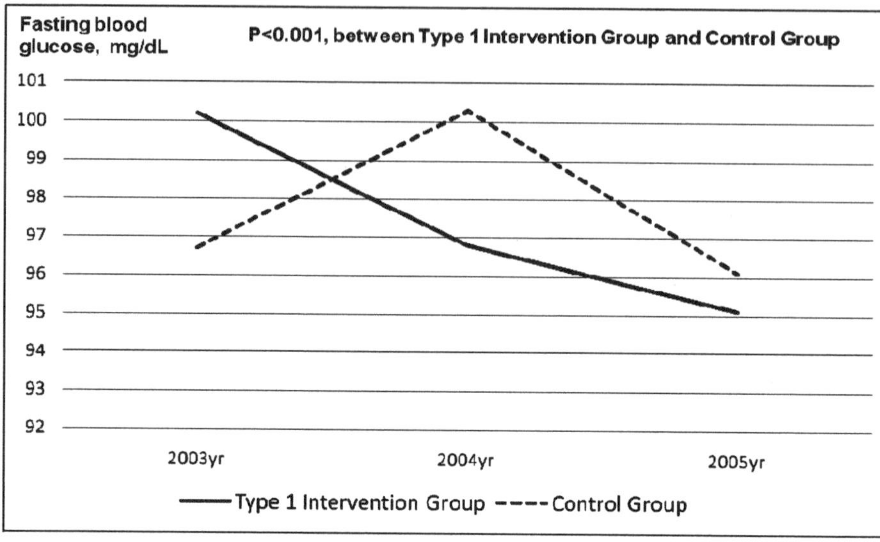

Fig. 5.1 Comparison of fasting blood glucose between the intervention group and the control group: Intervention Programme in 2003 (Data shown in Table 5.1)

5.3.2 Results of the Intervention Programme in 2004

The analysis of the linear mixed model (Table 5.2) revealed that BMI among the subjects of Type 1 and Type 3 intervention groups was significantly lower as compared to the subjects of control group. The differences are schematically shown in Fig. 5.2. Furthermore, FBG among the subjects of Type 3 intervention group was significantly lower as compared to the control group. The differences are schematically shown in Fig. 5.3. However, variables such as BW, BMI, SBP, DBP, T-cho, LDL-cho, HDL-cho, TG, and HbA1c were not significantly different among the groups in the analysis of the mixed linear model.

5.4 Discussion

According to our 6-month intervention study (either home-based or gym-based exercise, averages of BMI or FBP were significantly reduced after 2 years from starting point of interventions. These results were consistent with the results of several clinical trials in other countries. For example, Corpeleijn et al. (2006) reported that BMI and FBG were significantly reduced after a 12-month intervention composed of at least 30-min moderate physical activity per day for at least 5 days a week. Slentz et al. (2004) showed that there was a significant dose–response relationship between the amount of exercise and amount of weight

Table 5.2 Results of intervention programme in 2004: comparison between the intervention group and the control group by analysis of the linear mixed model

Items	Unit	Year	Type 1 intervention group	*P* value (comparison with the control group)	Type 2 intervention group	*P* value (comparison with the control group)	Type 3 intervention group	*P* value (comparison with the control group)	Control group
Body mass index (BMI)	kg/m^2		$n = 72$[a]		$n = 148$		$n = 95$		$n = 757$
		2004	26.2 ± 2.0[b]	< 0.01	26.0 ± 1.9	0.04	26.2 ± 2.1	< 0.01	26.5 ± 1.9
		2005	25.7 ± 2.1		25.8 ± 2.1		25.7 ± 2.0		26.4 ± 2.1
		2006	25.8 ± 2.3		25.7 ± 2.1		25.9 ± 2.0		26.4 ± 2.2
Systolic blood pressure (SBP)	mmHg		$n = 72$		$n = 148$		$n = 95$		$n = 756$
		2004	135.6 ± 15.2	0.71	134.6 ± 15.9	0.50	134.4 ± 16.9	0.30	136.5 ± 14.4
		2005	134.4 ± 14.0		132.7 ± 13.9		131.1 ± 14.7		135.6 ± 15.2
		2006	135.6 ± 13.7		133.8 ± 13.3		132.8 ± 15.7		135.0 ± 14.3
Diastolic blood pressure (DBP)	mmHg		$n = 72$		$n = 147$		$n = 95$		$n = 756$
		2004	80.3 ± 11.2	0.89	78.0 ± 10.4	0.12	78.7 ± 10.1	0.79	80.1 ± 9.5
		2005	79.4 ± 8.6		77.0 ± 9.7		77.4 ± 9.1		79.2 ± 9.7
		2006	78.5 ± 10.0		78.1 ± 9.4		77.5 ± 7.9		78.6 ± 9.4
Total cholesterol (T-cho)	mg/dL		$n = 72$		$n = 148$		$n = 95$		$n = 757$
		2004	213.5 ± 24.4	0.053	211.0 ± 33.3	0.83	215.1 ± 38.3	0.74	212.2 ± 30.6
		2005	204.2 ± 27.5		208.1 ± 30.7		212.4 ± 35.7		210.0 ± 30.0
		2006	207.7 ± 28.0		207.7 ± 28.0		211.1 ± 36.5		208.4 ± 29.4
LDL cholesterol (LDL-cho)	mg/dL		$n = 37$		$n = 81$		$n = 42$		$n = 349$
		2004	130.2 ± 33.4	0.16	129.7 ± 29.9	0.98	135.3 ± 35.6	0.67	133.3 ± 27.4
		2005	127.2 ± 23.5		128.2 ± 28.4		135.2 ± 34.6		130.4 ± 26.9
		2006	122.8 ± 24.2		128.4 ± 28.1		135.4 ± 32.1		128.9 ± 25.8
HDL cholesterol (HDL-cho)	mg/dL		$n = 72$		$n = 148$		$n = 95$		$n = 757$
		2004	56.6 ± 14.0	0.32	58.4 ± 17.0	0.20	57.2 ± 15.1	0.81	55.5 ± 13.0
		2005	55.6 ± 13.6		57.7 ± 12.1		57.6 ± 14.2		56.0 ± 13.3
		2006	55.5 ± 13.3		57.7 ± 12.4		57.2 ± 14.7		55.3 ± 12.5

(continued)

Table 5.2 (continued)

Items	Unit	Year	Type 1 intervention group	*P* value (comparison with the control group)	Type 2 intervention group	*P* value (comparison with the control group)	Type 3 intervention group	*P* value (comparison with the control group)	Control group
Triglyceride (TG)	mg/dL		$n = 36$		$n = 81$		$n = 42$		$n = 349$
		2004	137.6 ± 44.3	0.33	126.4 ± 71.8	0.76	133.9 ± 67.4	0.44	123.5 ± 57.2
		2005	129.2 ± 51.6		114.6 ± 58.4		117.9 ± 61.8		120.9 ± 65.7
		2006	137.8 ± 60.8		121.4 ± 63.2		116.1 ± 51.3		121.2 ± 56.1
Fasting blood glucose (FBG)	mg/dL		$n = 36$		$n = 81$		$n = 42$		$n = 349$
		2004	98.0 ± 14.7	0.55	99.0 ± 15.6	0.59	94.2 ± 11.7	0.04	98.3 ± 12.6
		2005	97.3 ± 15.9		96.5 ± 13.8		93.2 ± 7.8		99.6 ± 17.1
		2006	97.0 ± 12.7		97.7 ± 16.5		94.8 ± 11.0		101.1 ± 16.3
HbA_{1c}	%		$n = 62$		$n = 135$		$n = 88$		n = 717
		2004	5.2 ± 0.6	0.15	5.3 ± 0.6	0.99	5.2 ± 0.6	0.43	5.3 ± 0.6
		2005	5.3 ± 0.5		5.4 ± 0.7		5.3 ± 0.6		5.4 ± 0.8
		2006	5.2 ± 0.4		5.4 ± 0.7		5.3 ± 0.6		5.4 ± 0.7

[a]Number of study subjects in analysis
[b]Mean±standard deviation (SD)
#: Adapted from the article by Oura et al. (2008)

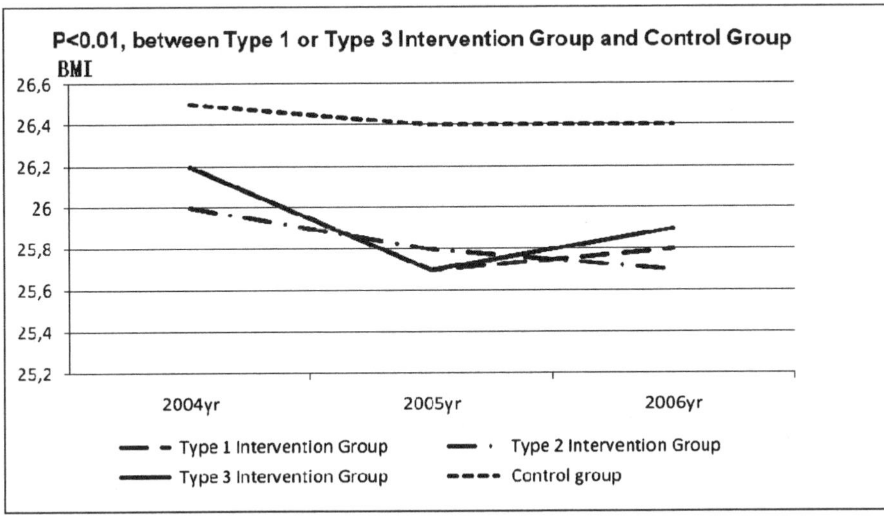

Fig. 5.2 Comparison of body mass index (BMI) between the intervention group and the control group: Intervention Programme in 2004 (Data shown in Table 5.2)

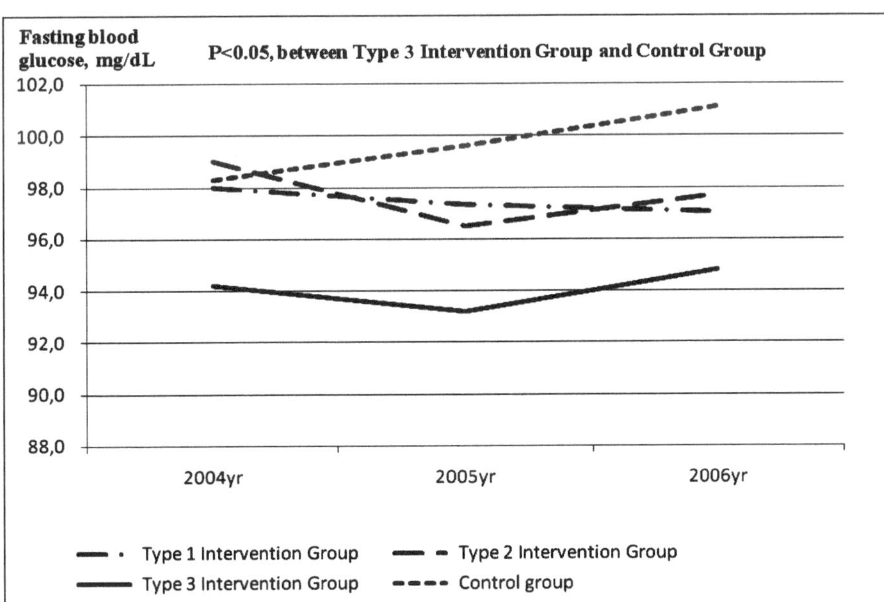

Fig. 5.3 Comparison of fasting blood glucose between the intervention group and the control group: Intervention Programme in 2004 (Data shown in Table 5.2)

loss and fat mass loss. Lindström et al. (2003) indicated that after a 3-year intervention of circuit-type moderate intensity resistance training, the intervention group showed significantly greater improvement in weight reduction and measure of glycemia. Jakicic et al. (2003) reported that after a 12-month intervention of exercise in addition to dietary intervention, significant weight loss was achieved. Irwin et al. (2003) suggested that after a 12-month intervention of moderate-intensity sports or recreational activity, significant reduction was observed in weight, total body fat, and subcutaneous abdominal fat. Miller et al. (2002) stated that after a 9-week intervention of supervised moderately intensive exercise 3 times per week, weight in the intervention group was significantly reduced.

The limitations of this study should also be mentioned. Selection bias might have occurred in both of the intervention programmes of 2003 and 2004 because the low participation rate from the target population, imperfect completion rate of the 6-months intervention programme of physical activities and imperfect completion of the blood chemical test in the 3 years of follow-up for the intervention and control groups.

5.5 Conclusion

The intervention programmes either home-based or gym-based exercises may be effective to decrease the BMI and/or blood glucose among people having sedentary lifestyle. However, careful generalisation is required as our findings were obtained from selected participants.

Acknowledgments This study was conducted in cooperation with staffs at the Sapporo City Bureau.

References

Anderssen SA, Carroll S, Urdal P, Holme I (2007) Combined diet and exercise intervention reverses the metabolic syndrome in middle-aged males: results from the Oslo Diet and Exercise Study. Scand J Med Sci Sports 17: 687–695

Blumenthal JA, Sherwood A, Gullette ECD, Babyak M, Waugh R, Georgiades A, Craighead LW, Tweedy D, Feinglos M, Appelbaum M, Hayano J, Hinderliter A (2000) Exercise and weight loss reduce blood pressure in men and women with mild hypertension. Effects on cardiovascular, metabolic, and hemodynamic functioning. Arch Intern Med 160: 1947–1958

Cooper AR, Moore LAR, McKenna J, Riddoch CJ (2000) What is the magnitude of blood pressure response to a programme of moderate intensity exercise? Randomized controlled trial among sedentary adults with unmedicated hypertension. Br J Gen Pract 50: 958–962

Corpeleijn E, Feskens EJM, Jansen EHJM, Mensink M, Saris WHM, de Bruin TWA, Blaak EE (2006) Improvements in glucose tolerance and insulin sensitivity after lifestyle intervention are related to changes in serum fatty acid profile and desaturase activities: the SLIM study. Diabetologia 49: 2392–2401

Elmer PJ, Obarzanek E, Vollmer WM, Simons-Morton D, Stevens VJ, Young DR, Lin P-H, Champagne C, Harsha DW, Svetkey LP, Ard J, Brantley PJ, Proschan MA, Erlinger TP, Appel LJ (2006) Effects of comprehensive lifestyle modification on diet, weight, physical fitness, and blood pressure control: 18-month results of a randomized trial. Ann Intern Med 2006; 144: 485–495

Green BB, McAfee T, Hindmarsh M, Madsen L, Caplow M, Buist D (2002) Effectiveness of telephone support in increasing physical activity levels in primary care patients. Am J Prev Med 22: 177–183

Higashi Y, Sasaki S, Sasaki N, Nakagawa K, Ueda T, Yoshimizu A, Kurisu S, Matsuura H, Kajiyama G, Oshima T (1999) Daily aerobic exercise improves reactive hyperemia in patients with essential hypertension. Hypertension 33: 591–597

Hinderliter A, Sherwood A, Gullette ECD, Babyak M, Waugh R, Georgiades A, Blumenthal JA (2002) Reduction of left ventricular hypertrophy after exercise and weight loss in overweight patients with mild hypertension. Arch Intern Med 162: 1333–1339

Irwin ML, Yasui Y, Ulrich CM, Bowen D, Rudolph RE, Schwartz RS, Yukawa M, Aiello E, Potter JD, McTiernan A (2003) Effect of exercise on total and intra-abdominal body fat in postmenopausal women: a randomized controlled trial. JAMA 289: 323–330

Jakicic JM, Marcus BH, Gallagher KI, Napolitano M, Lang W (2003) Effect of exercise duration and intensity on weight loss in overweight, sedentary women: a randomized trial. JAMA 290: 1323–1330

Jancey JM, Lee AH, Howat PA, Clarke A, Wang K, Shilton T (2008a) The effectiveness of a physical activity intervention for seniors. Am J Health Promot 22: 318–321

Jancey JM, Clarke A, Howat PA, Lee AH, Shilton T, Fisher J (2008b) A physical activity program to mobilize older people: a practical and sustainable approach. Gerontologist 48: 251–257

Kawamori R (2002) Diabetes trends in Japan. Diabet Metab Res Rev 2002; 18: S9-13

Knowler WC, Barrett-Connor E, Fowler SE, Hamman RF, Lachin JM, Walker EA, Nathan DM (2002) Reduction in the incidence of type 2 diabetes with lifestyle intervention or metformin. N Engl J Med 346: 393–403

Koba S, Sasaki J (2006) Treatment of hyperlipidemia from Japanese evidence. J Atheroscler Thromb 13: 267–280

Kraus WE, Houmard JA, Duscha BD, Knetzger KJ, Wharton MB, McCartney JS, Bales CW, Henes S, Samsa GP, Otvos JD, Kulkarni KR, Slentz CA (2002) Effects of the amount and intensity of exercise on plasma lipoproteins. N Engl J Med 347: 1483–1492

Lindström JL, Louheranta A, Mannelin M, Rastas M, Salminen V, Eriksson J, Uusitupa M, Tuomilehto J (2003) The Finnish diabetes prevention study: lifestyle intervention and 3-year results on diet and physical activity. Diabetes Care 26: 3230–3236

Mattila R, Malmivaara A, Kastarinen M, Kivelä S-L, Nissinen A (2003) Effectiveness of multidisciplinary lifestyle intervention for hypertension: a randomized controlled trial. J Hum Hypertens 17: 199–205

Miller ER 3rd, Erlinger TP, Young DR, Jehn M, Charleston J, Rhodes D, Wasan SK, Appel LJ (2002) Results of the diet, exercise, and weight loss intervention trial (DEW-IT). Hypertension 40: 612–618

Ohkubo T, Hozawa A, Nagatomi R, Fujita K, Sauvaget C, Watanabe Y, Anzai Y, Tamagawa A, Tsuji I, Imai Y, Ohmori H, Hisamichi S (2001) Effects of exercise training on home blood pressure values in older adults: a randomized controlled trial. J Hypertens 19: 1045–1052

Orchard TJ, Temprosa M, Goldberg R, Haffner S, Ratner R, Marcovina S, Fowler S (2005) The effect of metformin and intensive lifestyle intervention on the metabolic syndrome: the diabetes prevention program randomized trial. Ann Intern Med 142: 611–619

Oura A, Sakauchi F, Shang E, Mori M (2008) Effectiveness of a community-based health promotion by an exercise intervention program in middle aged and elderly people in Sapporo, Japan. Sapporo Med J 77: 23–23 (Japanese text)

Poston WSC, Haddock CK, Olvera NE, Suminski RR, Reeves RS, Dunn JK, Hanis CL, Foreyt JP (2001) Evaluation of a culturally appropriate intervention to increase physical activity. Am J Health Behav 25: 396–406

Ross R, Dagnone D, Jones PJH, Smith H, Paddags A, Hudson R, Janssen I (2000) Reduction in obesity and related comorbid conditions after diet-induced weight loss or exercise-induced weight loss in men: a randomized, controlled trial. Ann Intern Med 133: 92–103

Sakauchi F, Oura A, Shang E, Mori M (2008) Assessment of long effect in health promotion by an exercise intervention program: From results in Sapporo Health Promotion Program in 2003. J Hokkaido Public Health 22: 62–68 (Japanese text)

Simons-Morton DG, Blair SN, King AC, Morgan TM, Applegate WB, O'Toole M, Haskell WL, Albright CL, Cohen SJ, Ribisl PM, Shih JH (2001) Effects of physical activity counseling in primary care. The Activity Counseling Trial: a randomized controlled trial. JAMA 286: 677–687

Slentz CA, Duscha BD, Johnson JL, Ketchum K, Aiken LB, Samsa GP, Houmard JA, Bales CW, Kraus WE (2004) Effects of the amount of exercise on body weight, body composition, and measures of central obesity: STRRIDE-a randomized controlled study. Arch Intern Med 164: 31–39

Steptoe A, Kerry S, Rink E, Hilton S (2001) The impact of behavioral counseling on stage of change in fat intake, physical activity, and cigarette smoking in adults at increased risk of coronary heart disease. Am J Publ Health 91: 265–269

Stevens VJ, Obarzanek E, Cook NR, Lee I-M, Appel LJ, West DS, Milas NC, Mattfeldt-Beman M, Belden L, Bragg C, Millstone M, Raczynski J, Brewer A, Singh B, Cohen J (2001) Long-term weight loss and changes in blood pressure: Results of the trials of hypertension prevention, phase II. Ann Intern Med 134: 1–11

Ueshima H, Zhang X-H, Choudhury SR (2000) Epidemiology of hypertension in China and Japan. J Hum Hypertens 14; 765–769

Chapter 6
Measuring the Local Burden of Diarrhoeal Disease Among Slum Dwellers in the Megacity Chennai, South India

Patrick Sakdapolrak, Thomas Seyler, and Sanjeevi Prasad

6.1 Introduction

India is one of the focal points of the global megapolisation process. The country is facing urban poverty and the urban poor bear a large disease burden. In the South Indian metropolis of Chennai, one of India's seven megacities, an estimated 18.9% (Census of India 2001) to 40.9% (NFHS-3 in Gupta et al. 2009: 74) of the population lives in areas categorised as slums. Slums are characterised as areas with lack of access to basic services, substandard housing, overcrowding, insecure tenure, poverty as well as unhealthy living conditions (UN-Habitat 2003: 11). Consequently slum dwellers are not only more exposed to social and environmental health risks (e.g. lack of sanitation facilities), but also have less capacities to cope with them. The health status of slum dwellers is poor in comparison to other residents. The results of the third National Family Health Survey (NFHS-3, 2005-06) (Gupta et al. 2009) clearly indicates this intra-urban health inequality. The South Indian megacity Chennai is a case in point (NFHS-3, 2005-06): while the infant mortality rate for Chennai as a whole was 27.6, the rate in non-slum areas was 24.2 as compared to 38 in slum areas. A look at the disease-specific health burden shows that slum dwellers are suffering a higher burden of infectious diseases: tuberculosis, a widespread infectious disease in India, has a prevalence of 863 per 100,000 among male slum dwellers in Chennai. The prevalence in non-slum areas in contrast is 437 per 100,000. In addition, slum dwellers have, in certain areas, a higher burden of non-infectious diseases as well: the prevalence of diabetes among female slum dwellers was 3,901 per 100,000 in Chennai. It was slightly higher than the prevalence among non-slum female residents, which was 3,867 per 100,000.

P. Sakdapolrak (✉)
Department of Geography, Bonn University, Bonn, Germany
e-mail: Sakdapolrak@giub.uni-bonn.de

T. Seyler • S. Prasad
French Institute of Pondicherry, Pondicherry, India

A. Krämer et al. (eds.), *Health in Megacities and Urban Areas*,
Contributions to Statistics, DOI 10.1007/978-3-7908-2733-0_6,

What does this epidemiological profile and intra-urban inequality imply in terms of public health intervention? Effective health policy decision making requires a clear picture of disease burden in order to prioritise resource allocation. In the last decades several summary measures of population health (e.g. QUALY, DALY, HALE) have been developed to provide this information. Summary measures combine mortality and ill-health into a single index to measure overall population health (Murray et al. 2002). The functions of summary measures are manifold (Murray et al. 2000: 982): they allow comparing the health status of different populations and assessing the relative impact of different diseases of a given population. Furthermore, the changes in population health statuses can be monitored and inequalities can be identified and quantified. In addition, debates on priority settings for health service delivery and planning can be fuelled by such indicators. Moreover, they allow the analysis of benefits of health interventions using a common measure. To sum up, summary measures are tools that have the potential to guide policy makers in their decision to target diseases and allocate resources.

The Disability-Adjusted Life Years[1] (DALYs) is the most widespread summary measure (Malsch et al. 2006: 7). It was developed by Murray (1994) and adopted by large international organisations such as the World Bank and the World Health Organisation. The following study uses DALY as a method to assess disease burden. It seeks to illustrate how the burden of a particular disease can be empirically and locally measured and what difficulties can arise. The focus will be on the burden of diarrhoeal disease among slum dwellers in the megacity of Chennai.

Diarrheal disease remains a major public health issue in many developing countries – particularly India. According to UNICEF/WHO (2009: 5–7) 2.5 billion cases of diarrhoea occur each year among children under 5 years worldwide. A third of these cases occur in South Asia. Diarrhoea remains the leading cause of death among children. UNICEF/WHO (2009: 5–7) estimate that in 2004 20% of children's deaths – that are 1.5 million cases – were due to diarrhoea. Thirty eight percent of these deaths among children under 5 years occurred in South Asia: with 386,600 deaths, India was by far the country with the highest number. Diarrhoea is a common symptom of gastrointestinal infection, which can be caused by various pathogens (bacteria, viruses and protozoa) (Bern 2004; UNICEF/WHO 2009: 9). The leading cause of acute diarrhoea is rotavirus. The main bacterial pathogens are Shigella, Campylobacter, Salmonella and V. cholerae (ibid.). The main transmission route is faecal-oral transmission. It is estimated that 88% of deaths due to diarrhoea worldwide could be prevented through access to safe water, adequate sanitation and good hygiene practices (UNICEF/WHO 2009: 10–13). The following study aims at estimating the burden of diarrhoea disease among slum dwellers. In doing so the study wants to provide empirical evidence and input for the measurement of the global burden of disease.

[1]A critical discussion on the burden of disease approach and DALYs is provided by Pinheiro et al. in this volume.

Before presenting the results on the burden of disease among slum dwellers in Chennai, we will address different aspects of the study design, study population and the measurement methods.

6.2 Methods

The study was undertaken as a part of the research project called *Spatial Epidemiology* and *Health Vulnerability* of Slums Dwellers in the Megacity of Chennai.

6.2.1 Data Collection

6.2.1.1 Cohort Study in the Slums

We selected two slums in Chennai located along the river Cooum and the Buckingham canal. The estimated total population in the two slums in 2007 was 2,956. We randomly selected 219 households and included all household members in the cohort (1,041 individuals). After informed consent, each household representative was interviewed using a structured questionnaire to collect socio-demographic data on the household and its members. We followed the 219 households over time during a total of 15 weeks–17 weeks in May and June 2007 during the dry season and 8 weeks in October and November 2007 during the rainy season.

The two study areas are characterised by high population density, substandard housing and inadequate access to basic infrastructure (see Fig. 6.1). Three quarters of the households live in single room brick houses. The house rows are divided by narrow paths. A quarter of the households lives in thatched huts. The average size of the rooms, which are mostly without ventilation, is 10 m^2. The two study areas have rudimentary access to basic infrastructure. Water is supplied through public water points. One water point is shared by 50–75 households. Only a limited number of public toilets are available and open defecation is therefore common.

On average there are five members per household (see Table 6.1 and Fig. 6.2). The median age of the sample population was 23 years. The sex ratio of the sample was 1,031 females to 1,000 males. The sex ratio among the age group between 0 and 6 was 1,108 women to 1,000 men. The number of children under five was 113. Thirty-six percent of adults over 17 years never went to school and 35.4% have not completed primary school (8th grade). Forty-four percent of adult females never went to school against 27.2% of adult males. Sixty-one percent of the working age population is working. Most people are working in the informal sector as load carriers, construction workers and house maids. The median income is 1,500 INR (24 Euro) per month. The average per capita income in a household is 818 INR (13 Euro) per month.

Fig. 6.1 Slum in Chennai
Source: Sakdapolrak 2007

6.2.1.2 Syndromic Surveillance

The household representatives used a "health calendar"[2] to report health events among the household members. In particular, the following symptoms were

[2]The syndromic surveillance with a "health calendar" is based on a study on diarrhoea disease in Uzbekistan conducted by Herbst (2006) and Herbst et al. (2008).

Table 6.1 Basic characteristics of the sample population (n = 1,041)

Household size (persons)	4.8
Gender ratio[a] (0–6 years)	1,108
Number of children per household (total)	
Up to 4 years	0.5 (133)
Up to 14 years	1.6 (351)
Proportion of adults (>17) without school attendance (%)	
Total	36.7
Male	27.2
Female	44.4
Workforce participation[b](%)	
Total	61.1
Male	82.3
Female	39.7
Household income (INR) (per capita[c])	817
Poorest quartile	369
2. Quartile	644
3. Quartile	944
Richest quartile	1,400

[a]Number of females to 1,000 males

[b]Proportions of persons earning income among the total number of working age (15–64) population

[c]Age adjusted per capita income after Russell (2005: 1,398)

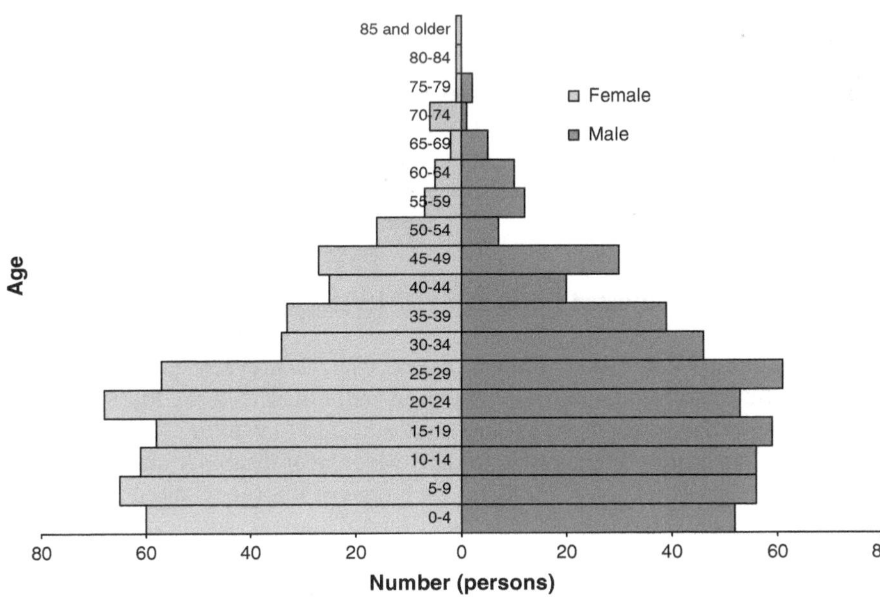

Fig. 6.2 Number of persons in the sample population by age and gender (n = 1,041)

reported on the calendar on a daily basis: diarrhoea, fever, joint pain, rash, headache, stomach-ache, cough and eye infection. For each household member, the date and duration of the symptom(s) were therefore recorded. Every week, we reviewed the health calendar with the household representative to ensure greater completeness and to collect additional information on morbidity and mortality.

In order to describe the disability associated with the reported symptom(s), we asked the household representative to describe the limitations caused by the symptom(s): (1) no limitation at all, (2) limitations in income earning activities, (3) limitations in social and recreational activities, (4) limitations in basic daily activities like preparing meals, house-keeping, (5) need assistance for eating and personal hygiene.

6.2.1.3 Case Definition

We defined a case of diarrhoeal disease as the occurrence of one or more loose stools in a 24 h period reported on the health calendar during a week of the surveillance period among a cohort member.

6.2.2 Data Analysis

6.2.2.1 Incidence Rate of Diarrhoeal Disease by Age Group, Gender and Season

We computed incidence rate by age group, gender and season by dividing the number of cases in each group/season by the total number of person-weeks followed in each group/season. Using incidence rates allows taking into account not only the number of people at risk but also the exact time of follow-up.

6.2.2.2 Disability-Adjusted Life Years (DALYs)

We estimated the burden of diarrhoeal disease using DALYs, which estimate the amount of time, ability or activity lost by an individual to disability (years lost to disability; YLD) or death (years lost to death; YLL) resulting from a disease. This loss is then adjusted to account for age, severity of disability and duration of disability. We estimated the DALYs for each case of diarrhoeal disease using the formula (Murray 1994: 441):

$$-\left[\frac{\mathrm{DC}e^{-\beta a}}{(\beta+\mathrm{r}^2)}\left[e^{-(\beta+\mathrm{r})(\mathrm{L})}(1+(\beta+\mathrm{r})(\mathrm{L}+\mathrm{a}))-(1+(\beta+\mathrm{r})\ \mathrm{a})\right]\right]$$

L represents the years lost to death or disability, and D is the disease-specific disability weight. C and β are positive constants, a is the age of the patient in years and r is the social discount rate. We assumed the disability weight for diarrhoeal disease to range from 0.086 to 0.119 – equal to the disability weight used in The Global Burden of Disease Report (Lopez et al. 2006: 119). For the disability weights, we adjusted for reported limitations by the household representative. To allow direct comparison with the DALYs from other diseases, we chose values for C, β and r equal to those used in the Global Burden of Disease Report (ibid.).

6.3 Results

6.3.1 General Morbidity and Mortality

Throughout the active surveillance of the health status of the 1,041 slum dwellers during 15 weeks, 2,600 cases of acute ill-health were reported. The incidence rate of any episode was 16.9 per 100 person-week. The most common reported symptoms were cough, headache, joint pain and fever. A quarter of the sample population did not report any ailment. 15.7% of the sample population reported a chronic illness.[3] The most common chronic conditions were migraine, cardio-vascular disease, chronic respiratory disease, diabetes and hypertension.

During the study year (2007) seven deaths were reported, including six men and one woman aged between 20 and 45 years. The causes of mortality were two cases of liver failure, two suicides, one myocardial infarction, one fatal accident and one death from unknown cause.

6.3.2 Incidence Rate of Diarrhoeal Disease

During the whole study period, we reported a total of 111 cases of diarrhoeal diseases in the two slums. Sixty-four diarrhoeal cases occurred during 7 weeks of the dry season, from May 1, 2007 to June 30, 2007 (Fig. 6.3). Forty-seven diarrhoeal cases occurred during 8 weeks of the monsoon season, from October 1, 2007 to November 30, 2007 (Fig. 6.3).

We followed 1,041 individuals during the dry season for a total of 7,231 person-weeks. The corresponding incidence rate of diarrhoeal disease per 100 person-weeks for the dry season was 0.899. The incidence rate among children aged

[3]Chronic illness is defined as a health problem that persists for more than 3 months preceding the time the survey was conducted.

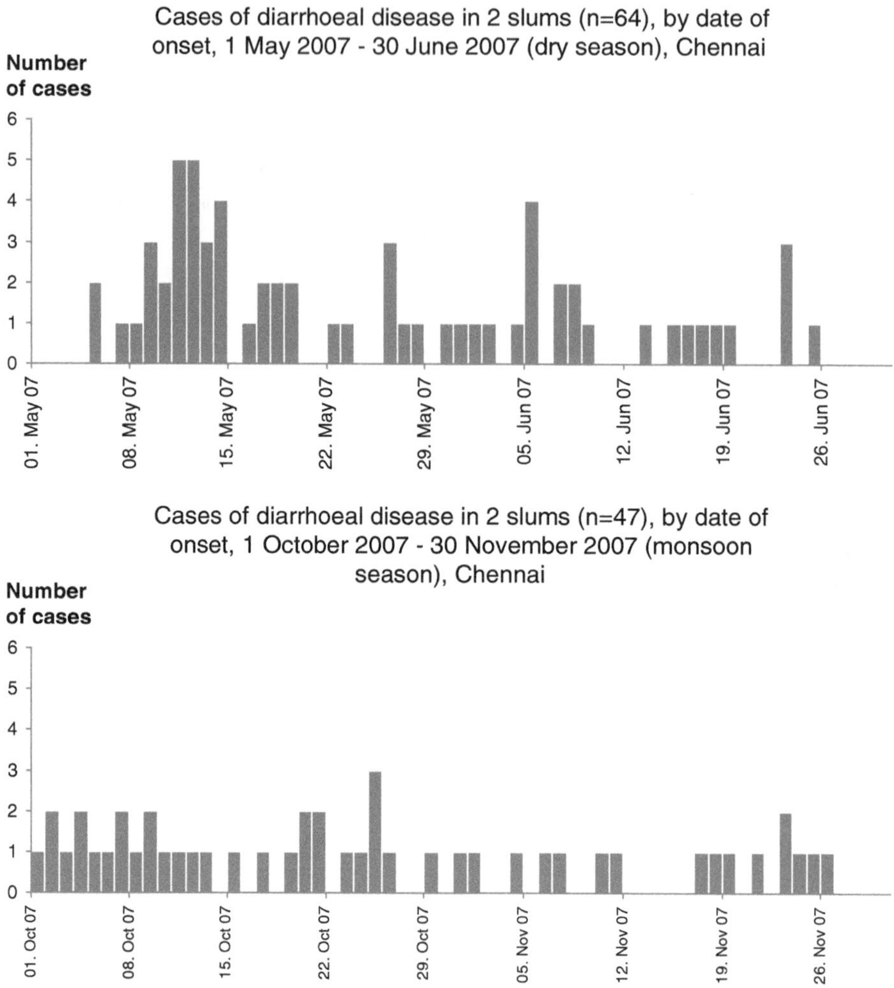

Fig. 6.3 Cases of diarrhoea disease in sample population during dry and monsoon season

0–4 years was 1.78 per 100 person-weeks. It was 0.784 and 0.773 per 100 person-weeks among individuals aged 5–14 and 15 years and above, respectively (Fig. 6.4a).

We followed 1,002 individuals during monsoon season for a total of 7,996 person-weeks. The corresponding incidence rate of diarrhoeal disease per 100 person-weeks for the monsoon season was 0.588. The incidence rate among children aged 0–4 years was 2.03 per 100 person-weeks. It was 0.546 and 0.360 per 100 person-weeks among individuals aged 5–14 and 15 years and above, respectively (Fig. 6.4b).

Overall, the incidence rate of diarrhoeal disease was 0.736 per 100 person-weeks. It was 0.508 per 100 person-weeks for males and 0.955 per 100 person-weeks for

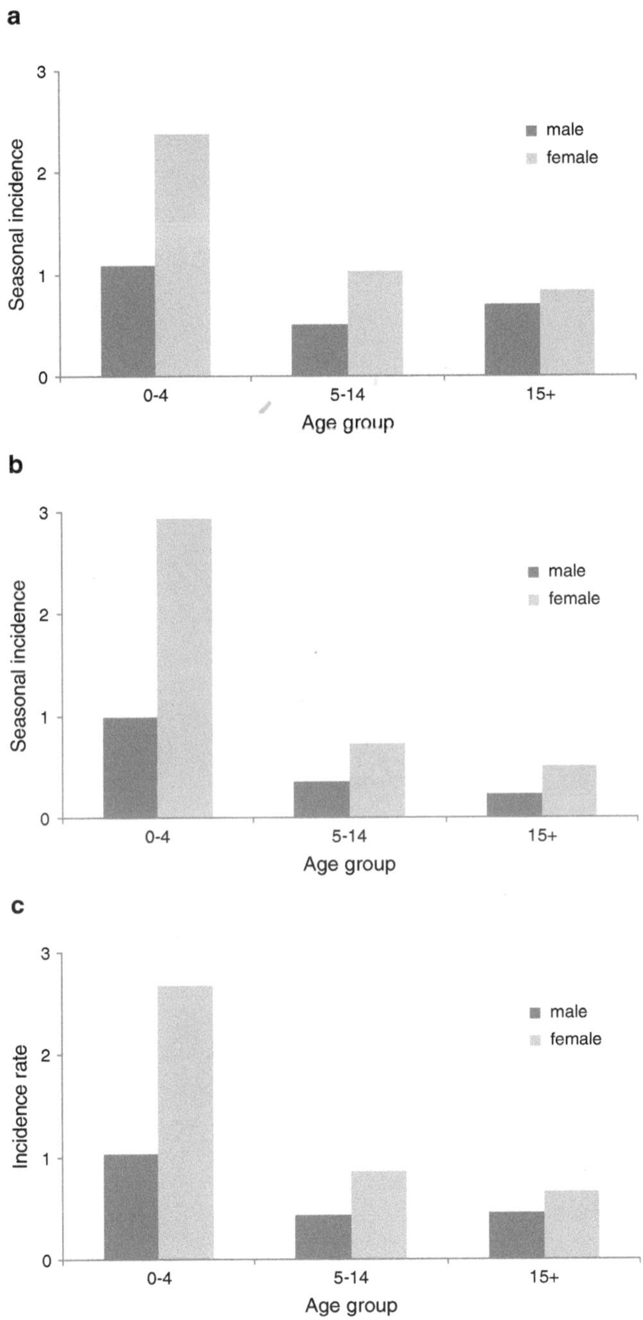

Fig. 6.4 Incidence rate of diarrhoea per 100 person-weeks (total, dry and monsoon season). (**a**) Incidence rate of diarrhoeal disease per 100 person-weeks, in 2 slums, dry season, by gender, Chennai, 2007. (**b**) Incidence rate of diarrhoeal disease per 100 person-weeks, in 2 slums, monsoon season, by gender, Chennai, 2007. (**c**) Incidence rate of diarrhoeal disease per 100 person-weeks, in 2 slums, by gender, Chennai, 2007

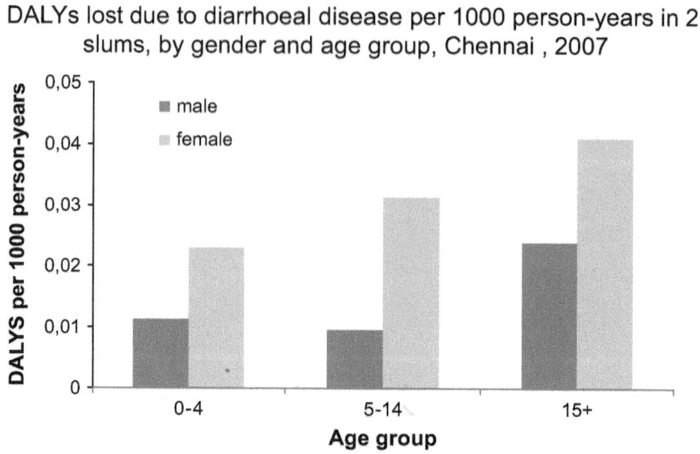

Fig. 6.5 DALYs lost due to diarrhoeal disease per 1,000 person-years

females. Among children aged 0–4 years, the incidence rate per 100 person-weeks was 1.03 for males against 2.68 for females (Fig. 6.4c). Among teenagers and adults aged 15 and above, the incidence rate per 100 person-weeks was 0.455 for males against 0.659 for females (Fig. 6.4c).

6.3.3 DALYs Lost to Diarrhoeal Disease

The mean duration of the 111 cases of diarrhoeal disease in the two slums was 2.4 days. No death due to diarrhoeal disease was reported. A total of 0.00825 DALYs were lost to diarrhoeal disease during the 15 weeks of follow-up in the two slums. This is equivalent to 0.0282 DALYs per 1,000 person-years. Among females, the DALYs lost to diarrhoeal disease per 1,000 person-years was 0.0231, 0.0315 and 0.0411 for the age groups 0–4, 5–14 and 15 and above, respectively (Fig. 6.5). Among males, it was 0.0113, 0.00958 and 0.0238 for the age groups 0–4, 5–14 and 15 and above, respectively (Fig. 6.5).

6.4 Discussion

Our study measured the local burden of diarrhoeal disease among residents of two slums in the megacity of Chennai. We described the processes of data gathering and data processing, which are necessary to empirically measure the burden of disease. In the following section we will critically evaluate our approach in the light of other published studies. In the health calendar we actively monitored nine symptoms, including diarrhoea. We identified diarrhoea as the occurrence of a particular symptom: at least one loose stool in a 24-hours period. This case definition is rather broad and unspecific. For example, Baqui et al. (1991) argues that three or more loose stools or any number of loose stools containing blood in a 24-hour period seemed to be the best definition of diarrhoea. They also pointed out that the end of an episode is best defined by three diarrhoea-free days. As stressed by Bern (2004), differences in the case definition, especially with regard to the end of episodes, make substantial difference on the estimates of incidence of diarrhoeal disease. Due to the broad focus of our study we did not differentiate between persistent diarrhoea, acute watery diarrhoea or dysentery. We were not able to perform laboratory confirmation in order to identify the pathogens causing the diarrhoeal episodes. The differentiation between different pathogens is important for the measurement of the disease burden and for control measures as the severity and the risk factors associated with different pathogens differ. In our study we approximate the differences in severity through the assessment of the limitations caused by the episode through the respondent. By comparing our results with other studies, it is important to consider the case definition used.

A community longitudinal study, as was done here, provides the most reliable data for diarrhoea incidence (Bern 2004). The frequent household visits (active surveillance) – weekly in our case – leads to a higher reported incidence rate (Bern 2004). The draw-back of the longitudinal community-based approach is that both the sample size and sample period are not large enough to make mortality estimates. The calculation of the burden of disease is therefore restricted to the years lived with disability (YLD). Another aspect of the surveillance method that might have an effect on the results is the characteristics of the person responsible for the "health calendar". With the health calendar we seek to monitor the health status of every household member. In practice, one member of the household – in most of the cases a female adult member – was responsible for the reporting within the household during the week and corresponded with the field assistant who checked and collected the calendar. It can be expected that the reported morbidity of the person who is responsible for the data collection is higher. Dilip (2007) estimated the bias due to a proxy respondent and stated that the morbidity rate of the respondent is 65% higher than of the person whose morbidity is indirectly reported.

6.4.1 Diarrhoea Incidence and Burden

The overall diarrhoea incidence rate among our population was 0.736 per 100 person-weeks (0.382 person-years). When we stratify by age, we observe a higher risk among children under five (1.913 per 100 person-weeks or 0.995 per person-years) compared to older children (0.659 per 100 person-weeks or 0.342 per person years) and adults (0.556 per 100 person-weeks or 0.289 per person-years). Compared to the incidence of diarrhoeal disease among young children reported in other studies, our results are comparable but slightly lower (see Table 6.2). The overall

Table 6.2 Incidences of diarrhoea per child (<5 years) per year. Global estimates and empirical results from India

Author/Reference	Year	Episodes/Child/Year
Global estimates		
Snyder and Merson (1982)	1982	2.2
Bern et al. (1992)	1992	2.6
Institute of Medicine (1986)	1986	3.5
Martines et al. (1993)	1990	3.5
Martines et al. (1993) estimates for India	1990	2.7
Longitudinal studies from India[a]		
Bhan et al. (1989) (Rural Uttar Pradesh)	1985–1986	0.7
Sircar et al. (1984) (Urban Calcutta)	1985–1986	1.1
Mathur et al. (1985) (Rural Andhra Pradesh)	Early 1980s	1.6
Kumar et al. (1987) (Rural Northern India)	Early 1980s	2.2
Bhandari et al. (1992) (Urban Uttar Pradesh)	1993	9.9

Source: Bern (2004)

[a]Community-based longitudinal studies of children diarrhoea incidence in developing countries with a 1 year follow-up and surveillance at least every 2 weeks

results correspond to the established higher vulnerability to diarrhoea of young children (UNICEF/WHO 2009: 10). According to the Institute of Medicine (1986) 60% of diarrhoea-related morbidity and 90% of mortality occur among children younger than 5 years.

When we stratify by gender, we see a considerable higher incidence among female persons compared to males. The difference is particularly strong among children under five and persists to a lesser extend in the older age groups. We do not have a clear explanation for this gender disparity. The reporting bias described above could have an impact in the age group 15 and above. But it does not explain the disparity among young children under five as the morbidity of male as well as female children is reported by proxy respondents. The established disparity could be an outcome of the prevalent gender discrimination in India as it has been described in various studies (see Bhan 2001, Sen 1992). Female children in the study area might not have equal access to various health related goods and services (e.g. food, health care), which then leads to a worse health condition expressed by a higher incidence of diarrhoea.

Whereas the incidence rate was higher among young children aged 0–4 years than among older children and adults, the burden in DALYs per person was higher among the older age group. This reflects the age-weighting function in the DALYs calculation that assigns less weight to cases among the very young and the very old. The age-weighting function is based on a social preference to value a year lived by young adults more highly than a year lived by children and older adults (Murray 1994: 435). Mathers et al. (2006) stated that age-weighting is one of the most controversial issues regarding DALYs calculation. The criticism comes from normative judgment that every year of life must have an equal value. Some critics also pointed out that the age weighting function is not based on empirical evidence while others argued that it makes the burden of disease analysis more complex.

In the burden of disease study (WHO 2004), WHO estimated that 0.0148 DALYS per person were lost to diarrhoeal disease in India in 2004. In 2005, the National Commission on Macroeconomic and Health (NCMH), India estimated that 0.0217 DALYS per person were lost to diarrhoeal disease in India (NCMH 2005). In our study we found that in the two slums of Chennai, 0.0000282 DALYS per person per year were lost due to diarrhoeal disease. Our estimate is much lower as no death due to diarrhoeal disease was reported in our cohort. In the NCMH estimates, premature mortality (Years Life Lost) accounted for 98.2% of DALYS lost to diarrhoeal disease. Taking only into account the morbidity-related burden of diarrhoeal disease (excluding deaths) for urban areas in India, the DALYs estimate of the NCMH study was 0.000384 per person per year. This is still more than ten times higher than our estimate. This difference can partly be explained by the higher duration of the diarrhoeal episode assumed in the 2005 study (4 days) and the higher incidence of diarrhoeal disease. The difference between our results and the national estimates for urban India shows that the local burden of disease in specific areas and time periods among different social groups might vary. We share the same conclusion as Würthwein et al. (2001) in their study of local burden of disease in Nanou District, Burkina Faso. Global or national estimates of the burden of disease should be complemented by local estimates to guide local policy making.

Acknowledgements We thank the German Research Foundation (DFG) and the French Institute of Pondicherry for their financial support of the project. At the time of the study, Mr. Seyler and Mr. Prasad were funded by the French Institute of Pondicherry, Mr. Sakdapolrak was funded by the German Research Foundation (DFG).

References

Baqui AH, Black RE, Yunus M, Hoque AR, Chowdhury HR, Sack RB. Methodological issues in diarrhoeal disease epidemiology. Definition of diarrhoea episodes. International Journal of Epidemiology 1991;20:1057-63

Bern C. Diarrhoeal diseases. In: Murray CJL, Lopez AD, Mathers C, eds. The global epidemiology of infectious diseases. Geneva: World Health Organisation; 2004.

Bern C, Martines J, de Zoysa I, Glass RI. The magnitude of the global problem of diarrhoeal disease. A 10-year update. Bulletin of the World Health Organization 1992;70:705–14

Bhan G. India gender profile. Brighton: Institute of Development Studies; 2001.

Bhan MK, Bhandari N, Sazawal S, et al. Descriptive epidemiology of persistent diarrhoea among young children in rural northern India. Bulletin of the World Health Organization 1989;67: 281-8.

Bhandari N, Bhan MK, Sazawal S. Mortality associated with acute watery diarrhea, dysentery, and persistent diarrhea in rural north India. Acta Paediatrica Supplements 1992;381:3-6.

Census of India. Census of India 2001. Total Slum population. In: Census of India; 2001.

Dilip TR. Age-specific analysis of reported morbidity in Kerala, India. World Health & Population 2007;9(4):98-108.

Gupta T, Arnold F, Lhugdim H. Health and living conditions in eight Indian cities. National Family Health Survey (NFHS-3). Mumbai: International Institute for Population Science; 2009.

Herbst S. Water, sanitation, hygiene and diarrheal diseases in the Aral Sea area (Khorezm, Uzbekistan) Göttingen: Cuvillier Verlag; 2006

Herbst S, Fayzieva D, Kistemann T. Risk factor analysis of diarrhoeal diseases in the Aral Sea area. (Khorezm, Uzbekistan). International Journal of Environmental Health Research 2008; 18(5):305-21

Institute of Medicine. The burden of disease resulting from various diarrheal pathogens. In: Institute of Medicine, ed. New Vaccine Development: Establishing Priorities. Washington DC: National Academy Press; 1986:C1–13

Kumar V, Kumar R, Datta N. Oral rehydration therapy in reducing diarrhoea-related mortality in rural India. Journal of Diarrhoeal Diseases Research 1987;5:159-64.

Lopez AD, Mathers CD, Ezzati M, Jamison DT, Murray CJL. Global Burden of Disease. Washington: The World Bank; 2006.

Malsch AKF, Pinheiro P, Krämer A, Hornberg C. Zur Bestimmung von "Environmental / Burden of Disease" (BoD / EBD) in Deutschland. Expertise für das Landesinstitut für den Öffentlichen Gesundheitsdienst (lögd) NRW. Bielefeld: Landesinstitut für den Öffentlichen Gesundheitsdienst 2006.

Martines J, Phillips M, Feachem RG. Diarrheal diseases. In: Jamison DT, Mosely WH, Measham AR, Bobdadillia JL, eds. Disease Control Priorities in Developing Countries. New York: Oxford University Press for the World Bank; 1993:91-116.

Mathers CD, Salomon JA, Ezzati M, Begg S, Vander Hoorn S, Lopez AD. Sensitivity and uncertainty analyses for burden of disease and risk factor estimates. In: Lopez AD, Mathers CD, Ezzati M, Jamison DT, Murray CJL, eds. Global Burden of Disease and Risk Factors. New York: Oxford University Press; 2006:399-426

Mathur R, Reddy V, Naidu AN, Ravikumar, Krishnamachari KA. Nutritional status and diarrhoeal morbidity. A longitudinal study in rural Indian preschool children. Human Nutrition–Clinical Nutrition 1985;39:447-54.

Murray CJL. Quantifying the burden of disease: the technical basis for disability-adjusted life years. Bulletin of the World Health Organization 1994;72(3):429-45.

Murray CJL, Salomon JA, Mathers C. Policy and Practice - A critical examination of summary measures of population health. Bulletin of the World Health Organization 2000;78(8):981-94.

Murray CJL, Salomon JA, Mathers CD, Lopez AD. Summary measures of population health. Concepts, ethics, measurement and applications. Geneva: World Health Organisation; 2002

NCMH. Estimation of the burden of diarrhoeal diseases in India. In: NCMH, ed. Burden of disease in India. Delhi: National Commission on Macroeconomic and Health, Ministry of Health and Family Welfare, Government of India; 2005:182-7

Russell S. Treatment-seeking behaviour in urban Sri Lanka: Trusting the state, trusting private providers. Social Science & Medicine 2005;61:1396-407.

Sen A. Missing Women: Social inequality outweighs women's survival advantage in Asia and North Africa. British Medical Journal 1992;304(6827):587-8.

Sircar BK, Deb BC, Sengupta PG, et al. A longitudinal study of diarrhoea among children in Calcutta communities. Indian Journal of Medical Research 1984;80:546–50.

Snyder JD, Merson MH. The magnitude of the global problem of acute diarrhoeal disease: a review of active surveillance data. Bulletin of the World Health Organization 1982;60: 605–13.

UN-Habitat. The challenge of slums. Nairobi: Earthscan; 2003.

UNICEF/WHO. Diarrhoea: Why children are still dying and what can be done. Gevena: WHO Press; 2009

WHO. Global burden of disease 2002 estimates. World Health Organization, 2004. (Accessed 29 May 2009, at http://www.who.int/healthinfo/global%20burden%20disease/estimates%202000%202002/en/index.html.)

Würthwein R, Gbangou A, Sauerborn R, Schmidt C. Measuring the local burden of disease. A study of years of life lost in sub-Saharan Africa. International Journal of Epidemiology 2001; 30:501-8

Chapter 7
Urban Health in North Rhine-Westphalia

Rainer Fehr, Rolf Annuss, and Claudia Terschüren

7.1 North Rhine-Westphalia: The Most Populous State in Germany

North Rhine-Westphalia (*Nordrhein-Westfalen*, NRW) is a large **federal state** (*Bundesland*) of Germany (Fig. 7.1). Reaching from 50° 19′ to 52° 32′ North (distance 242 km) and from 5° 52′ to 9° 28′ East (distance 252 km), NRW is situated in the western part of the country, sharing borders with Belgium and the Netherlands.

Among the 16 German federal states, with respect to area NRW ranks third (34,086 km^2 = 9.5% of Germany), but it is the largest state in terms of population (17,996,621 = 21.9% of Germany, Dec. 2007). With 528 persons/km^2 the population density is much higher than the German average (230 persons/km^2); it is surpassed only by the city states of Berlin, Hamburg and Bremen. Just as in Germany as a whole, the population in NRW currently decreases.

Among the states, NRW also holds rank highest in terms of **economic output**, contributing c. 22% of Germany's gross domestic product. NRW can be regarded the world's 16th largest economy. The state represents Europe's largest industrial concentration. In NRW, rural and industrial regions are packed closely together. In June 2009, the unemployment rate in Germany was 8.1% (June 2008: 7.5%). For Western Germany, an average rate of 6.9% (6.2%) was reported. The average unemployment rate in North Rhine-Westphalia was 9.0% (8.4% in June 2008) (Bundesagentur für Arbeit 2009).

R. Fehr (✉) • R. Annuss • C. Terschüren
NRW Institute of Health and Work (LIGA.NRW), Department of Prevention and Innovation, Ulenbergstr. 127-131, 40225 Düsseldorf, Germany
e-mail: rainer.fehr@liga.nrw.de

A. Krämer et al. (eds.), *Health in Megacities and Urban Areas*, Contributions to Statistics, DOI 10.1007/978-3-7908-2733-0_7,

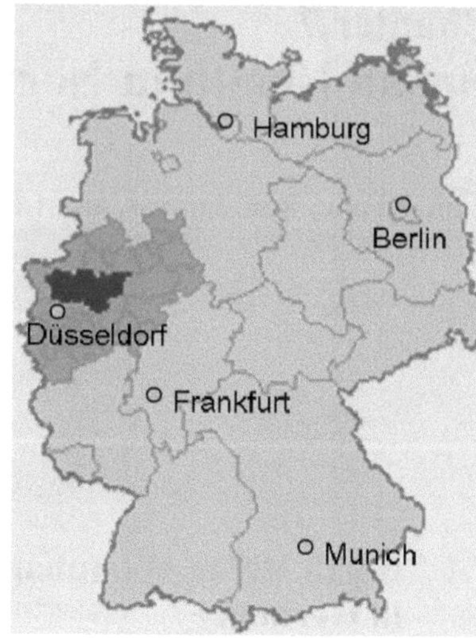

Fig. 7.1 Map of North Rhine-Westphalia (NRW) in Germany; with Ruhr area highlighted

7.2 "Urban Health" in Germany

In Germany, the concept of "**urban health**" conveys different meanings to different persons, including the following: health status of urban populations as compared to rural populations; distribution of health determinants in urban vs. rural areas; and distribution of health care needs in urban vs. rural areas. All these issues constantly undergo changes. The study of urban health, therefore, requires a dynamic approach, prepared to deal with significant changes even within short periods of time.

Rooted in ancient traditions, e.g. Vitruv (first century B.C.), city hygiene in Germany developed rapidly since the eighteenth century and for some time provided the leading paradigm for urban planning and city development. Later on, scholars applied the approaches of hygiene (e.g. Eikmann 1993; Akbar 2005) and of public health promotion (e.g. Stumm and Trojan 1994; Trojan 2001; Mossakowski et al. 2007; Horstkotte and Zimmermann 2008) to urban health. The former WHO health program "Health for all" included a target directed specifically towards "Urban health". One landmark book publication posed the question, "Does the city make us ill?" (Machule et al. 1996). More recently, green spaces attract attention from a public health perspective (Brei and Hornberg 2009). Meanwhile, the European Public Health Association (EUPHA) has started dealing intensely with urban health issues. In 2005, their president stated: "Studies on the health impact of urbanisation reveal that urbanisation can have both positive and negative effects on health. Urban life can be rich and fulfilling since it is more diverse, stimulating, and full of new opportunities. (. . .) Cities are sources of ideas, energy, creativity, and technology" (Tellness 2005). Recently, Trojan and Nickel (2008) developed a standardised instrument to measure empowerment by capacity building in urban quarters.

Each county and city (>100,000 inhabitants) of NRW features a public health department (*Gesundheitsamt*). Based on the Public Health Service act of NRW (1997 and revisions), the tasks include the following: child and youth health service, health reporting, epidemiology, health promotion, infectious disease protection, hygiene, environmental health, social-psychiatric services, pregnancy advice, and dental care especially for school children (Ministry for Internal Affairs North Rhine-Westphalia 2005).

In North Rhine-Westphalia, about 4 million patients need to be treated in **hospitals** each year. In total, 432 hospitals and university clinics take care of the patients. In 2007, 31,069 physicians were registered as working at the hospitals in NRW. Out of these, 16,738 were registered specialists at the hospitals and 14,331 were physicians in advanced training to become a medical specialist. Almost 95,000 nurses and other personnel took care of in-patients at hospitals (IT.NRW 2009).

Concerning **ambulatory care**, 24,191 physicians in NRW were involved based on the German statutory health insurance system in 2007. On average, each of these physicians provided out-patient care for 744 inhabitants. Within this group of physicians, 10,763 (44.5%) worked as general practitioners. On average, each general practitioner was responsible for 1,672 inhabitants. A total of 13,428 (55.5%) physicians offered specialized ambulatory care, e.g. as pediatrician, gynecologist, internal or eye specialist, surgeon, or dermatologist (1,340 inhabitants per medical specialist) (LIGA.NRW 2007a).

As for **regional need and supply of physicians**, throughout Germany the regional Associations of Statutory Health Insurance Physicians (*Kassenärztliche Vereinigung*), in coordination with the health insurers, have to propose a plan for guaranteeing the provision of services (*Sicherstellungsauftrag*). In doing so, they have to consider the goals of spatial and regional planning as well as of hospital planning. The need of physicians in ambulatory care is calculated according to a specific algorithm based on population numbers. The plans specify the required numbers of physicians by medical discipline, type of region, etc. If in a region the number of physicians (in any subgroup) is higher than 110% of the calculated need, then no new accreditations are being approved. Currently, neither counties nor cities in NRW are undersupplied in relation to these calculations. It is expected, however, that within the next 10 years, the Eastern region of the state (i.e. Westphalia) might become short-staffed if the physicians older than 65 years of age continue having difficulties in finding successors (Ärzte Zeitung 2008).

In Germany, the **health economy** now is a highly important economic sector, with 4.4 million employees. Health expenses were 8.8% of gross domestic product in 1980; in 2003, they represented 11.1% (higher fractions existing only in USA and Switzerland). In North Rhine-Westphalia, more than 1 million persons earn their living in health care, medical engineering, pharmaceutical industry, or other health-related sectors. The number of persons working in health care and health-related industries continues to grow. More than 330,000 employees worked at hospitals and other in-patient institutions. More than 260,000 persons were engaged in the treatment and care of out-patients. Additionally, more than 165,000 were employed at nursing homes and at ambulatory services for the elderly, and more than 41,000 employees worked as optometrists or in the production of assistive health technology, e.g. wheel chairs or orthopedic shoes.[1]

7.3 NRW Cities and Health/Overview

NRW comprises 5 regional districts and 54 local administrative units,[2] i.e. 31 counties and 23 cities. The counties, featuring a rural character, are mostly located in the south and the east of the state. The large cities of Cologne and Düsseldorf are located in the west by the river Rhine. The metropolitan area of "Ruhr City" in the state's center consists of 11 cities and 4 counties (cf. Sect. 7.5).

In addition to the administrative status of "city" vs. "county", it can be useful to include other dimensions of the **urban – rural** polarity, e.g. on population density. Using the value of 1,000 inhabitants per km^2 as a cut-off point, then a total of 22 areas in NRW is above this limit and these are categorized as "urban", with 32 "rural" areas below this limit. Not surprisingly, the bulk of the members of the urban (high population density) group are cities, but also one county (Mettmann) falls into this group. Likewise, most of the members of the rural (low population density) group are counties, but the two cities of Münster and Hamm also belong to this group.

The LIGA institute developed and maintains an online database with approx. **300 indicators** describing the health situation in NRW as well as health determinants and health system parameters.[3] These indicators were agreed by the Conference of the German Ministers of Health in 1991. They cover a wide range of topics (demography, life expectancy, mortality, morbidity, health care institutions, health related behaviour, environmental risk factors, etc.), and often represent time series. Data for more than 70 indicators are also available for NRW's cities and counties.

In NRW - and other German states - the cities and counties are legally obliged to produce local health reports, reflecting the health situation in cities and rural areas, the distribution of health determinants and relevant parameters of the health care system (Stockmann et al. 2008). Many cities produce local health reports, some of them being reports dedicated to specific topics, including young and old people, families, female and male health, migration, social situation, handicaps, nursing care, hygiene, environmental health, addiction, or psychiatry. For further analyses, an online tool is available at the LIGA website. The "**Health Atlas NRW**" interactively produces a number of different views on the indicator data, including trends, rankings, profiles, and comparisons between counties and/or cities.

---

[1]www.gesundheitswirtschaft-nrw.de, download: July, 20, 2009

[2]Status: October, 21, 2009

[3]www.liga.nrw.de/themen/gesundheit_berichte_daten/gesundheitsindikatoren/indikatoren_laender/index.html?PISESSION=4cfc41c83ccf3e6da8666decfe4512a2, download: October, 14, .2009

7.4 Urban vs. Rural Health in North Rhine-Westphalia: "Gaps of Trends and Trends of Gaps" – Current Analyses

On a world-wide scale, health (in)equity issues move up the political agenda (Marmot et al. 2008). In order to (1) better identify and understand health inequities in NRW, and (2) take measures to reduce or overcome such inequities, an approach developed by WHO Centre for Health Development in Kobe, Japan, can be utilized: the Urban HEART (Health Equity Assessment and Response) tool (WHO Centre for Health Development 2008). Concerning "assessment", this implies two main approaches. The "**Urban health equity monitor**" is a diagram showing time trends of selected indicators, including the values of the most advantaged and the most disadvantaged performers. The difference between these extremes is called equity yardstick, it is a gauge of how effectively inequity factors have been responded to. This tool can be used either within or across cities.

For NRW, we adapted the health equity monitor approach to include, for selected indicators, the time trends for cities and counties, maintaining the focus on the difference between the extremes ("**gap**"). In this chapter, we look at arrays of time trend curves of three different variables: life expectancy at birth; rate of live births; and fraction of live births with underweight. Life expectancy tables for all 54 administrative units were calculated based on death probabilities according to Farr's death rate method. Infant mortality rate was calculated as deaths in the first year of life per 1,000 live births. The rate of live births with underweight (<2,500 g) was calculated as the number of resident live births in the specified area with a birth weight of less than 2,500 g per 1,000 live births (AOLG 2003).

For each variable, we use the time trends observed in each locality ($n = 54$), i.e. city or county. In order to characterize the distinctive features of the urban situation, we distinguish between "urban" ($n = 22$) and "rural" ($n = 32$) localities, based on population density as described above. Since this is not based on random "sampling", we do not use statistical testing. In order to decrease "noise" and to better identify "signals", we used 3-year moving averages instead.

7.4.1 *Life Expectancy at Birth*

During the study period of 15 years, life expectancy at birth for **females in rural areas** increased from a mean value of 79.4 years to 82.1 years, implying an increase by 2.7 years. The range of variation (max – min between the different rural areas) decreased slightly from an initial value of 2.2 years to a final value of 1.8 years. A peak variation was observed for 1998–2000 (2.8 years). Throughout the period, life expectancy was highest in Münster (i.e. the one city, which was reclassified as "rural" due to relatively low population density, light blue line in Fig. 7.2). As for **males in rural areas**, life expectancy increased from an initial mean value of 72.3 years to a final value of 76.2 years. Compared to females, this increase

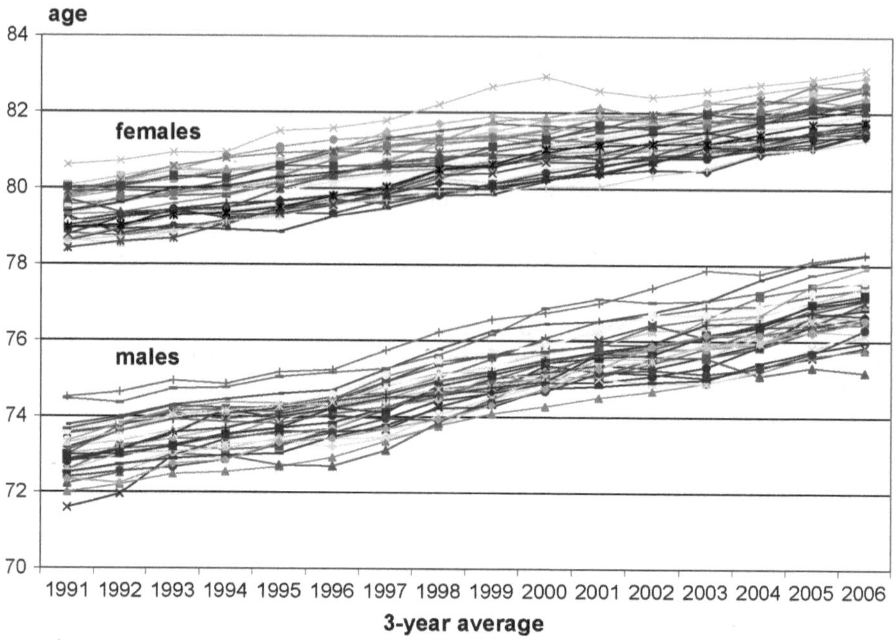

Fig. 7.2 Life expectancy in rural areas: females, males (*Source*: LIGA.NRW)

of 3.9 years is slightly higher than in females. The initial range for males was 2.9 years, which over time increased to a value of 3.1 years, being somewhat larger than for females and with a trend in opposite direction (Fig. 7.2).

Concerning the situation of **females in urban areas**, life expectancy at birth increased from a mean of 79.1 years (1990–1992) to 81.6 years (2005–2007). This positive trend results in an increase of 2.5 years. Between the different urban areas of NRW, the range of variation (max - min) varied between 2.4 years and 3.2 years, showing no clear trend.

Life expectancy of **males in urban areas** also increased from 72.3 years to 76.1 years. The range of variation increased from an initial value of 3.1 years to a distinctly larger value of 4.5 years (45% increase), i.e. there is a widening gap of life expectancy for males between the different urban areas. Throughout the study period, the life expectancy for females and even more so for males was lowest in Gelsenkirchen (females, 1990–1992: 77.7 years, 2005–2007: 80.1 years; males, 1990–1992: 70.7 years, 2005–2007: 74.0 years; females: turquoise line, males light green line in Fig. 7.3).

Comparing life expectancy for **females in rural vs. urban areas in NRW**, the average life expectancy is higher in rural areas. The gap between life expectancy for women in rural vs. urban areas was widening during the investigated period (1990–1992: 3 months; 2005–2007: 6 months). As in females, the comparison of life expectancy of males **in rural vs. urban areas** revealed higher estimates for men in rural areas. However, in contrast to the females the gap between rural

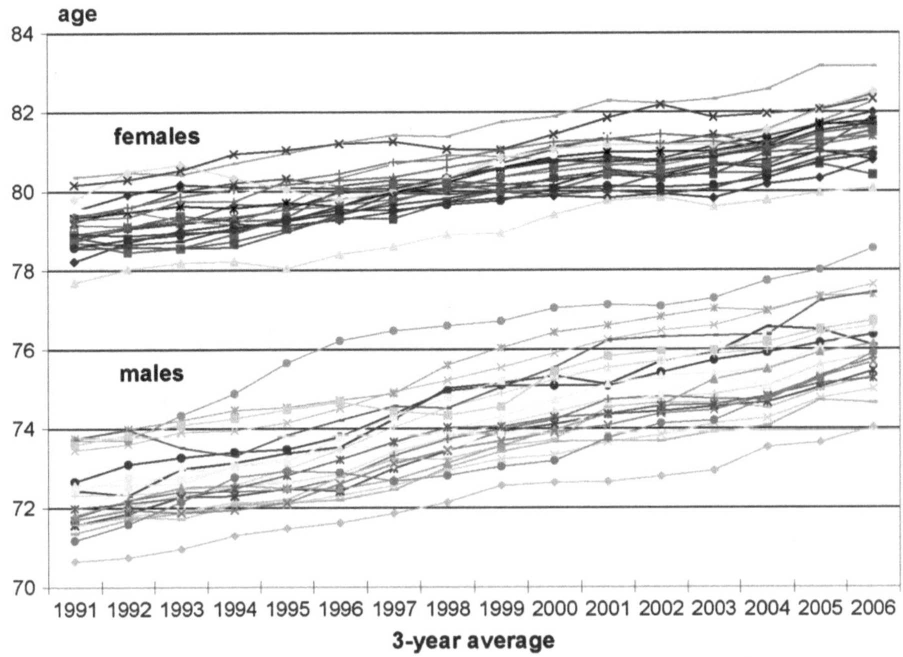

Fig. 7.3 Life expectancy in urban areas: females, males (*Source*: LIGA.NRW)

and urban remained almost constant over the investigated period, varying around 9 month. Combining rural and urban areas, the overall gap of life expectancy between **females and males** in North Rhine Westphalia is narrowing. The difference in average life expectancy at birth between boys and girls slightly decreased from 1990 to 1992 (6.6 years) to 2005–2007 (5.3 years).

A recent study looking into healthy life expectancy found that besides the quantity of life years also the quality of life years of the NRW population increased (Pinheiro and Krämer 2009). The Severe-Disability-Free Life Expectancy (SDFLE) at birth was 69.9 years in 1999 and rose to 71.7 years in 2005. For Long-Term-Care-Free Life Expectancy (LTCFLE) at birth, the authors calculated an increase from 75.3 years (1999) to 76.6 years (2005).

7.4.2 Infant Mortality

Within the last two decades, the infant mortality in North Rhine-Westphalia decreased considerably from 8.1 per 1,000 live births (3-year moving average 1988–1990) to 4.7 per 1,000 live births (2005–2007).

Initially, in 1988–1990 in urban as well as in rural areas, extremes of infant mortality rates higher than 10 per 1,000 live births were observed. In 1988–1990 the

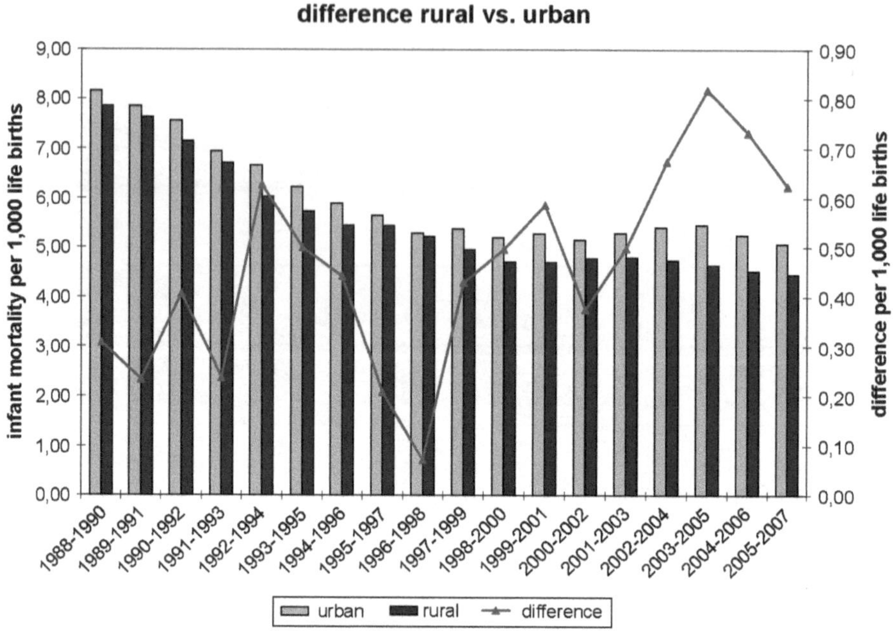

Fig. 7.4 Rate of infant mortality, urban and rural areas

average infant mortality was 7.9 per 1,000 live births in rural areas and 8.2 per 1,000 live births in urban areas. In the end of the study (2005–2007), the average infant mortality rate for the rural areas decreased to 4.5 and for urban areas to 5.1 per 1,000 live births. The **difference between rural and urban** areas in average infant mortality was initially small (0.3 per 1,000 live births). As shown in Fig. 7.4 there is no clear trend over the period. We found **peaks of wide gaps** in 1992–1994 (0.63), in 1999–2001 (0.59), and in 2003–2005 (0.82) between rural and urban areas, whereas in 1996–1998 the gap between the average rates was almost closed (0.07 per 1,000 live births). Since 2003–2005, we can observe the gap narrowing again (Fig. 7.4). Despite the variations in difference, the average rural infant mortality rate was continuously lower than the urban average rate. The range of variation (max-min) within the group of rural areas (average range 3.8 per 1,000 live births) is similar to the range found within the urban areas (4.2 per 1,000 live births; data not shown).

In 2007 in Germany, the average infant mortality was 3.9 per 1,000 live births (Statistisches Bundesamt 2009). To evaluate the causes of the elevated infant mortality in cities of North Rhine-Westphalia, Danke et al. (2008) investigated causes of deaths in infants with special consideration of the migration status of the parents. They found an association of infant mortality and social status of the family. Among infants of migrants in comparison to non-migrants, malformations were more often the cause of death.

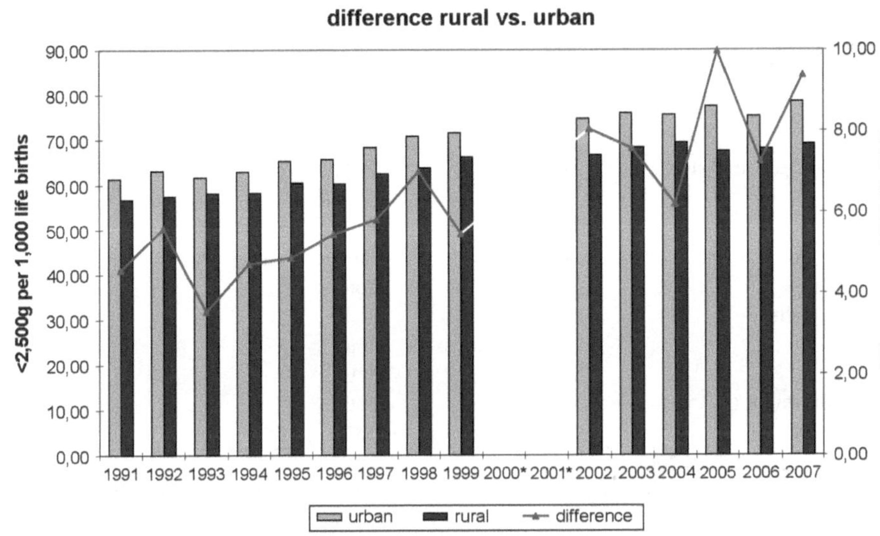

Fig. 7.5 Rate of low weight newborns, rural and urban areas

7.4.3 Live Births with Birth Weight Lower Than 2,500 g

In North Rhine-Westphalia, each year about 150,000 children are born. In 2007, more than 10,000 of these 150,000 babies born alive weighed less than 2,500 g.

The average rate of newborns weighing less than 2,500 g increased from 58 per 1,000 live births in 1991 to 72 per 1,000 live births in 2007. In 2007, the rate of underweight newborns ranged from 99 per 1,000 live births (maximum, city of Mülheim) to 50 per 1,000 live births (minimum, county of Olpe). The rate of newborns weighing less than 2,500 g at birth per 1,000 live births was constantly higher in urban areas of North Rhine-Westphalia than in rural areas (1991–2007). In **rural areas**, the average **rate of low weight newborns increased** from 57 (1991) to 69 (2007) per 1,000 live births (Fig. 5). In **urban areas**, the **average rate increased** from 61 (1991) to 78 (2007) per 1,000 live births. The **gap between the average rates** of the rural and urban areas **is widening**. The difference between the average rural and urban rate initially was 5.0 per 1,000 live birth. In the end of the study, a difference of 9.0 per 1,000 live births was registered. Extreme maximum rates were observed in 1999 (92 low weight newborns per 1,000 live birth in Herne), in 2004 (97 per 1,000 live birth in Gelsenkirchen) and in 2007 (99 per 1,000 in Mülheim).

We observed an increasing trend in live birth <2,500 g in both the urban and the rural areas. However, the increase is higher in the urban areas. The trend might be influenced by the fact, that the university clinics are located in the urbanized areas of North Rhine-Westphalia, especially in the metropolitan area of the Ruhr area. In highly specialized neonatology centres affiliated to these university clinics, more children born preterm and with extreme low birth weight are able to survive.

Table 7.1 Summary of results

Health indicator	Region	Sex	Trend[a]	Gap[b]
		Female	↗	↘
	Rural	Male	↗	↗
		Female	↗	→
Life expectancy at birth (LE)	Urban	Male	↗	↗
		Female		↗
	Rural vs. urban	Male		→
Comparison	Both	Female vs. male		↘
	Rural	Both	↘	
Infant mortality	Urban	Both	↘	
Comparison	Rural vs. urban	Both		→
	Rural	Both	↗	
Low birth weight (<2,500 g)	Urban	Both	↗	
Comparison	Rural vs. urban	Both		↗

[a]Trend over time (1990–2007)
[b]Difference between extremes of cities or counties for LE by sex, difference between regions or sexes
↗ increasing trend; widening gab, ↘ decreasing trend; narrowing gap, → no distinct trend

The study of Danke et al. (2008) found an association between proportion of low birth-weight infants and migration background of the parents. This finding also adds to the possible explanations of the rates observed for cities located in the Ruhr area. The results reported in this section 7.4 are summarized in Table 7.1

7.5 "Ruhr Cities" Metropolitan Area

The **Ruhr area** was a highly industrialized area within NRW including coal mining, steel production, and chemical plants. However, like in many other developed industrial regions within the last few decades, the main employers changed from metalworking into service industry (Bosch and Nordhause-Janz 2005). Within the Ruhr area, a subset of cities and counties developed a specific identity as "Ruhr City". This metropolitan area is a close approximation of "megacity". In German, this conglomerate is called *Regionalverband Ruhr;* it includes the 11 cities of Bochum, Bottrop, Dortmund, Duisburg, Essen, Gelsenkirchen, Hagen, Hamm, Herne, Mülheim a.d. Ruhr, Oberhausen, and the four counties of Ennepe-Ruhr, Recklinghausen, Unna, and Wesel. This association was founded in 1920 as Settlement Union of the Ruhr Mining Area (*Siedlungsverband Ruhrkohlenbezirk).* The area is 4,435 km^2, i.e. 13% of NRW, the population was 5.2 million (1 Jan 2009), i.e. 29% of NRW. In "Ruhr City", the population density on average is 1,199 inhabitants per km^2, and approx. 630,000 of the inhabitants have a migration background.[4] The population forecast for Ruhr-City in 2030 projected a population loss of 8.1% in average (range: −2.3% to −14.7%) in comparison to 2009.

[4]www.rvr-online.de; accessed 20 July 2009

Characteristics of "Ruhr City", compared to NRW as a whole, include the age distribution of inhabitants. The ratio of inhabitants older than 65 years to inhabitants aged 18 to 64 years (i.e. working age) is called **old-age dependency ratio** (*Altenquotient*). In North Rhine-Westphalia in 2007, the average old-age dependency ratio was 0.32. In "Ruhr City", many cities report higher ratios. The highest ratio of NRW was reported for the city of Mülheim a.d. Ruhr, where live nearly 40 persons older than 65 years per 100 persons at working age (18–64 years). The lowest ratio of NRW (0.26) was ascertained for Paderborn, a predominantly rural county located in the eastern part of NRW. – "Ruhr-City" is aging faster than other regions of North Rhine-Westphalia (Terschüren et al. 2009).

In the area of "Ruhr City", we find the highest **unemployment** rates of North Rhine-Westphalia and of Western Germany. In June 2009, the unemployment rates surpassed 12% in Dortmund (12.8%), Duisburg (13.2%), Essen (12.4%), and Gelsenkirchen (15.1%). Rates as high as this, do not exist elsewhere in the West of Germany; they can only be observed in former East Germany (Bundesagentur für Arbeit 2009).

Out of the about 4 million patients per year in NRW (cf. above), roughly 35% are treated in "Ruhr City". Altogether 122 **hospitals**, i.e. 28% o NRW hospitals, are located in "Ruhr-City". The 11 cities of "Ruhr-City" host 78 of these hospitals and university clinics (18% of all hospitals in NRW) (IT.NRW 2009). The city featuring the largest number of hospitals (i.e. 15) is Essen.

Within this densely populated area, each general **medical practitioner** is statistically responsible for more inhabitants than on NRW average (1,672 inhabitants per 1 general practitioner in 2007). In the cities of Oberhausen, Dortmund, and Hamm one general practitioner has to take care of more than 1,900 inhabitants. In Bochum, Bottrop, Duisburg, and Herne, the rate still is more than 1,800 inhabitants per general practitioner (LIGA.NRW 2007a).

Based on social structure indicators which clearly reflect differences in living conditions and on factors to describe the economic prosperity and population characteristics (proportion of poorer, elderly, unemployed and foreign groups and density) Strohmeier et al. (2007) identified the Ruhr area as the poverty cluster of North Rhine-Westphalia. Morbidity and mortality of the inhabitants of the most urbanized area of NRW are strongly influenced by social status (Strohmeier et al. 2007; Klapper et al. 2007).

From public health perspective, one particularly important parameter is the so-called **avoidable mortality**. Avoidable deaths are those which probably would have occurred at older age or didn't have occurred at all, if adequate prevention, screening, and treatment would have been provided. Taking the average standardized mortality rate (SMR) of North Rhine-Westphalia as reference, three out of 11 cities of the Ruhr area show a statistically significant higher SMR for lung cancer (ICD 10 C33-34) than the NRW average (LIGA.NRW 2007b). In Duisburg, Gelsenkirchen, and Oberhausen about 30% of the lung cancer deaths are avoidable in comparison to the NRW average SMR. For only three of these cities SMRs are equal to the SMR of NRW in total (Table 7.2). SMRs for breast cancer are statistically inconspicuous. Looking at ischemic heart disease the majority of the Ruhr area cities (8 of 11 cities)

Table 7.2 Avoidable mortality in Ruhr-City due to selected chronic diseases

	Avoidable mortality					
	Cancer of the lung, trachea, and bronchus (ICD C33-34)		Breast cancer (ICD C50)		Ischemic heart disease (ICD I20-25)	
	Both sexes (15–64 years)		Females (20–64 years)		Both sexes (35–64 years)	
Administrative units of Ruhr-City	*N*[a]	SMR[b]	*N*[a]	SMR[b]	*N*[a]	SMR[b]
Bochum	71	0.99	25	0.89	73	0.99
Bottrop	26	1.17	11	1.22	29	1.24
Dortmund	120	1.10	44	1.05	141	1.26*
Duisburg	118	1.29*	34	0.96	137	1.46*
Essen	131	1.17	42	0.97	144	1.25*
Gelsenkirchen	63	1.29*	20	1.04	75	1.49*
Hagen	40	1.09	16	1.12	51	1.33*
Hamm	34	1.02	8	0.60	43	1.28*
Herne	42	1.30	10	0.85	42	1.28*
Mülheim/Ruhr	34	1.00	12	0.89	38	1.09
Oberhausen	54	1.33*	17	1.06	56	1.34*
Ennepe-Ruhr	62	0.93	25	0.96	73	1.06
Recklinghausen	139	1.14*	48	1.02	161	1.29*
Unna	75	0.95	34	1.09	92	1.12
Wesel	91	1.01	40	1.11	79	0.85
NRW	**3,283**	**1.00**	**1,285**	**1.00**	**3,377**	**1.00**

Source: Health reporting and surveillance system, LIGA.NRW
*Statistically significant elevation above NRW reference
[a]5-year average: 2003–2007
[b]SMR (standardized mortality ratio, reference: NRW mortality)

show statistically significant SMRs of 30% and higher than the NRW average. One epidemiological study currently investigates the causes for this striking negative outcome in the metropolitan area (e.g. Dragano et al. 2009). Among the counties included in the Ruhr area, the county of Recklinghausen with the city Recklinghausen (c. 120,000 inhabitants) showed significantly elevated SMRs for lung cancer and ischemic heart disease.

International studies support these findings on geographical (Shaw et al. 2000) and deprivation-dependent variation in life expectancy at birth and morbidity-related health indicators (e.g. Woods et al. 2005; Lang et al. 2008a, b). Breeze et al. (2005) showed that health-related quality of life (QoL) among older people is significantly reduced by living in deprived areas. In this British study, in the most deprived areas about 30% of the excess risk of poor home management and self-care was accounted on health symptoms. A systematic review of Galobardes et al. (2008) found that mortality risk for all causes was higher among those who experienced poorer socioeconomic circumstances during childhood.

Using the burden of disease, Terschüren et al. (2009) predicted disease burden for the metropolitan area of the Ruhr area in 2025 by calculating disability adjusted life years (DALY) as the sum of life years lost due to premature death and years lived with disability due to selected diseases. The projection included selected tumour site, myocardial infarction (MI), and dementia. For the metropolitan area, increases in DALYs are expected for all diseases included, i.e. selected tumours (20%), MI (17%), and dementia (36%).

7.6 Conclusion: Continued Need of Prevention and Health Promotion Programmes Especially in NRW Cities

These findings of the analysis based on the health indicators are contributing to the demand for prevention and health promotion in the cities, if possible, tailored for different sub-groups within the setting city.

Health policy in NRW already covers a wide scope of approaches and activities, ranging from NRW health targets and prevention programs all the way to infectious disease control, medical drugs safety and hospital planning, to name but a few items. On state level as well as in the cities and counties, multi-stakeholder **health conferences** are regularly being held, bringing together multiple actors of health care and health policy. There is an initiative "**Healthy state NRW**", maintaining an annual contest for health-related projects which consecutively are documented in a specialised database.[5] The NRW **prevention concept** includes topics such as non-smoking, maternal & child health, assistance for obese children, and prevention of falls of the elderly.

In the Ruhr area, more inhabitants are disadvantaged by lower education, lower income and long-term unemployment. In comparison to rural areas, more families have many children, or live as single parents. An epidemiological cohort study, the Heinz-Nixdorf Recall Study, including more than 4,800 middle aged urban inhabitants of the Ruhr area revealed that low socio-economic position was associated with poor social networks and social support (Weyers et al. 2008). A federal program "**Health promotion for socially disadvantaged persons**" was started in Germany in order to strengthen and disseminate good practice concerning projects and activities of health promotion for socially disadvantaged persons. Components include: an Internet platform,[6] practice examples database (>1,700 projects), "Good practice" certification (>70 projects), and "Regional hubs" (*Regionale Knoten*) in all 16 federal states.

In this context, a Regional hub was located at LIGA.NRW, which pursues the following main targets: Sensitizing for health inequity; supporting knowledge transfer; connecting actors from different sectors. While one focus is on health promotion for unemployed persons, the other one refers to establishing close cooperation with the **Federal and *Länder* Program "Social City"** (*Soziale*

[5]www.mags.nrw.de/03_Gesundheit/2_Versorgung/gesundheitspreis/index.php (in German), accessed 31 July 2009

[6]www.gesundheitliche-chancengleichheit.de (in German), accessed 31 July 2009

Stadt).[7] A first regional conference on "Health promotion in the socially integrative city" took place in November 2008 (Dickersbach and Weth 2008).

North Rhine-Westphalia is aware of the specific health status of the population in its megacity. In rural and urban areas of NRW the trends in life expectancy and infant mortality are developing positively – also in cities of the Ruhr area. Public health programs and local projects in the cities of the Ruhr area aim to push this further and to narrow the gaps.

References

Akbar, O. (2005): Selbstheilungskräfte der Stadt - Zivile Qualitäten des Urbanen [The Self Healing Power of Cities - Quality of Urban Life]. Das Gesundheitswesen 67: 827–831

Arbeitsgemeinschaft der Obersten Landesgesundheitsbehörden (AOLG) (2003): Indikatorensatz für die Gesundheitsberichterstattung der Länder [Indicators for Health Reporting and Surveillance of the Federal States of Germany]. Volumes 1 and 2, third edition; lögd, Bielefeld

Ärzte Zeitung (2008): Die Arztpraxis in Fußnähe wird es nicht mehr überall geben [In the future the physicians practice will not be in walking distance for everyone]. Ärzte Zeitung Verlags GmbH Neu-Isenburg, 3 November 2008, online: www.aerztezeitung.de/extras/druckansicht/ ? sid=518006&pid=524072, accessed: 6 August 2009

Bosch, G., Nordhause-Janz, J. (2005): Arbeitsmarkt NRW: Entwicklungen und Herausforderungen [Job market NRW: developments and challenges]. In: Institut Arbeit und Technik (IAT, ed.), Jahrbuch 2005 [Annual Report 2005], Chapter 2, pp. 47–64

Breeze, E., Jones, D. A., Wilkinson, P., Bulpitt, C. J., Grundy, C., Latif, A. M., Fletcher, A. E. (2005): Area deprivation, social class, and quality of life among people aged 75 years and over in Britain. Int J Epidemiol 34 (2): 276–283

Brei, B., Hornberg, C (2009): Die Bedeutung von Stadtgrün aus gesundheitswissenschaftlicher Sicht [The meaning of neighbourhood greenness in cities - the public health perspective]. Public Health Forum 17 (62): 11.e1-11.e3

Bundesagentur für Arbeit [Federal agency for employment] (2009): Arbeitsmarkt in Deutschland nach Regionen [Job market in Germany by region]. www.pub.arbeitsagentur.de/hst/services/statistik/000000/html/start/karten/aloq_aa.html; accessed 7 July 2009

Danke, K., Blecher, C., Bardehle, D., Cremer, D., Razum, O. (2008): Kleinräumige Analyse der Säuglingssterblichkeit in Bielefeld unter besonderer Berücksichtigung des Migrationshintergrundes, 2000–2006 [Small Area Analysis of Infant Mortality in Bielefeld with Special Consideration of the Migration Status od Parents, 2000–2006]. Gesundheitswesen 70 (11): 624–630

Dickersbach, M., Weth, C. (2008): Dokumentation der ersten Regionalkonferenz Nordrhein-Westfalen "Gesundheitsförderung in der Sozialen Stadt" vom 25.11.2008 in Düsseldorf [Report of the first regional conference in NRW "Health Promotion in the Socially Integrated City"]. Landesinstitut für Gesundheit und Arbeit NRW (LIGA.NRW) (eds.). 66S

Dragano, N., Hoffmann, B., Stang, A., Moebus, S., Verde, P.E., Weyers, S., Möhlenkamp, S., Schmermund, A., Mann, K., Jöckel, K.-H., Erbel, R., Siegrist, J., Heinz Nixdorf Recall Study Investigative Group (2009): Subclinical coronary atherosclerosis and neighbourhood deprivation in an urban region. Eur J Epidemiol 24: 25–35

Eikmann, Th. (1993): Gesundheit. In Sukopp, H. Wittig, R. (Hrsg.): Stadtökologie, S. 70–96, Springer Verlag, Heidelberg -New York

[7]www.sozialestadt.de/programm/ (in German), accessed 31 July 2009

Galobardes, B., Lynch, J. W., Smith, G. D. (2008): Is the association between childhood socio-economic circumstances and cause-specific mortality established? Update of a systematic review. J Epidemiol Community Health 62 (5): 387–390

Horstkotte, E., Zimmermann, E. (2008): Ungleiche Lebensbedingungen und Gesundheitschancen bei Kindern im Vorschulalter: Schutzfaktoren fördern - Risikofaktoren begrenzen! [Unequal Living Conditions and Health Chances in Preschool Children: Support Protective Factors - Limit Risk Factors]. Das Gesundheitswesen 70: 662–666

Information und Technik Nordrhein-Westfalen (IT.NRW) (2009): Landesdatenbank, Krankenhausstatistik: Grunddaten [hospital statistics: basic data, (number of hospitals, beds, physicians, and nurses)]. https://www.landesdatenbank.nrw.de; accessed: 3 July 2009

Klapper, A., Bardehle, D., Razum, O. (2007): Alters- und geschlechtsspezifische Mortalität im Ruhrgebiet von 1994 bis 2004 [Age- and sex-specific mortality in the Ruhr region from 1994 to 2004]. Das Gesundheitswesen 69 (10): 521–526

Landesinstitut für Gesundheit und Arbeit Nordrhein-Westfalen [Institute for Health and Work NRW] (LIGA.NRW) (2007a): Indikator 8.8 – Ärztinnen und Ärzte in ambulanten Einrichtungen, Nordrhein-Westfalen nach Verwaltungsbezirken, 2007 [Health indicator 8.8 - Physician in out-patient care, North Rhine-Westphalia by administrative districts, 2007]. www.loegd.nrw.de/gesundheitberichterstattung/gesundheitsindikatoren/kommunale_gesundheitsindikatoren/frameset.html, accessed 3 July 2009

Landesinstitut für Gesundheit und Arbeit Nordrhein-Westfalen [Institute for Health and Work NRW] (LIGA.NRW) (2007b): Indikator 3.14 – Vermeidbare Sterbefälle nach ausgewählten Diagnosen, Nordrhein-Westfalen nach Verwaltungsbezirken, Mittelwert 2003–2007 [Health indicator 3.14 – Avoidable deaths of selected diagnosis, North Rhine-Westphalia by administrative districts, average of 2003–2007]. www.loegd.nrw.de/gesundheitberichterstatung/gesundheitsindikatoren/kommunale_gesundheitsindikatoren/frame set.html, accessed 29 June 2009

Lang, I. A., Llewellyn, D. J., Langa, K. M., Wallace, R. B., Huppert, F. A., Melzer, D. (2008a): Neighborhood deprivation, individual socioeconomic status, and cognitive function in older people: analyses from the English Longitudinal Study of Ageing. J Am Geriatr Soc 56 (2): 191–198

Lang, I. A., Llewellyn, D. J., Langa, K. M., Wallace, R. B., Melzer, D. (2008b): Neighbourhood deprivation and incident mobility disability in older adults. Age Aging 37 (4): 403–410

Machule D., Mischer O., Sywottek, A. (eds.) (1996): Macht Stadt krank? Vom Umgang mit Gesundheit und Krankheit. Hamburg: Verlag Dölling und Galitz; pp. 320

Marmot, M., Baum, F., Bégin, M., et al. (2008): Closing the gap in a generation. Health equity through action on the social determinants of health. Final Report. WHO/World Health Organization, Geneva, pp. 246

Mossakowski, K., Nickel, S., Schäfer, I., Süß, W., Trojan, A., Werner, S. (2007): Die Quartiersdiagnose: Daten und Ansätze für ein stadtteilorientiertes Präventionsprogramm des Öffentlichen Gesundheitsdienstes - erste Ergebnisse eines Forschungsprojektes. Prävention und Gesundheitsförderung 2: 82–89

Ministry for Internal Affairs North Rhine-Westphalia (2005): Gesetz über den Öffentlichen Gesundheitsdienst (ÖGDG) vom 25 November 1997 [Act on Public Health Services, 25 November 1997]. Düsseldorf, 18pp. www.loegd.nrw.de/1pdf_dokumente/2_gesundheitspolitik_gesundheitsmanagement/gesetze/oegdg.pdf, accessed 5 August 2009

Pinheiro, P., Krämer, A. (2009): Calculation of health expectancies with administrative data for North Rhine-Westphalia, a Federal State of Germany, 1999–2005. Population Health Metrics 7 (4): doi:10.1186/1478-7954-7-4

Shaw, M., Orford, S., Brimblecombe, N., Dorling, D. (2000): Widening inequality in mortality between 160 regions of 15 European countries in the early 1990s. Soc. Sci. Med. 50 (7–8): 1047–1058

Statistisches Bundesamt [Federal Agency for Statistics Germany] (2009): Säuglingssterbefälle je 1.000 Lebendgeborene und durchschnittliches Sterbealter [Infant mortality per 1,000 live births and average age of deaths], Wiesbaden. www.gbe-bund.de/oowa921-istall/servlet/oowa/aw92/dboowasys921. xwdevkit/xwd_init?gbe.isgbetol/xs_start_neu/i17379374/26424677#SOURCES; accessed 12 August 2009

Stockmann, S., Kuhn, J., Zirngibl, A., Mansmann, U. (2008). Kommunale Gesundheitsberichterstattung in Deutschland: eine empirische Erhebung [Health Reporting at the Community Level in Germany: Results of a Survey]. Das Gesundheitswesen 70: 679–683

Strohmeier, K. P., Schulz, A., Bardehle, D., Annuß, R., Lenz, A. (2007): Sozialräumliche Clusteranalyse der Kreise und kreisfreien Städte und Gesundheitsindikatoren in NRW [Health Indicator-based Cluster Anlysis of Districts and Urban Districts in North Rhine-Westphalia]. Das Gesundheitswesen 69 26–33.

Stumm, B.; Trojan, A. (eds.) (1994): Gesundheit in der Stadt. Modelle, Erfahrungen, Perspektiven [Health in the City. Models – Experiences – Perspectives]. Fischer Verlag, Frankfurt/M., Germany

Tellness, G. (2005): President's column: positive and negative health effects of urbanisation. Eur J Public Health 15 (5): 552

Terschüren, C., Mekel, O. C. L., Samson, R., Classen, T. K., Hornberg, C., Fehr, R. (2009): Health status of 'Ruhr-City' in 2025 - predicted disease burden for the metropolitan Ruhr area in North Rhine-Westphalia. Eur J Public Health (advanced access published May 21), doi:10.1093/eurpub/ckp060

Trojan A. (2001). Soziale Stadtentwicklung und Armutsbekämpfung als Gesundheitsförderung [Urban Development and the Fight Against Poverty as Health Promotion Measures]. Das Gesundheitswesen 63 (Sonderheft 1): S43-S47

Trojan, A., Nickel, S. (2008): Empowerment durch Kapazitätsentwicklung im Quartier - erste Ergebnisse und Einschätzung eines Erhebungsinstruments. Gesundheitswesen (70): 771–778

Weyers, S., Dragano, N., Möbus, S., Beck, E-M., Stang, A., Möhlenkamp, S., Jöckel, K. H., Erbel, R., Siegrist, J. (2008): Low socio-economic position is associated with social poor social networks and social support: results from the Heinz Nixdorf Recall Study. Int. J. Equity in Health 7 (13): doi:10.1186/1475-9276-7-13

WHO Centre for Health Development (2008): Urban HEART. Health Equity Assessment & Response Tool. WHO Kobe, Japan. pp. 54, www.who.or.jp/2008/urbanh/URBAN_Health_Equity_Assessment_and_Response_Tool_(HEART).pdf, accessed: 29 October 2009

Woods, L.M., Rachet, B., Riga, M., Stone, N., Shah, A., Coleman, M.P. (2005): Geographical variation in life expectancy at birth in England and Wales is largely explained by deprivation. J Epidemiol Community Health 59: 115–120

Part III
Environmental Health Risks

Chapter 8
Health Effects of Air Pollution and Air Temperature

Alexandra Schneider, Susanne Breitner, Irene Brüske, Kathrin Wolf, Regina Rückerl, and Annette Peters

8.1 Introduction

The aim of environmental epidemiology is to detect a possible risk or to investigate the exposure-response relation with time, duration, location and amount of exposure being the major determinants for that relationship. The assessment of health effects in environmental epidemiology can for example be done by using routine data such as emergency room visits or death certificates. It is very well known that the health of a population is very dependent on a stable and functioning ecosystem. Air as well as climate has a major impact on the function and procedures within the ecosystem.

8.1.1 Air Pollution

Air pollution is one of the most serious environmental problems in all countries and societies regardless their economic development. In developing countries, millions of people are exposed to high levels of indoor air pollution by smoke from open fires or poorly designed stoves. In industrial countries on the other hand, millions of people live in urban areas with elevated levels of air pollution due to burning fossil fuels for energy and transportation in industrial processes or traffic. Although successful efforts for emission control have been undertaken in the developed world, there is existing epidemiological evidence that air pollution remains a health risk even under current regulations. Rapid expansion of industry, increased automobile and truck traffic and high demands for powering homes, especially in large urban areas (megacities), result in severe air pollution problems.

A. Schneider (✉) • S. Breitner • I. Brüske • K. Wolf • R. Rückerl • A. Peters
Helmholtz Zentrum München Deutsches Forschungszentrum für Gesundheit und Umwelt (GmbH), Institut für Epidemiologie II, Ingolstädter Landstr. 1 85764 Neuherberg
e-mail: alexandra.schneider@helmholtz-muenchen.de

A. Krämer et al. (eds.), *Health in Megacities and Urban Areas*, Contributions to Statistics, DOI 10.1007/978-3-7908-2733-0_8,

The size distribution of total suspended particles in the ambient air can be divided in several subclasses according to their aerodynamic diameter. Coarse particles with an aerodynamic diameter between 2.5 and 10 μm and fine particles with an aerodynamic diameter smaller than 2.5 μm ($PM_{2.5}$) add up to the inhalable fraction of particles with an aerodynamic diameter less than 10 μm (PM_{10}). These particles are the main contributors to measured particle mass. In addition, there exist ultrafine particles (UFP) with an aerodynamic diameter below 0.1 μm which do not add much to particle mass due to their small size and weight, but are the main contributors in particle number counts. Particles can additionally be distinguished by their density, their shape or surface area and their chemical composition. For particles, the primary mode of entry into the body is the respiratory system and it is the size that mainly determines the location of the deposition in the respiratory tract with larger particles remaining already in the oral cavity, the fauces and the throat. Smaller particles are deposited in the large and small bronchial tubes and mainly the ultrafine fraction reaches the alveoli. Particles and particle components may cross the pulmonary epithelium into the circulation and interact directly with the cardiovascular system (Nemmar et al. 2002; Oberdorster et al. 2004). The primary mode of entry into the body is through the respiratory system; the greatest population attributable risk from air pollution is due to cardiovascular disease. Epidemiologists have been able to show that the inhalation of ambient particles not only causes local effects in the lung but also systemic effects in the cardiovascular system (Pope and Dockery 2006). In general, there exist two main types of health effects studies in air pollution epidemiology. Long-term studies examine the annual average exposure to air pollution in association with (cause-specific) morbidity and mortality, whereas short-term studies are based on the day-to-day fluctuations in exposure in association with various possible health outcomes of the respiratory and cardiovascular system.

First clear evidence for health effects of air pollution was documented in the London smog episode in December 1952. This smog episode was responsible for an estimated two- to three-fold increase in mortality and showed beyond doubt that episodes of high air pollution have a detrimental effect on respiratory and cardiovascular health. In the greater London area, the number of excess deaths was in the order of 4,000 and the demand for hospital beds far exceeded supply. Other "natural experiments" like closing the city center for private cars during the Olympic Games in Atlanta 1996 showed that a significant decrease in air pollution can relatively quickly improve the health situation of the population as seen by inspecting Medicaid and hospital discharge files before/after versus during the intervention (Friedman et al. 2001). Furthermore, an intervention study by Clancy et al. (2002) compared age-standardized death rates for 72 months before and after the ban of coal sales in Dublin. After the ban, average black smoke concentrations in Dublin declined by 70%. Adjusted non-trauma death rates decreased highly significantly, and cause-specific death rates, such as respiratory and cardiovascular, dropped even more.

8.1.2 Weather and Climate

In addition to air pollution, expected climate changes in the near future will pose another environmental problem with a strong impact on population health worldwide. Clear climate changes in the last 50 years have already been well documented, but the influence on population health has not been sufficiently examined so far. It has been shown that anthropogenic greenhouse gas emissions started and accelerated climate change. The last report of the "Intergovernmental Panel on Climate Change" (IPCC fourth assessment report; http://www.ipcc.ch) revealed for the first time for Europe already strong influences of climate change on the biosphere as well as on the cryosphere. As the causal interrelations are very complex in the climatic system, it will be very difficult to find clear associations between long-term environmental changes and certain health outcomes. Therefore, the IPCC suggests starting research in this field by analyzing the association of the mortality or morbidity incidences with short-term changes in weather and climate and to use short-term temperature fluctuations as one of the main markers. Therefore, the interest in studies examining the temperature-mortality association increased in the last years. It was found that particularly hot days as well as particularly cold days are related to an increase in mortality. Heat effects were mostly restricted to a shorter time-period than cold effects (Braga et al. 2002). Besides age and disease status, gender, socioeconomic status and access to air conditioning were observed to be effect modifiers for the association of temperature with mortality (Donaldson et al. 2001; O'Neill et al. 2003). Medical supply and preventive measures also play an important role. Cities like Rome, Italy, and Paris, France, for example, have established an early heat-warning system (Michelozzi et al. 2004; Pascal et al. 2006) to inform the susceptible parts of the population about expected heat waves. Effective public health strategies and the right preventive measures for the most vulnerable parts of the population at the right time are the necessary consequences from the research described above.

8.2 Methods

For the analysis of associations between short-term changes in exposure and health data special epidemiological designs and statistical methods are needed.

Time-series studies associate time-varying exposures to time-varying event counts. Data of these kinds of studies is often intensive and rely on routinely collected environmental data (e.g. PM_{10} or temperature) that are usually used for the surveillance of air quality or for the recording of weather patterns as well as on health data (e.g. cause-specific mortality). Time series studies are a type of ecologic study because they analyze population-averaged health outcomes and exposure levels. However, due to the temporal nature of the design, confounding concerns that usually come up with ecological studies such as reverse causation fallacy are avoided in time-series studies. As risk factors usually do not change very quickly over time the population at risk is used as its own control.

The objective of a time-series design is to assess short-term changes in the health outcome series which follow changes in exposure. The response of the health outcome might occur immediately after exposure but it is also possible that it needs a certain lag time to become apparent. An effect at "lag 0" thus means an effect on the same day, an effect at "lag 1" means an effect 1 day later, etc. The time-series of both, the exposure and the outcome variable, mostly display periodic patterns such as annual or seasonal patterns which need to be filtered out by statistical methods; otherwise these time-dependent factors will confound the association of interest. Further common issues of time-series studies are adjustment for time-varying factors such as influenza epidemics, temporal autocorrelation within the data, overdispersion, the shape of the exposure-response function and mortality displacement ("harvesting effect"). For air pollution, the exposure-response curve is often assumed to be linear, but for air temperature, other shapes like V, U or J are commonly accepted.

A so-called **panel study** is a small prospective cohort study consisting of individual time-series of repeated measurements. The studies usually follow a small group of individuals intensively over a short time period (e.g. several months). During this time-period, repeated observations (e.g. every day, every 4 weeks) are made on the exposure, the health outcome variables as well as on potential confounders. Panel studies share the common objective of investigating short-term effects with time-series studies and they both use similar analytical methods (e.g. specific fixed or random effects regression models dependent on the distribution of the outcome variable). However, panel studies have the advantage of individual measurements that can be considered for analysis; this can include time-activity tracking, personal monitoring of multiple environmental agents, personal behavior characteristics or medication usage. Primary outcomes of interest in panel studies are clinical or subclinical symptoms, physiological indicators of health status, or measures of minor morbidity such as certain ECG-parameters or markers for inflammation or coagulation in the blood. The aim of these analyses is often to explore certain biological pathways that might explain the observed links between exposure measures such as air pollution or air temperature and more severe health outcomes such as mortality or hospital admissions.

A third design for short-term health effect studies is the **case-crossover design**. It can be viewed heuristically as a modification of the matched case-control design where each case acts as his/her own control. As in the above mentioned panel studies, time-independent confounders, such as personal characteristics, are therefore controlled for by design. The distinction from a case-control study is that the case-crossover design is based on making time-varying exposure comparisons. The so-called index-time, which is the exposure just prior to the event, is compared with exposure at comparable control or referent times. Moreover, if the referent times are matched to the index time with respect to time-dependent confounders (e.g. season, day of the week), these effects are also controlled for by design. Therefore, the referent selection strategy is a key issue of that design. Janes et al. (2005) recommend using several referent periods to increase efficiency. Restricting referent periods bidirectionally to the same day of the week, month and year as the index day controls for season and day of the week and conditional logistic regression can be used to obtain estimates free of overlap bias.

Studying chronic or long-term health effects requires a different design than studying acute or short-term health effect as the latter focuses on the exacerbation rather than the induction of a disease. Long-term studies also have ecological character as the level of exposure is classified to be the same for all study participants from the same area. However, non-exposure variables can be gathered individually for each participant. Thus it can be assured that the observed difference in the health outcome can reliably be attributed to the difference in exposure rather than to some other confounding factor. For health outcome data that are certain events (e.g. death, myocardial infarction), some form of Cox proportional hazards regression modeling is often used. For continuous outcome data like markers that reflect a certain stage of atherosclerosis a multiple linear or logistic regression model is applicable. New approaches try also to individualize the exposure level information by using person-specific measures such as distance of residence to major roads or by using modeled exposure maps for certain areas together with geographic information system data.

Health impact assessment has been defined by the WHO as a combination of procedures or methods by which a policy may be judged as to the effects it may have on the health of a population (WHO 1999). For air pollution, health impacts have been assessed and provide now estimates for the burden of disease attributable to air pollution but additionally they estimate the potential benefits from policies that are aimed to improve air quality. The WHO-recommended methodology for health impact assessment (WHO 2001) suggests calculating the attributable fraction to estimate the impact of changing the exposure from the actually measured concentration to a concentration taken from a possible scenario assuming that all the population is exposed to the mean concentration of a certain area (e.g. a city). This attributable fraction is based on a relative risk (RR) calculated by using an effect estimator from a long-term study (e.g. from the ACS study) and the difference in the actual and the scenario concentration. From the attributable fraction the expected number of "saved" attributable deaths based on the total number of deaths in that area can be estimated due to an exposure reduction from the actual level to a reduced scenario level. It can be speculated that theses deaths would have occurred earlier than they would normally have occurred because of the actual measured exposure concentration. The choice of the exposure-response functions is very influential in this process. Guidelines for health impact assessment suggest using estimates from cohort studies to capture long-term effects.

8.3 Results

8.3.1 Long-Term Studies of Air Pollution Health Effects

In a prospective cohort study, Dockery et al. (1993) estimated the effects of air pollution on mortality with data from a 14- to 16-year mortality follow-up of

8,111 adults in six U.S. cities, while controlling for individual risk factors. The adjusted mortality-rate ratio for the most polluted of the cities as compared with the least polluted was 1.26 (95%-confidence interval (CI): 1.08–1.47). Air pollution was positively associated with death from lung cancer and cardiopulmonary disease but not with death from other causes. A follow-up of this study showed that an overall reduction in $PM_{2.5}$ levels resulted in reduced long-term mortality risk (Laden et al. 2006).

In 1982 the American Society Study (ACS) enrolled approximately 1.2 million adults as part of the Cancer Prevention II study. Vital status and cause of death data were linked with air pollution data for metropolitan areas throughout the U.S. Each 10 $\mu g/m^3$ elevation in fine particulate air pollution was associated with an increased risk in mortality of approximately 6% (95%-CI: 2–11%) for all-cause, 9% (95%-CI: 3–16%) for cardiopulmonary, and 14% (95%-CI: 4–23%) for lung cancer mortality, respectively, calculated as an average of the two air pollution measurement periods 1979–1983 and 1999–2000 (Pope et al. 2002).

Hoek et al. (2002) investigated a random sample of 5,000 people from the full cohort of the Netherlands Cohort Study on Diet and Cancer from 1986 to 1994. Long-term exposure to traffic-related air pollutants (black smoke and nitrogen dioxide) was estimated for the 1986 home addresses. Cardiopulmonary mortality was associated with living near a major road (RR 1.95, 95%-CI: 1.09–3.52) and, less consistently, with the estimated ambient background concentration (1.34, 95%-CI: 0.68–2.64). A further analysis with 120,852 individuals of the same original cohort study (Beelen et al. 2008), followed from 1987 to 1996, showed that traffic intensity on the nearest road was independently associated with mortality.

Based on the results of the ACS study on all-cause mortality, the CAFE (Clean Air for Europe) project calculated the decrease in life-expectancy due to anthropogenic $PM_{2.5}$ impact on premature deaths in 25 European Union (EU) countries. For the year 2000 the decrease in life-expectancy was an estimated 8.1 months in the EU on average, in Germany even 9.2 months. The CAFE-scenario for 2020 was modeled under 2005 discussed EU-regulations and shows an improvement (the decrease in life-expectancy was about 5.9 months in the EU on average and 6.8 months in Germany), but still a loss in life expectancy (CAFE scenario final report 2005).

Also based on the ACS estimates, the APHEIS (Air Pollution and Health: A European Information System) network tried to estimate the number of premature deaths from all causes that could be prevented by reducing $PM_{2.5}$ annual levels in 26 European cities with altogether 40 million inhabitants. The authors stated that the reduction to 15 $\mu g/m^3$ could lead to a reduction in mortality among people aged 30 years or older that would be four times greater than the reduction in mortality that could be achieved by reducing the $PM_{2.5}$-level to 25 $\mu g/m^3$ (about 0.4%) and two times greater than a reduction to 20 $\mu g/m^3$. The mortality reduction could grow by more than seven times if fine PM levels were reduced to 10 $\mu g/m^3$ instead of 25 $\mu g/m^3$ (Ballester et al. 2008).

8.3.2 Short-Term Studies of Air Pollution Health Effects

From 1988 to 1993, the averages of the annual mean PM_{10} concentrations at 799 sites monitored by the U.S. Environmental Protection Agency (EPA) declined by 20%. Despite these improvements in air quality, Samet et al. (2000) reported associations between particle concentrations and the number of deaths per day in 20 of the largest cities and metropolitan areas in the U.S. from 1987 to 1994 with mean 24 h PM_{10} concentrations well below the standard. Analyses of the daily number of deaths occurring within an urban region have shown that an increase of each 10 $\mu g/m^3$ PM_{10} was associated with an increase in mortality of 0.21% (posterior standard error: 0.06) with a lag time of 1 day. The result is based on recent re-analyses of the National Morbidity, Mortality, and Air Pollution Study that included 90 urban areas of U.S. (Dominici et al. 2005).

The APHEA project (Air Pollution and Health: A European Approach) was a large multicenter study investigating the short-term effects of air pollution on health in 29 European cities (Katsouyanni et al. 2001). An increase of PM_{10} by 10 $\mu g/m^3$ was associated with increases of 0.76% (95%-CI: 0.47–1.05%) in cardiovascular deaths and 0.58% (95%-CI: 0.35–0.90%) in respiratory deaths (Analitis et al. 2006).

A study on traffic exposure and the onset of heart attack used data from non-fatal heart attacks from the Myocardial Infarction (MI) Registry in Augsburg, Germany, with 691 cases between 1999 and mid 2001 and a detailed recollection of activities during the 4 days before the event (bedside interviews). The analysis showed an odds ratio of 2.7 (95%-CI: 2.1–3.6) for times spent in traffic (car, bike or public transport) in association with having a heart attack 1 h later when adjusting for strenuous exercise, being outdoors and getting up in the morning (Peters et al. 2004, 2005).

8.3.3 Health Effects of Weather and Climate: Cold Effects

In 1998 it was estimated that there are up to 250,000 excess deaths in Western Europe due to cold weather (Keatinge 1998). The project PHEWE (Assessment and Prevention of Acute Health Effects and Weather Conditions in Europe) used mortality and climate data of almost every climatic region of the European continent between 1990 and 2000, and found that a 1°C decrease in minimum apparent temperature, a combined measure for temperature and humidity, during the cold season was associated with a 1.35% (95%-CI: 1.16–1.53%) increase in the daily number of total natural deaths (Analitis et al. 2008). The authors observed even higher percentages for cardiovascular mortality with 1.72% (95%-CI: 1.44–2.01%) and respiratory mortality with 3.30% (95%-CI: 2.61–3.99%). The increase was greater for older age groups and also in on average warmer (more southern) cities. The effect persisted up to 23 days with no evidence for mortality displacement. In addition, an increase in cardiovascular events in general was observed in winter months (Barnett et al. 2005). Danet et al. (1999) found a risk increase of 11% for a first MI, 26% for re-infarction and 11% for fatal infarctions in association with a 10°C decrease in temperature. A similar study of Wolf et al. (2009) using data between 1995 and 2004 of the Myocardial Infarction (MI) Registry in Augsburg, Germany, also showed an increased risk for MIs which were survived longer than 28 days, for fatal MIs as well as for first MIs. In both studies the effects of cold did not only occur during the cold season but were observed throughout the whole year. Wolf et al. (2009) were also able to show that the effects of cold were stronger during warmer summers and warmer winters. Like the PHEWE project other studies also showed that the effects of cold last up to 4 weeks (Braga et al. 2002). In a review, Nayha (2002) comes to the conclusion that there is an U-shaped association between coronary heart disease and air temperature with an increase in the event rate of 1% per 1°C temperature decrease on the cold side. The so-called "thermal optimum" (lowest mortality rates) was estimated to lie between 15°C and 20°C.

8.3.4 Health Effects of Weather and Climate: Heat Effects

In his review Nayha (2002) estimates a 4% increase in coronary heart disease event rate with a 1°C temperature increase above a temperature of 25°C, which means that deaths related to hot weather cannot only occur during heat waves. During the heat wave of 2003, up to 70,000 excess deaths (Robine et al. 2008) were estimated all over Europe, particularly in France. Heat wave effects occur after a very short time lag (same day or 1 day lag) (Basu and Samet 2002) and are not only pronounced in cardiovascular but also in respiratory mortality (Hajat et al. 2002). The effects are often modified by age, disease status, gender, socio-economic status, behavior, air condition, and prevention measures. Sometimes mortality displacement is observed and after a short increase in mortality a following compensatory decline in the number of deaths occurs a few days later. However, this only explains a small percentage of the observed increase in mortality during heat episodes (Le Tertre et al. 2006). In the above mentioned PHEWE project, a 1°C increase in maximum apparent temperature during the warm season was associated with a 3.12% (95%-CI: 0.60–5.72%) increase in the daily number of total natural deaths in Mediterranean cities and with a 1.84% (95%-CI: 0.06–5.72%) increase in north-continental cities (Baccini et al. 2008). The effects were stronger for respiratory deaths and for the elderly. The effect was limited to the 1 week following temperature excess, with evidence for mortality displacement. The temperature effects were also found for hospital admissions due to respiratory causes, but not for cardiovascular causes (Michelozzi et al. 2009). In a study during the California heat wave 2006, Knowlton et al. (2009) published a RR of 10.15 (95%-CI: 7.79–13.43) for hospitalization and a RR of 6.30 (95%-CI: 5.67–7.01) for emergency department visits for heat-related causes.

8.4 Discussion

8.4.1 Air Pollution

Despite important gaps in mechanistic knowledge, a comprehensive evaluation of all research findings on the health impact of air pollution provides persuasive evidence that exposure to particulate air pollution has adverse health effects, especially on the cardiopulmonary system. Overall, it was found that the effect estimates of long-term studies looking at the association between PM and mortality are larger than those from the daily time-series or case-crossover studies that evaluated daily changes in exposure.

The mechanisms responsible for promoting these adverse health effects are strongly debated. Repeated exposures to elevated ambient air pollution concentrations might not only transiently deteriorate risk factor profiles. Several putative pathways have been hypothesized to contribute to deaths from cardiovascular diseases (Brook et al. 2003). They can be summarized in three pathophysiological pathways: (1) Particles deposited in the bronchial tree can alter systematic autonomic balance, either indirectly, by provoking oxidative stress and inflammation in the lung or directly, by stimulating pulmonary neural reflexes, or a combination of both. Alterations in autonomic tone might contribute to the instability of a vascular plaque or initiate cardiac arrhythmias. Exposure to air pollution has been linked to ventricular arrhythmias (Berger et al. 2006), alteration in heart rate and heart rate variability (Gold et al. 2000), repolarization abnormalities (Henneberger et al. 2005), ST-segment depression (Pekkanen et al. 2002) and increased blood pressure (Ibald-Mulli et al. 2001). These studies show that besides the autonomic tone also ECG parameters reflecting myocardial substrate and vulnerability are affected by air pollution. All three are key factors in the mechanisms of cardiac death (Zareba et al. 2001). (2) Pulmonary oxidative stress and inflammation induce a systemic chain reaction by the release of circulating pro-oxidative and pro-inflammatory mediators from the lungs. The mediators include cytokines (e.g. interleukin-6), acute-phase reactants (e.g. C-reactive protein or fibrinogen), vasoconstrictive hormones (e.g. endothelins) and activated leukocytes, which may trigger various cardiovascular reactions (Seaton et al. 1995; Mills et al. 2007; Ruckerl et al. 2007; Schneider et al. 2008a; Peters et al. 1997; Baccarelli et al. 2007). (3) UFP or soluble particle constituents may rapidly cross the pulmonary epithelium into the circulation and interact directly with the cardiovascular system. These small particles might affect the vascular endothelium and atherosclerotic plaques, but also provoke local inflammation and oxidative stress (Oberdorster et al. 2002; Nemmar et al. 2004). Once in the circulation, UFP might have direct effects on the heart and other organs.

Direct pollutant effects are hypothesized to trigger acute cardiovascular events occurring within a few hours after exposure. This includes direct effects on hemostasis and the cardiovascular system by particles translocated into the circulation (pathway 3), but also alterations in autonomic tone by activation of pulmonary neural reflexes (pathway 1). Indirect air pollutant effects are supposed to rather evoke delayed and chronic cardiovascular responses. Pulmonary oxidative stress and inflammation may contribute to systemic oxidative stress and inflammation (pathway 2) which again may activate hemostatic pathways, impair vascular function, and accelerate atherosclerosis.

It is speculated that the contribution of air pollution to the development and exacerbation of atherosclerosis is an underlying factor in the worldwide observed associations between long-term air pollution and cardiovascular morbidity and mortality. Künzli et al. (2005) were able to demonstrate an increase in carotid intima-media thickness, a subclinical measure of atherosclerosis, with an increase of $PM_{2.5}$ in a cross-sectional study. This can be interpreted as the first epidemiological evidence for an association between long-term residential exposure and atherosclerosis.

8.4.2 Weather and Climate

It is widely discussed if global warming might lead to a stronger reduction in winter mortality compared to the increase in summer mortality due to more frequent, longer and more intense heat waves. But since we do not have widely accepted criteria for the definition of a heat wave and also do not have criteria for determining heat-related death (heat is often not given as causing or contributing reason for death on the death certificate), the actual magnitude of heat-related mortality could be greater than estimated so far. Persons living in urban environments may suffer from the so-called "urban heat island effect" which leads to higher temperature and humidity during the day and to more heat retain during the night. As in industrialized countries more and more of the population becomes urbanized and the percentage of people with higher age increases continuously, the threat of temperature-related mortality will probably become more severe over the next decades. Models of the relation between temperature and mortality are needed to predict the consequences of global warming, particularly for those most vulnerable and least able to adapt (Basu and Samet 2002). The PESETA project (Projection of economic impacts of climate change in sectors of the European Union based on bottom-up analysis: http://peseta.jrc.ec.europa.eu/) estimated 86,000 extra deaths per year with a global mean increase of 3°C in 2071–2100. But cold weather and cold spells could still affect Europe as climate change also includes more temperature variability and temperature extremes. At the moment, most European countries suffer from 5% to 30% excess winter mortality. Moreover, one should take into account more complex weather indicators such as air mass types as they might explain some of the geographic variation as well as variation in effect magnitude observed in different studies.

Potential mechanisms to explain the increased risk for coronary events in association with decreasing temperature include the stimulation of cold receptors in the skin and therefore the sympathetic nervous system, leading to a rise in the catecholamine level. The consequences are vasoconstriction, increased heart rate and blood pressure (Alpérovitch et al. 2009). An increased blood pressure decreases the ratio of myocardial oxygen supply to demand and may lead to myocardial ischemia, particularly in the vulnerable myocardium. Moreover, a drop in temperature could be related to an increase in fibrinogen and C-reactive protein (Schneider et al. 2008b). In cold conditions, the plasma concentration of certain clotting factors, blood lipids and platelet count and their in vitro aggregation are all increased and promote clotting (Keatinge et al. 1984; Elwood et al. 1993). Furthermore, reduced plasma volume and increased blood viscosity during cold exposure also tend to promote thrombosis (Keatinge et al. 1984). Hence, well-known risk factors are elevated during colder periods and recurrent changes in markers of atherothrombosis may contribute to the risk of triggering acute coronary events. The seasonal variation of cardiovascular events however could also be affected by the frequency of respiratory infections in winter, less physical activity and possibly changed nutrition habits. In the elderly, thermoregulation efficiency is reduced and the fibrinolytic system changes with increasing age which might explain the often higher effects of cold in the higher age groups.

The mechanisms of heat exposure have been less frequently studied. Keatinge et al. (1986) reported increased heart rate, blood viscosity, platelet and red blood cell count as well as dehydration and endothelial cell damage after the exposure of volunteers to air with 41°C. It is conceivable that heat exposure leads to an overload for the thermoregulation, which results in vasodilatation, a drop in blood pressure and less evaporation of perspiration from the skin. Dehydration leads to a loss in electrolytes, a significant increase in blood viscosity and changes in hemostasis. Elevated night time temperature does not allow recovery from severe heat stress experienced during the day. In addition, high levels of ozone might have a contributory effect (Bell et al. 2004).

8.5 Conclusion

8.5.1 Air Pollution

In summary, both fine and ultrafine particles are associated with respiratory and cardiovascular morbidity and mortality and appear to do so independently. There is also epidemiological evidence of similar responses to fine and ultrafine particles, although the size of the effects is often larger for ultrafine than for fine particles (at least on a per mass basis). In general, the relative effects of particulate air pollution are greater for respiratory than cardiovascular mortality. Nevertheless, due to the higher background rate of cardiovascular mortality, the absolute number of deaths attributable to particulate air pollution is much higher for cardiovascular than for respiratory deaths (Dockery 2001; Pope et al. 2004). Studies on particles mass concentration indicate that there is a linear relationship between PM_{10} and $PM_{2.5}$

and various health indicators for concentration levels between 0 and 200 $\mu g/m^3$, and no threshold in particle concentrations below which health would not be jeopardized. In 2007 the U.S. White House Office of Management and Budget found that for every dollar that was spent on air pollution regulations between five and ten dollars were saved with regard to reductions in hospitalizations and emergency room visits, lost work and school days, and premature deaths. Across the huge number of regulations from all federal agencies 63–88% of the estimated benefits were due to air pollution regulations between 1996 and 2006 (http://www.whitehouse.gov/omb/inforeg/2007_cb/2007_draft_cb_report.pdf).

8.5.2 Weather and Climate

Scientific consensus exists that climate change is anthropogenically forced, with effects on the ecological system and human health already in evidence (IPCC fourth assessment report). The effects include more frequent natural disasters such as storms, floods, heat waves, droughts, and wildfires resulting in injury, disease and mortality. The projected temperature increase for Europe by the end of the twenty-first century is 2.3–6.0°C. Climate change will also affect air quality, particularly ground-level ozone and allergenic pollens adding to the burden of chronic illnesses. The setup of targeted warning systems for cold events (e.g. in United Kingdom) or heat events (e.g. in Rome) are certainly measures that will need more and more attention in the next decades.

References

Alperovitch A, Lacombe JM, Hanon O et al (2009) Relationship between blood pressure and outdoor temperature in a large sample of elderly individuals: the Three-City study. Arch Intern Med; 169(1):75–80

Analitis A, Katsouyanni K, Biggeri A et al (2008) Effects of cold weather on mortality: results from 15 European cities within the PHEWE project. Am J Epidemiol; 168(12):1397–1408.

Analitis A, Katsouyanni K, Dimakopoulou K et al (2006) Short-term effects of ambient particles on cardiovascular and respiratory mortality. Epidemiology; 17(2):230–233.

Baccarelli A, Zanobetti A, Martinelli I et al (2007) Effects of exposure to air pollution on blood coagulation. J Thromb Haemost; 5(2):252–260

Baccini M, Biggeri A, Accetta G et al (2008) Heat Effects on Mortality in 15 European Cities. Epidemiology; 19(5):711–719

Ballester F, Medina S, Boldo E et al (2008) Reducing ambient levels of fine particulates could substantially improve health: a mortality impact assessment for 26 European cities. J Epidemiol Community Health; 62(2):98–105

Barnett AG, Dobson AJ, McElduff P, Salomaa V, Kuulasmaa K, Sans S (2005) Cold periods and coronary events: an analysis of populations worldwide. J Epidemiol Community Health; 59(7): 551–557

Basu R, Samet JM (2002) Relation between elevated ambient temperature and mortality: a review of the epidemiologic evidence. Epidemiol Rev; 24(2):190–202

Beelen R, Hoek G, van den Brandt PA et al (2008) Long-term effects of traffic-related air pollution on mortality in a Dutch cohort (NLCS-AIR study). Environ Health Perspect; 116(2):196–202
Bell ML, McDermott A, Zeger SL, Samet JM, Dominici F (2004) Ozone and short-term mortality in 95 US urban communities, 1987–2000. JAMA; 292(19):2372–2378
Berger A, Zareba W, Schneider A et al (2006) Runs of ventricular and supraventricular tachycardia triggered by air pollution in patients with coronary heart disease. J Occup Environ Med; 48(11):1149–1158
Braga AL, Zanobetti A, Schwartz J (2002) The effect of weather on respiratory and cardiovascular deaths in 12 U.S. cities. Environ Health Perspect; 110(9):859–863
Brook RD, Brook JR, Rajagopalan S (2003) Air pollution: the "Heart" of the problem. Curr Hypertens Rep; 5(1):32–39
CAFE (2005) A final set of scenarios for the Clean Air For Europe program. Technical report for the European Commission. International Institute for Applied Systems Analysis. Laxenburg, Austria, Clean Air For Europe Program.
Clancy L, Goodman P, Sinclair H, Dockery DW (2002) Effect of air-pollution control on death rates in Dublin, Ireland: an intervention study. Lancet; 360(9341):1210–1214
Danet S, Richard F, Montaye M et al (1999) Unhealthy effects of atmospheric temperature and pressure on the occurrence of myocardial infarction and coronary deaths. A 10-year survey: the Lille-World Health Organization MONICA project (Monitoring trends and determinants in cardiovascular disease). Circulation; 100(1):E1-E7
Dockery DW, Pope CA, III, Xu X et al (1993) An association between air pollution and mortality in six U.S. cities. N Engl J Med; 329(24):1753–1759
Dockery DW (2001) Epidemiologic evidence of cardiovascular effects of particulate air pollution. Environ Health Perspect; 109 Suppl 4:483–486
Dominici F, McDermott A, Daniels M, Zeger SL, Samet JM (2005) Revised Analyses of the National Morbidity, Mortality, and Air Pollution Study (NMMAPS), Part II: Mortality Among Residents of 90 Cities. Revised Analyses of Time-series Studies of Air Pollution and Health. Boston: Health Effects Institute: 9–24
Donaldson GC, Rintamaki H, Nayha S (2001) Outdoor clothing: its relationship to geography, climate, behaviour and cold-related mortality in Europe. Int J Biometeorol; 45(1):45–51
Elwood PC, Beswick A, O'Brien JR et al (1993) Temperature and risk factors for ischaemic heart disease in the Caerphilly prospective study. Br Heart J; 70:520–523
Friedman MS, Powell KE, Hutwagner L, Graham LM, Teague WG (2001) Impact of changes in transportation and commuting behaviors during the 1996 Summer Olympic Games in Atlanta on air quality and childhood asthma. JAMA; 285(7):897–905
Gold DR, Litonjua A, Schwartz J et al (2000) Ambient pollution and heart rate variability. Circulation; 101(11):1267–1273
Hajat S, Kovats RS, Atkinson RW, Haines A (2002) Impact of hot temperatures on death in London: a time series approach. J Epidemiol Community Health; 56(5):367–372
Henneberger A, Zareba W, Ibald-Mulli A et al (2005) Repolarization changes induced by air pollution in ischemic heart disease patients. Environ; 113(4):440–446
Hoek G, Brunekreef B, Goldbohm S, Fischer P, van den Brandt PA (2002) Association between mortality and indicators of traffic-related air pollution in the Netherlands: a cohort study. Lancet 360(9341):1203–1209
Ibald-Mulli A, Stieber J, Wichmann HE, Koenig W, Peters A (2001) Effects of air pollution on blood pressure: a population-based approach. Am J Public Health; 91(4):571–577
Intergovernmental Panel on Climate Change"(IPCC) fourth assessment report (2007) (Accessed October 16, 2009, at http://www.ipcc.ch.)
Janes H, Sheppard L, Lumley T (2005) Case-crossover analyses of air pollution exposure data: referent selection strategies and their implications for bias. Epidemiology; 16(6):717–726
Katsouyanni K, Touloumi G, Samoli E et al (2001) Confounding and effect modification in the short-term effects of ambient particles on total mortality: results from 29 European cities within the APHEA2 project. Epidemiology; 12(5):521–531

Keatinge WR, Coleshaw SR, Cotter F, Mattock M, Murphy M, Chelliah R (1984) Increases in platelet and red cell counts, blood viscosity, and arterial pressure during mild surface cooling: factors in mortality from coronary and cerebral thrombosis in winter. Br Med J (Clin Res Ed); 289(6456):1405–1408

Keatinge WR, Coleshaw SR, Easton JC, Cotter F, Mattock MB, Chelliah R (1986) Increased platelet and red cell counts, blood viscosity, and plasma cholesterol levels during heat stress, and mortality from coronary and cerebral thrombosis. Am J Med; 81(5):795–800

Keatinge WR (1998) Mortality in winter. Eur Heart J; 19(3):361–362

Knowlton K, Rotkin-Ellman M, King G et al (2009) The 2006 California heat wave: impacts on hospitalizations and emergency department visits. Environ Health Perspect; 117(1):61–67

Künzli N, Jerrett M, Mack WJ et al (2005) Ambient Air Pollution and Atherosclerosis in Los Angeles. Environ Health Perspect; 113(2):201–206

Laden F, Schwartz J, Speizer FE, Dockery DW (2006) Reduction in fine particulate air pollution and mortality: Extended follow-up of the Harvard Six Cities study. Am J Respir Crit Care Med; 173(6):667–672

Le Tertre A, Lefranc A, Eilstein D et al (2006) Impact of the 2003 heatwave on all-cause mortality in 9 French cities. Epidemiology; 17(1):75–79

Michelozzi P, Accetta G, De Sario M et al (2009) High temperature and hospitalizations for cardiovascular and respiratory causes in 12 European cities. Am J Respir Crit Care Med; 179(5): 383–389

Michelozzi P, de Donato F, Accetta G et al (2004) Impact of Heat Waves on Mortality - Rome, Italy, June-August 2003. JAMA; 291(21):2537–2538

Mills NL, Tornqvist H, Gonzalez MC et al (2007) Ischemic and thrombotic effects of dilute diesel-exhaust inhalation in men with coronary heart disease. N Engl J Med; 357(11):1075–1082

Nayha S (2002) Cold and the risk of cardiovascular diseases. A review. Int J Circumpolar Health; 61(4):373–380

Nemmar A, Hoet PH, Vanquickenborne B et al (2002) Passage of inhaled particles into the blood circulation in humans. Circulation; 105(4):411–414

Nemmar A, Hoylaerts MF, Hoet PH, Nemery B (2004) Possible mechanisms of the cardiovascular effects of inhaled particles: systemic translocation and prothrombotic effects. Toxicol Lett; 149(1–3):243–253

Oberdorster G, Sharp Z, Atudorei V et al (2002) Extrapulmonary translocation of ultrafine carbon particles following whole-body inhalation exposure of rats. J Toxicol Environ Health A; 65(20): 1531–1543

Oberdorster G, Sharp Z, Atudorei V et al (2004) Translocation of inhaled ultrafine particles to the brain. Inhal Toxicol; 16(6–7):437–445

O'Neill MS, Zanobetti A, Schwartz J (2003) Modifiers of the temperature and mortality association in seven US cities. Am J Epidemiol; 157(12):1074–1082

Pascal M, Laaidi K, Ledrans M et al (2006) France's heat health watch warning system. Int J Biometeorol; 50(3):144–153

Pekkanen J, Peters A, Hoek G et al (2002) Particulate air pollution and risk of ST-segment depression during repeated submaximal exercise tests among subjects with coronary heart disease: the Exposure and Risk Assessment for Fine and Ultrafine Particles in Ambient Air (ULTRA) study. Circulation; 106(8):933–938

PESETA project - Projection of economic impacts of climate change in sectors of the European Union based on bottom-up analysis (2007) (Accessed October 16, 2009, at http://peseta.jrc.ec.europa.eu/.)

Peters A, Doring A, Wichmann HE, Koenig W (1997) Increased plasma viscosity during an air pollution episode: a link to mortality? Lancet; 349(9065):1582–1587

Peters A, von Klot S, Heier M et al (2004) Exposure to traffic and the onset of myocardial infarction. N Engl J Med; 351(17):1721–1730

Peters A, von Klot S, Heier M et al (2005) Particulate air pollution and nonfatal cardiac events. Part I. Air pollution, personal activities, and onset of myocardial infarction in a case-crossover study. Res Rep Health Eff Inst (124):1–66

Pope CA, III, Burnett RT, Thun MJ et al (2002) Lung cancer, cardiopulmonary mortality, and long-term exposure to fine particulate air pollution. JAMA; 287(9):1132–1141

Pope CA, III, Burnett RT, Thurston GD et al (2004) Cardiovascular mortality and long-term exposure to particulate air pollution: epidemiological evidence of general pathophysiological pathways of disease. Circulation; 109(1):71–77

Pope CA, III, Dockery DW (2006) Health effects of fine particulate air pollution: lines that connect. J Air Waste Manag Assoc; 56(6):709–742

Robine JM, Cheung SL, Le Roy S et al (2008) Death toll exceeded 70,000 in Europe during the summer of 2003. C R Biol; 331(2):171–178

Ruckerl R, Greven S, Ljungman P et al (2007) Air Pollution and Inflammation (IL-6, CRP, Fibrinogen) in Myocardial Infarction Survivors. Environ; 115(7):1072–1080

Samet JM, Zeger SL, Dominici F et al (2000) The National Morbidity, Mortality, and Air Pollution Study. Part II: Morbidity and mortality from air pollution in the United States. Res Rep Health Eff Inst; 94(Pt 2):5–70

Schneider A, Neas L, Herbst MC et al (2008a) Endothelial dysfunction: associations with exposure to ambient fine particles in diabetic individuals. Environ Health Perspect; 116(12):1666–1674

Schneider A, Panagiotakos D, Piccioto S et al (2008b) Air temperature and inflammatory responses of myocardial infarction survivors in a European panel study. Epidemiology; 19(3): 391–400

Seaton A, MacNee W, Donaldson K, Godden D (1995) Particulate air pollution and acute health effects. Lancet; 345(8943):176–178

White House Office of Management and Budget (OMB) Report (2007) (Accessed October 16, 2009, at http://www.whitehouse.gov/omb/inforeg/2007_cb/2007_draft_cb_report.pdf.)

WHO (1999) European Center for Health Policy. Health impact assessment: main concept and suggested approach. The Gothenburg Consensus Paper. Copenhagen, Regional Office for Europe

WHO (2001) Quantification of health effects of exposure to air pollution. Copenhagen, Regional Office for Europe: EUR/01/5026342

Wolf K, Schneider A, Breitner S et al (2009) Air temperature and the occurrence of myocardial infarction in Augsburg, Germany. Circulation; 120(9):735–742

Zareba W, Nomura A, Couderc JP (2001) Cardiovascular effects of air pollution: what to measure in ECG? Environ Health Perspect; 109 Suppl 4:533–538

Chapter 9
Climate Change and Infectious Diseases in Megacities of the Indian Subcontinent: A Literature Review

Md. Mobarak Hossain Khan, Alexander Krämer, and Luise Prüfer-Krämer

9.1 Introduction

Global environmental change or climate change is a growing and challenging area of multidisciplinary research. It poses an emerging threat to global public health as well as to the wellbeing of many populations (Costello et al. 2009; Campbell-Lendrum and Corvalan 2007). It inhibits the progress of poverty reduction and the reaching of the Millennium Development Goals (Mitchell and Tanner 2006). Annually, over 150,000 deaths and 5 million disability-adjusted life years (DALYs) losses occur due to such change (Patz and Olson 2006). According to the recent report of the UCL Lancet Commission, the health effects of climate change will be even stronger in the next decades and will place the lives and wellbeing of billions of people at increased risk (Costello et al. 2009).

The major aspects of environmental change which adversely affect health outcomes are changing patterns of disease, water and food insecurity, vulnerable shelter and human settlements, extreme weather events, and increasing population growth and migration (Costello et al. 2009). The health risks attributable to climate change are inequitable (Campbell-Lendrum and Corvalan 2007; Bigio 2002) and depend on the level of urbanisation, prevailing socio-economic conditions, preventive behaviours, and the adaptive capacity of the human populations (Bhattacharya et al. 2006; McGeehin and Mirabelli 2001). The poorest populations with limited access to health care are the most vulnerable to the impact of global environmental change (Costello et al. 2009; Campbell-Lendrum and Corvalan 2007; WHO 2003). For instance, about 99% of all extreme climate/weather-related global deaths in 1990 occurred in developing countries (WHO 2003). The lack of necessary

M.M.H. Khan (✉) • A. Krämer
Department of Public Health Medicine, School of Public Health, Bielefeld University, Bielefeld, Germany
e-mail: mobarak.khan@uni-bielefeld.de

L. Prüfer-Krämer
Travel Clinic, Bielefeld, Germany

A. Krämer et al. (eds.), *Health in Megacities and Urban Areas*,
Contributions to Statistics, DOI 10.1007/978-3-7908-2733-0_9,

institutional, economical and financial capacity, as well as the ability to rebuild the infrastructure damaged by the natural disasters makes poor countries more vulnerable to the impact of climate change (Costello et al. 2009; Campbell-Lendrum and Corvalan 2007; Huq et al. 2003)

Climate effects are mostly area-specific (Hess et al. 2008). Some of the hot spots for environmental change are cities/megacities, coastal regions, low lying lands prone to river flooding, areas experiencing weather extremes, areas with a high endemicity of diseases sensitive to climate change, and areas currently experiencing food insecurity (Hess et al. 2008; IPCC 2007; Patz and Kovats 2002). Rapidly growing cities and megacities of developing countries are particularly vulnerable to climate change mainly because of their geographical locations (e.g. >50% of all megacities are situated at sea level), high population density, severe water and air pollution, poor sanitation, inadequate drainage systems, poor solid waste management, heat waves, ecological imbalance due to unplanned urbanisation and deforestation, and slum development in climate-prone areas (Kovats and Akhtar 2008; Alam and Rabbani 2007; Campbell-Lendrum and Corvalan 2007; Patz and Kovats 2002). City-based commercial, industrial and transport activities contribute to significant amounts of greenhouse gases (Grimm et al. 2008; Alam and Rabbani 2007) and increase the surface temperature (e.g. Dhaka megacity in Bangladesh).

Most of the world largest cities including the megacities of the Indian subcontinent (Fig. 9.1) are coastal cities (Bigio 2002) and are directly disturbed by

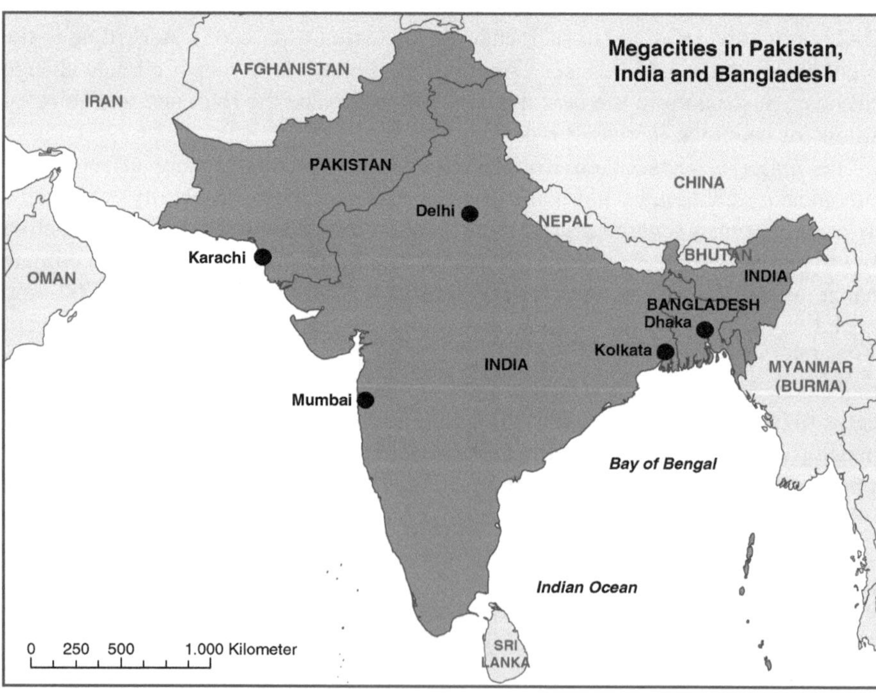

Fig. 9.1 Five megacities in the Indian sub-continent

climate change (Kovats and Akhtar 2008; Bigio 2002; Patz and Kovats 2002). Particularly the sea level rise, which is often associated with changing storm frequency and intensity, intensified rainfall and flooding, and changing patterns of the flow of water mainly from rain, snowmelt, or over land (also called run-off) (Kovats and Akhtar 2008; Nicholls 1995) can increase the vulnerability remarkably. Already many megacities are exposed to the threat of hurricanes or tropical storms, and flooding due to storm surges or both (Nicholls 1995).

9.2 Linkages Between Climate Change and Infectious Diseases

Numerous studies already documented the linkages between climate change and infectious diseases (e.g. Greer et al. 2008; Zhang and Hiller 2008; Patz and Olson 2006; Khasnis and Nettleman 2005; Sutherst 2004; Hunter 2003; Lipp et al. 2002; Epstein 2001). For instance, the risk of diarrhoea could be up to 10% higher in 2030 in regions experiencing climate change as compared to the regions without such change. Similarly about 6% of malaria cases in some middle income countries are attributed to climate change (WHO 2003). Although epidemiological studies regarding infectious diseases in megacities are very limited, we assumed that the burden of infectious diseases among megacity populations will be comparatively higher as compared to rural areas. There are some potential reasons behind our assumption. For instance, the breeding sites for vectors can be extended in urban areas through e.g. decreasing water supply, increasing construction of overhead water storage tanks in most of the houses and increasing water storage practices, and availability of discarded tyres and bottles especially in rainy seasons (Ratho et al. 2005). The spread of potential vectors may expand to areas of higher altitudes or adjacent latitudes (Bhattacharya et al. 2006; Hunter 2003). Deforestation and new habitation due to e.g. urbanisation can influence malaria through creation of new breeding areas and vector varieties as well as immigration of susceptible populations (WHO 2003). Inundation and flooding due to sea level rise, storms and heavy rainfall may result in higher probabilities for water-borne diseases such as cholera and other diarrheal diseases. High population density and higher contact rate in urban areas will increase the likelihood of transmission of infections with the possibility of outbreaks and epidemics. Various infections can be imported into cities due to national and international migration. Special human host characteristics like impaired immunity or immunological deficits due to malnutrition or chronic infections may increase the likelihood for the acquisition of infections on the individual level. Poor sanitation and sewage disposal in marginal settlements can enhance the risk.

In this review we summarise the epidemiological findings of three most climate-sensitive infectious diseases namely diarrhoea/cholera (as water-borne disease), dengue and malaria (as vector-borne diseases) in five megacities of the Indian sub-continent. These diseases are generally more sensitive to climate change than others (Zhang and Hiller 2008; Khasnis and Nettleman 2005; Sutherst 2004;

Curriero et al. 2001). Moreover, these diseases are common in both Indian subcontinent (Kovats and Akhtar 2008). Some environmental characteristics of these megacities are also reported separately. Another section is included to discuss some multi-level strategies in order to reduce the impact of climate change. In this review, most of the references were obtained from "Pubmed" and "Google scholar". We also checked the reference sections of the selected articles/reports and included some of them after review. Such type of assessment in megacities may be crucial not only due to the scarcity of information but also for getting some useful hints about the burden of these infectious diseases in other large cities or megacities. Moreover, these five megacities will be the leading megacities in the world (except Tokyo) by 2025. For instance, the rank of Mumbai in the world megacities will shift from position 4 in 2007 to position 2 in 2025. Similarly, Delhi will move from 6 to 3, Dhaka from 9 to 4, and Karachi from 12 to 10 (UN 2008).

9.3 Results

In this section, first the general environmental characteristics of each megacity are summarised, followed by the epidemiological findings of selected infectious diseases. A brief comparison of the megacities is also added at the end of this section.

9.3.1 Mumbai Megacity

Mumbai is the financial and commercial centre of India and a major industrial port (Khan et al. 2004). It is densely populated and is ecologically wet and dry (Tikar et al. 2008). It is a rapidly growing megacity, with a projected population of 26.4 million in 2025 (UN 2008). Annually over 250,000 rural migrants come to the city. The mean surface temperature has increased by 0.32°C per decade (Khan et al. 2004). This megacity could face profound consequences from climate change due to a high population density, and its major industrial and financial installations. The major proportion of its land is in low-lying areas. Therefore this megacity is vulnerable to the impact of frequent floods due to increasing rainfall and rising sea level. The impact of flooding is often exacerbated by blocked canals and drains (Kovats and Akhtar 2008). A majority of the population lives in slums, characterised by unhygienic living conditions, overcrowding, poor housing, and lack of basic amenities (Kothari 1987). These poor people have limited capacity to cope with the consequences of climate change (Kovats and Akhtar 2008).

Publications regarding diarrhoeal diseases and cholera in Mumbai megacity are very scarce. To our knowledge, no study explicitly examined the association of these infectious diseases in relation to climate variables. Although malaria was well contained in Mumbai through control of mosquito breeding sites and legal provisions

(Kumat 2000), unfortunately it has re-emerged in 1992. During 1992–1997, the city witnessed a manifold increase in the number of malaria cases diagnosed and treated by the public health system (Kumat 2000). It is reported that increasing resistance to chloroquine is one of the causes of resurgence of malaria in this city (Garg et al. 1999). Two studies published after 1995 reported outbreaks of dengue in Mumbai (Karande et al. 2005; Shah et al. 2004). According to these studies, dengue fever is emerging (Karande et al. 2005) or rising in Mumbai with increased incidence among children during the post-monsoon season (Shah et al. 2004).

9.3.2 Delhi Megacity

Delhi, the capital city of India, is located in the semi arid zone of northern India (Tikar et al. 2008). The population of Delhi grew rapidly from 1.44 to 12.82 million during 1951–2001. The densely populated (e.g. 9,294 persons per square km in 2001) and rapidly growing Delhi megacity experiences very high level of pollution. Vehicular traffic is the most important source of air pollution. The transport demand increased from 37.4 thousands in 1961 to 2,629.6 thousands in 1996 due to increasing population, urbanisation, and industrialisation. Major sources of water pollution are domestic, sewage and industrial effluents. The quantity of sewage and liquid wastes from human settlements and uncontrolled industries far exceeds both the city's wastewater management and carrying capacity of its sewers. The water quality is affected by inadequate availability of basic facilities and a rapidly increasing population. Exposure to environmental pollution is now almost an inescapable part of urban life (Nagdeve 2004).

Dengue seems to be common in Delhi since several decades. Many articles are available for this city in this respect. The epidemiology of dengue infection is rapidly changing in the city. Delhi has experienced six outbreaks of dengue virus infection namely in 1967, 1970, 1982, 1988, 1996 and 2003 (Gupta et al. 2005). However, the largest outbreak of dengue in Delhi occurred in 1996 during August-November and indicated a serious resurgence of dengue in this country. A total of 8,900 cases were reported and the death rate was 4.2%. The analysis of dengue outbreaks in Delhi indicated a seasonal trend. All outbreaks occurred during the monsoon (rainy season from August to November) and subsided with the onset of winter (Dar et al. 1999). Malaria seems to be uncommon in Delhi megacity. Although we checked about 500 abstracts, none of them was explicitly related to climate factors.

Cholera caused by either Vibrio cholerae O1 or O139 is endemic in Delhi and its peripheral areas (Sharma et al. 2007) with an increasing trend (Datta et al. 1993). The endemicity of cholera was almost constant in Delhi since 1992 (Sharma et al. 2007). It is found to be highly seasonal (Singh et al. 1995). A large scale cholera outbreak occurred in 1988 since its first detection in 1965 (Datta et al. 1993). For instance, the number of cholera cases in July–August in 1988 was five to ten times higher as compared to the same period in previous years (Khanna et al. 1990).

Children under the age of 4 years, irrespective of sex, were most affected by that outbreak. Lower socio-economic status, poor personal hygiene, absence of sanitary latrines, drinking water and food storage practices were the major risk factors (Singh et al. 1995; Datta et al. 1993). Recently Delhi experienced two outbreaks of cholera (in 2003 and 2004) with peaks in August and April respectively. One possible reason was the ability of vibrios' to grow rapidly in warm environmental temperatures (Sharma et al. 2007).

9.3.3 Calcutta Megacity

Calcutta is the third largest megacities (population 14.8 million) in India with a density of around 9,000 per sq km (Hasan and Khan 1999). Rampant land filling in low lying areas, conversion of wetlands to satellite townships, shrinkage of the drainage outfall basin, disturbed ecosystems, homelessness, congestions and degraded living conditions are the characteristics of Calcutta. The insufficient supply of urban services contributes to the slum development, illegal construction and undesirable land-use changes, deterioration in air and water quality, and poor health and hygiene. The fast growth of Calcutta has generated many environmental problems in the city. About 50% of the total population lives in slums and squatter settlements. About 44% of the population lives in very poor quality houses with low level of urban services. Huge amount of the uncollected waste remains on the roads and is scavenged by rag-pickers, animals and birds and can deteriorate environmental conditions. The uncollected waste flows into the gullies and open drains during the rainfall. Only 50% of its population and 27% of its area have sewage and drainage facilities. The total city sewage is discharged into the nearest water body and nearest open surface drains. Due to unplanned growth of the city, characterised by poor drainage and resultant water-logging, a simple rain submerges many parts of the city especially in the low-lying areas. The logged water favours malaria transmission and outbreaks. Around 700 t of air pollutants are emitted everyday, of which 240 t are created by vehicles. Poorly maintained vehicles adversely deteriorate the air pollution in the city. The flood water increases the chance of surface water contamination by sewage and waste water. The contaminated water then enters the distribution systems from stand points and other entry points. The high mobility of the slum dwellers is also a major source of disease transmission to all over the city (Hasan and Khan 1999).

Several studies reported the resurgence of malaria in Calcutta in the 1990s (e.g. Basu et al. 1998; Mandal et al. 1998). Malaria cases steadily increased from around 8,000 in 1984 to more than 23,000 in 1996 (Chattopadhyay and Sengupta 2000). The incidence of *P. falciparum* malaria increased more than eleven folds in 1996 as compared to 1990 (Mukhopadhyay et al. 1997). The occurrence of malaria in Calcutta also varied seasonally (Mandal et al. 1998). Calcutta faced a malaria epidemic in 1995 characterised by an increased occurrence of both *P. falciparum* malaria and *P. vivax* (Chattopadhyay and Sengupta 2000). *P. falciparum* accounts for approximately 60%

of malaria cases in Calcutta (Kim et al. 2006). Chloroquine resistance may be the reason for the increase of malaria in this city (Nandy et al. 2003).

Dengue fever (DF) and dengue haemorrhage fever (DHF) are recurring in Calcutta (Pramanik et al. 2007). This city had a long experience of recurring epidemics of dengue fever (Banik et al. 1994). Although the outbreak of dengue in Calcutta was first documented during the 1960s (Tandon and Raychoudhury 1998; Banik et al. 1994), it is reported increasingly in recent times (Hati 2006; Bhattacharjee et al. 1993). However, none of these studies reported the association between climate factors and dengue.

Both diarrhoea and cholera are prevalent in Calcutta (Deen et al. 2008; Sur et al. 2006; Sur et al. 2005; Banerjee et al. 2004; Dutta et al. 2003; Basu et al. 2000; Bhattacharya et al. 1994). The city is known as 'homeland of cholera'. A substantial burden of cholera (Sur et al. 2005) including several outbreaks of diarrhoea was reported in Calcutta (Sur et al. 2006). The overall incidence of treated diarrhoea and cholera episodes was 57.7 cases and 2.2 cases per thousand/year respectively (Sur et al. 2005). The burden of cholera was greatest among those less than 2 years of age (Sur et al. 2005; Deen et al. 2008). The prevalence of diarrhoeal diseases was also highest in the people of lower classes as compared to the upper classes. Acute watery diarrhoea was the commonest type, followed by dysentery and persistent types (Banerjee et al. 2004). Most of the outbreaks were reported among people living in urban slums. Unsafe water supply, poor environmental sanitation, indiscriminate defecation, and lack of personal hygiene are mainly responsible for the continued transmission of these diseases (Sur et al. 2006). These diseases are also reported to be seasonal (Sur et al. 2005; Basu et al. 2000). Improvement of living conditions and sanitation, dissemination of health education, and the supply of safe drinking water are some of the effective ways to reduce the impact of these diseases (Deen et al. 2008; Dutta et al. 2003). Although many studies are available, no study highlighted the association of climate factors with diarrhoeal diseases.

9.3.4 Dhaka Megacity

Dhaka, the capital city of Bangladesh, is one of the fastest growing megacities in the world (Burkart et al. 2008; UN 2008). It is now the ninth largest megacity in the world with about 13.5 million inhabitants (UN 2008). Every year 300,000–400,000 new migrants, mainly the rural poor, move to Dhaka and most of them initially concentrate in the slums and squatter settlements. The slum population increased from 20% in 1996 to 37% in 2005. Slum formations in the climate-prone or low-lying areas, poor housing, traffic congestions, water shortage, garbage mismanagement, higher temperature due to increasing green house gas emissions, and higher pollution are very much common in Dhaka (Khan and Kraemer 2008). Dhaka is generally warmer as compared to other parts of the country (Quadir et al. 2004). All these factors make this city vulnerable to the impact of climate change. For instance, this city experienced three severe floods during the last 20 years namely in

1988, 1998 and 2004 (Schwartz et al. 2006). Floods affect water resources and sanitary conditions and increase the susceptibility to infectious disease.

Data show that dengue fever and dengue hemorrhagic fever re-emerged in the megacity of Dhaka and other large cities in 2000 (Yunus et al. 2001). The city experienced some recent outbreaks particularly during the monsoon and rainy season from July to October (Islam et al. 2006; Podder et al. 2006; Wagatsuma et al. 2004; Rahman et al. 2002; Yunus et al. 2001). For instance, a total of 3,383 cases of DF and 581 cases of DHF were reported in Dhaka in 2000, of which 51 died (Yunus et al. 2001). The dengue outbreak in this city was reported to be associated with local factors (Podder et al. 2007). It affected all age groups including children (Chowdhury et al. 2004). Males seemed to be more affected than females, as male comprised 82.2% of dengue patients (74 out of 90) admitted to a hospital in Dhaka. Cities were more vulnerable because 77% of the patients came from the city of Dhaka (Alam et al. 2004). Month-wise data indicated that the seropositivity rate of dengue fever was 8.5% in July, 50.1% in August, and 10.0% in September (Chowdhury et al. 2004). Unfortunately, no data was available regarding the association of climate variables such as rainfall and temperature with dengue. As most of the dengue outbreaks occurred in Dhaka with increasing trends, it can be concluded that megacity inhabitants will be highly affected by dengue in the future (Islam et al. 2006; Podder et al. 2006; Chowdhury et al. 2004; Wagatsuma et al. 2004).

Although malaria is a public health problem in some of the forests and forest fringe areas of the north eastern and south eastern borders of Bangladesh (Alam 2008), Dhaka has not yet been found to be affected by this disease. No study regarding malaria and climate factors was found for Dhaka in our review.

Cholera is a major public health problem in Dhaka (Hashizume et al. 2008a; Lobitz et al. 2000; Pascual et al. 2000). Between March and April 2002, a resurgence of Vibrio cholerae O139 occurred in Dhaka and adjacent areas with an estimated 30,000 cases of cholera (Faruque et al. 2003). High temperature, river level, and floods have been invoked to explain the seasonality of cholera since the early times. Heavy rain leads to flooding, which may affect water and sanitation systems and thereby promote the use of contaminated water e.g. in bathing and washing (Hashizume et al. 2008). Three studies based on time series data indicated that climate change acts as a driver in the dynamics of disease (Pascual et al. 2000) and that the cholera epidemic is climate-linked (Rodo et al. 2002; Lobitz et al. 2000; Pascual et al. 2000). The ENSO system is the primary driver of inter-annual variability in global climate and clearly associated with cholera during the last two decades (Rodo et al. 2002). The seasonality of cholera incidences in Dhaka suggests that weather factors play a role through multiple pathways (Hashizume et al. 2008; Pascual et al. 2000). Cholera incidences were higher before the monsoon (high rainfall period) and at the end of the period with a trough in the middle of the monsoon. For a 10 mm increase above the rainfall threshold (45 mm), the number of cholera cases increased by 14% after controlling for temperature and other factors (Hashizume et al. 2008). High temperature and a rising river level was associated with increased rotavirus-associated diarrhoea in Dhaka (Hashizume et al. 2008b; Hashizume et al. 2007).

9.3.5 *Karachi Megacity*

Karachi is the largest city in Pakistan with a population of 5.2 million in 1981, 9.2 million in 1998 (ADB 2005) and 12.1 million in 2007 (UN 2008). Annually about 300,000 new migrants to the city create a serious shortage of housing and overburden the adequate water supply, public transport and city infrastructure. Up to 40% of Karachi's population live in squatter settlements. The rapid growth of the city is deteriorating the living conditions and environment continuously. Poverty, poor quality and overcrowded housing, inadequate access to public services, infrastructure and health care are some features of this city (ADB 2005).

The development of Karachi in terms of infrastructure, residential areas, new industries, increasing vehicles and rapid growth of the urban population produced remarkable effects on the urban temperature. During the period of 1947 to 2005, the mean maximum and annual temperatures increased by 4.60°C and 2.25°C respectively. Air pollution level in Karachi is among the highest in the world (Sajjad et al. 2009). The average rainfall is 7.71 in., of which 6.65 in. are received during the monsoon period lasting from June to September. The hottest month is June with a mean monthly temperature of about 97°F. The winter season is very short lasting from November to January. The strong coastal winds are characteristic for this region (Perveen et al. 2007).

The first outbreak of dengue occurred in 1994 in Pakistan, mostly affecting children (Jamil et al. 2007; Akram and Ahmed 2005; Paul et al. 1998). Some studies reported dengue outbreaks in Karachi after this period particularly in 2006 (Ahmed et al. 2008; Khan et al. 2007; Akram and Ahmed 2005; Qureshi et al. 1997). During 2005–2006, there was an unprecedented increase in epidemic DHF with a large number from Karachi during the period of August and October (Daily Times Monitor 2007). Unfortunately none of these articles focused on climate variables. In Pakistan, malaria was mainly concentrated among Afghan refugees (Kazmi and Pandit 2001; Suleman 1988; Nalin et al. 1985). According to our search, no study focused on climate factors and malaria in this megacity, although increasing trends of plasmodium falciparum infection (Khan et al. 2005) due to emerging chloroquine and quinine resistance in Pakistan (Khan et al. 2006) are reported.

A re-emergence of Vibrio cholerae O139 in 2000–2001 from a tertiary care hospital in Karachi was reported (Jabeen and Hasan 2003) although one study had reported its disappearance by 1996 (Sheikh et al. 1997). The bacterial pathogens showed a distinct seasonal variation with summer predilection (Alam et al. 2003; Sheikh et al. 1997) peaking in July and August (Sheikh et al. 1997). The examination of 818 stools collected from Karachi during 1990 and 1997 revealed that rotavirus was identified among 14% stools (Nishio et al. 2000). Another study found about 12.3% children with rotavirus in 1990 and 24.4% in 1991 (Agboatwalla et al. 1995). A rapidly expanding population in Karachi combined with civil unrest and a crumbling infrastructure experience no basic sanitation and no clean water to many of its inhabitants. In such situations, enteric infections continue to take their toll (Sheikh et al. 1997). We examined the abstracts of about 100 articles, but none of them focused on the association between climate variables and cholera/diarrhoeal diseases.

9.4 Comparative Analysis of the Five Megacities

The following information (Table 9.1) shows the similarities and differences of the five megacities with respect to some selected factors related to climate change and infectious diseases. The overall infrastructure of these megacities is poor and they are experiencing rapid urbanisation. A large portion of the population lives in slum areas. High density, lack of infrastructure, poor housing, social inequality, air and water pollution, water shortage, poor garbage management and sewage systems, poor health systems are some of the common characteristics in these megacities. Temperature is rising in all megacities and dengue is resurging with an increasing trend. Malaria seems to be very uncommon in Mumbai, Dhaka and Karachi. Diarrhoeal diseases are more common in Dhaka and Calcutta than in other megacities. Few studies are available regarding climate factors and infectious diseases in Dhaka, mainly published by the International Centre for Diarrhoeal Disease Research, Bangladesh (ICDDR,B).

9.5 Strategies to Reduce the Impact of Climate Change

Multi-level prevention strategies (Table 9.2) are needed to reduce the impact of climate change, which work at national, city and neighbourhood levels and bring together the stakeholders such as private ones and the civil society (Revi 2008).

At the micro level, increasing awareness, education, personal hygiene, capacity building and risk management are necessary. Cleaning of water coolers, storage tanks, and tyres are required to destroy the vector breeding sites (Ratho et al. 2005). Individual behaviours such as avoiding intensive car use, use of the community bus rather than the individual car, less use of energy, quitting smoking and walking rather than using vehicles for a small distance may be worthy to reduce the greenhouse gas emissions.

At the meso-level, strengthening of the community capacity to reduce the risk of infectious diseases through water, waste, garbage, and ecology management might be useful. Such strategies are particularly important during natural disasters and outbreaks. For instance, community involvement for providing information on mosquito control during the disease transmission period is necessary for sustainable

Table 9.1 Similarities and differences between the five megacities

Characteristics/infectious diseases	Mumbai	Delhi	Calcutta	Dhaka	Karachi
Coastal location	Yes	No	Yes	Yes	Yes
Population (million)	19.0	15.9	14.8	13.5	12.1
Overall infrastructure	Poor	Poor	Poor	Poor	Poor
Urbanisation	Rapid	Rapid	Rapid	Rapid	Rapid
Slum population	Huge	Huge	Huge	Huge	Huge
Population density	High	High	High	High	High
Poverty rate	High	High	High	High	High
Water pollution	Increasing	Increasing	Increasing	Increasing	Increasing
Drainage	Poor	Poor	Poor	Poor	Poor
Flood affected	Yes	No	Yes	Yes	–
Temperature	Increasing	Increasing	Increasing	Increasing	Increasing
Vulnerability level	High	High	High	High	High
Diarrhoea/cholera	Scarce information	Increasing (some recent outbreaks)	Highly prevalent and some recent outbreaks	Highly prevalent (mainly among poor people)	Emerging (O139) and showing seasonal trend
Malaria	Increasing recently	Scarce information	Resurging recently (perhaps due to resistance)	Scarce information	Scarce information
Dengue	Emerging (recent outbreak)	Emerging (many recent outbreaks)	Recurring and increasing	Emerging and some recent outbreaks	Emerging and some recent outbreaks
Studies explicitly focusing on infectious diseases and climate factors	Not available	Not available	Not available	Partially available (only related to diarrhoea and cholera)	Not available

–, no information

Table 9.2 Multi-level prevention strategies to reduce the impact of climate change on infectious diseases in megacities

Micro-level strategies
▪ Maintenance of personal hygiene and improved lifestyles
▪ Increasing awareness and education
▪ Increasing capacity for risk management (e.g. during flood, outbreaks)
Meso-level strategies
▪ Strengthening community capacity building e.g. through health education
▪ Increasing community waste, garbage and drainage management
▪ Increasing community involvement during natural disasters
▪ Increasing community involvement in policy decision
▪ Strengthening social organisational systems (e.g. social cohesion and support)
▪ Strengthening local ecosystem management through area-based development programs
Macro-level strategies
▪ Maintaining eco-parks, decreasing deforestation and increasing vegetation
▪ Destruction of breeding sites for vectors
▪ Mapping risk areas and vulnerable groups
▪ Improving health sectors and supply of adequate medicine during disasters
▪ Strengthening media to disseminate information regarding outbreak and disaster management
▪ Strengthening surveillance for infectious diseases
▪ Strengthening data collection and management information system (e.g. for time series analysis)
▪ Manpower development for research and outbreak management during disasters
▪ Increasing research for typing strains and molecular epidemiology
▪ Strengthening laboratory facilities for investigating pathogenicity, virulence and resistance
▪ Strengthening governance and city-based capacity building
▪ Strengthening urban planning for housing, water, drainage and garbage management
▪ Ecosystem management through national policy and monitoring
▪ Improved warning and forecasting systems
▪ Strengthening research collaboration and public-private partnerships
▪ Increasing accessibility to the health and laboratory facilities

control (Tikar et al. 2008). Drainage and sewerage management and reducing the space for log water by the community can improve the situation.

At the macro-level, good governance and proper urban planning are extremely important. Particularly housing and slum development and building regulations may help. Identifying the risk zones and vulnerable groups in the megacities, restructuring and developing primary health care systems to handle the epidemic situations in natural disasters more effectively are needed. More doctors who are capable to deal effectively with infectious diseases should be trained. Improvement in sewerage and drainage facilities and health education are necessary (Karande et al. 2005). Capacity building for natural disasters, involving civil societies, public-private partnerships, and strengthening collaborations among different health stakeholders are important. Improved periodic surveillance for infectious diseases to understand the impact of climate change is also important. Among others, information dissemination through mass media regarding disaster and outbreak management, early warning systems, maintaining eco-parks and enlarging green space within the city are necessary. River and water management should be improved. Encroachment of rivers, deforestation and unplanned urbanisation should be strictly controlled.

For controlling infectious diseases, early diagnosis, appropriate investigations, strict monitoring, and prompt supportive management are necessary (Shah et al. 2004). More research is needed to study emerging diseases, virulence, resurgence, and pathogenesis. Setting up of new laboratories with adequate equipments, easy accessibility to these facilities for the affected people and manpower development are necessary. Finally, improved forecasting technologies to combat the exacerbated impacts of climate change are required (Bhattacharya et al. 2006). To combat the impact of climate change on infectious diseases in megacities, interdisciplinary and integrated approaches are extremely necessary.

9.6 Concluding Remarks

Changing climate and growing megacities have drawn considerable public health attentions worldwide. Although megacities are more vulnerable to the impact of climate change, overall research in megacities in relation to climate change is very scarce. Both climate change and megacities can adversely affect all ecosystems and hence have the potential to influence water-borne and vector-borne diseases by expanding and creating conducive environments. Diarrhoea/cholera, dengue, and malaria are sensitive to climate factors. Unfortunately, few studies are available in five megacities of the Indian sub-continent which assessed the impact of climate change on infectious diseases. Particularly, such information in Karachi is very scarce.

Evidence showed that all these megacities are very similar and almost equally vulnerable in terms of coastal location (except Delhi), rapid urbanisation, poor infrastructure, high population densities, high poverty, huge slum development, flooding, and poor ecological management.

All megacities experienced recent dengue outbreaks. Dengue has re-emerged after 1990 in all megacities with an increasing trend. This vector-borne disease is found to be highly seasonal with higher number of outbreaks during the monsoon. Information about and burden of diarrhoea/cholera varied remarkably among the megacities. For instance, diarrhoea/cholera was found to be more common in Dhaka and Kolkata than other megacities. Information about malaria is scarce particularly in Dhaka and Karachi. None of the studies explicitly assessed the long-term association of these diseases (except cholera/diarrhoea in Dhaka) with climate factors by using time series data.

In spite of these limitations, the available data regarding climate change and infectious diseases suggest that the megacities will be increasingly affected by these diseases in the future particularly in the absence of adequate interventions. The varying burden of climate-sensitive infectious diseases indicates that every megacity should be investigated separately. Time series data (yearly and seasonally) are required for better understanding. The scarcity of burden of disease information emphasises the need for epidemiological studies of common infectious diseases and their trends over time. Epidemiological studies regarding non-communicable diseases and climate change in megacities are also needed (e.g. association between climate change, air pollution, air pollutions sensitive diseases like cardiopulmonary illness). The capacities of the health sector in megacities to cope with the impact of climate change should be assessed. Health sectors should be restructured to cope with the multidimensional impacts of a changing climate. Interdisciplinary approaches are highly warranted. Multi-level prevention strategies might be useful to control the outbreak of diseases especially dengue and diarrhoea/cholera in the megacities of the Indian sub-continent and other developing world regions.

References

Agboatwalla M, Isomura S, Akram DS, Isihara Y, Sakae K, Yamashita T, Nishuo O. Enteric viral infections in pre-school children in Karachi, Pakistan. Indian J Pediatr 1995; 62(3):345–51.

Ahmed S, Arif F, Yahya Y, Rehman A, Abbas K, Ashraf S, Akram DS. Dengue fever outbreak in Karachi 2006 – a study of profile and outcome of children under 15 years of age. J Pak Med Assoc 2008; 58(1):4–8.

Alam M, Akhtar YN, Ali SS, Ahmed M, Atiq M, Ansari A, Chaudhry FA, Bashir H, Bangash MA, Awais A, Safdar A, Hasnain SAF, Zafar A. Seasonal variation in bacterial pathogens isolated from stool samples in Karachi, Pakistan. J Pak Med Assoc 2003; 53(3):125–9.

Alam P. Malaria country report Bangladesh-2008. Dhaka: Directorate General of Health Services, 2008.

Alam R, Siddiqui FM, Rahman S, Haque AK, Sarker CB, Siddiqui NI, Das SN, Kahhar A, Zafar SA, Khandaker AK, Ahmed T, Haque MF. Management of dengue by the WHO guided national guidelines. Mymensingh Med J 2004;13(1):43–7.

Alam M, Rabbani MG. Vulnerabilities and responses to climate change for Dhaka. Environment and Urbanization 2007; 19(1):81–97.

Akram DS, Ahmed S. Dengue fever. Infect Dis J Pakistan 2005; 124–125.

Asian Development Bank. TA 4578-Pakistan: Karachi megacities preparation project. Final report. August 2005.

Banerjee B, Hazra S, Bandyopadhyay D. Diarrhoea management among under fives. Indian Pediatr 2004; 41: 255–260.

Banik GB, Pal TK, Mandal A, Chakraborty MS, Chakravarti SK. Dengue haemorrhagic fever in Calcutta. Indian Pediatr 1994; 31(6):685–687.

Basu K, Das PK, Mondal T, Nandy A, Bhowmick PK, Addy M. Resurgence of malaria in Calcutta in 1995: a hospital based study. Indian J Public Health 1998; 42(2):50–52.

Basu A, Garg P, Datta S, Chakraborty S, Bhattacharya T, Khan A, et al. Vibrio cholerae O139 in Calcutta, 1992–1998: incidence, antibiograms, and genotypes. Emerg Infect Dis 2000; 6(2):139–147.

Bhattacharjee N, Mukherjee KK, Chakravarti SK, Mukherjee MK, De PN, Sengupta M, et al. Dengue haemorrhagic fever (DHF) outbreak in Calcutta-1990. J Commun Dis 1993; 25 (1):10–14.

Bhattacharya S, Sharma C, Dhiman RC, Mitra AP. Climate change and malaria in India. Current Science 2006; 90(3): 369–375.

Bhattacharya SK, Bhattacharya MK, Dutta D, Garg S, Mukhopadhyay AK, Deb M, Moitra A, Nair GB. Vibrio cholerae O139 in Calcutta. Arch Dis Child 1994; 71:161–162.

Bigio AG. Cities and climate change. A background paper for World Development Report, 2002.

Burkart K, Gruebner O, Khan MMH, Staffeld R (2008). Megacity Dhaka: urban environment, informal settlements and public health. Geographische Rundschau International Edition 4, 4–10.

Campbell-Lendrum D, Corvalan C. Climate change and developing-country cities: implications for environmental health and equity. Journal of Urban Health 2007; 84(1):i109-i117.

Chattopadhyay O, Sengupta G. Review of trends of malaria in an urban community of Calcutta during 1984–1997. J Indian Med Assoc 2000; 98(10):615–618.

Chowdhury AQ, Miah RA, Akhtar N, Jubayer SM. Seroprevalence of dengue infections amongst the children. Mymensingh Med J. 2004;13(1):59–62.

Costello A, Abbas M, Allen A et al. Managing the health effects of climate change. Lancet 2009; 373:1693–1733.

Curriero FC, Patz JA, Rose JB, Lele S. The association between extreme precipitation and waterborne disease outbreaks in the United States, 1984–1994. Am J Public Health 2001; 91(8):1194–1199.

Daily Times Monitor. First report of Karachi's largest dengue epidemic published by AKU. Daily Times, Monday, August 20, 2007.

Dar L, Broor S, Sengupta S, Xess I, Seth P. The first major outbreak of dengue hemorrhagic fever in Delhi, India. Emerg Infec Dis 1999; 5(4):589–590.

Datta KK, Bandyopadhyay S, Khanna KK, Banerjee K. Epidemiological features of cholera outbreak in Delhi in 1988. J Commun Dis 1993; 25(2):57–61.

Deen JL, Seidlein L, Sur D, Agtini M, Lucas MES, Lopez AL, et al. The high burden of cholera in children: comparison of incidence from endemic areas in Asia and Africa. PloS Negl Trop Dis 2(2): e173. doi:10.1371/journal.pntd.0000173

Dutta S, Dutta D, Dutta P, Matsushita S, Bhattacharya SK, Yoshida SI. Shigella dysenteriae serotype 1. Calcutta, India. Emerg Infect Dis 2003; 9(8):1471–1474.

Epstein PR. Climate change and emerging infectious diseases. Microbes Infect 2001; 3(9):747–54.

Faruque SM, Chowdhury N, Kamruzzaman M, Ahmad QS, Faruque ASG, Salam MA, Ramamurthy R, Nair GB, Weintraub A, Sack DA. Reemergence of epidemic vibrio cholerae O139, Bangladesh. Emerg Infect Dis 2003; 9(9):1116–1122.

Garg MR, Gogtay NJ, Kotwani RN, Bodhe PV, Kshirsagar NA. Resurgence of malaria in Mumbai- is escalating chloroquine resistance a cause? J Assoc Physicians India 1999; 47(4): 377–379.

Greer A, Ng V, Fisman D. Climate change and infectious diseases in North America: the road ahead. CMAJ 2008; 178(6):715–22.

Grimm NB, Faeth SH, Golubiewski NE, Redman CL, Wu J, Bai X, Briggs JM. Global change and the ecology of cities. Science 2008; 319:756–760.

Gupta E, Dar L, Narang P, Srivastava VK, Broor S. Serodiagnosis of dengue during an outbreak at a tertiary care hospital in Delhi. Indian J Med Res 2005; 121:36–38.

Hasan S, Khan MA. Community-based environmental management in a megacity: considering Calcutta. Cities 1999; 16(2):103–110.

Hashizume M, Armstrong B, Hajat S, Wagatsuma Y, Faruque ASG, Hayashi T, Sack DA. Association between climate variability and hospital visits for non-cholera diarrhoea in Bangladesh: effects and vulnerable groups. Int J Epidemiol 2007; 36:1030–1037.

Hashizume M, Armstrong B, Hajat S, Wagatsuma Y, Faruque AS, Hayashi T, Sack DA. The effect of rainfall on the incidence of cholera in Bangladesh. Epidemiology 2008a; 19:103–111.

Hashizume M, Armstrong B, Wagatsuma Y, Faruque ASG, Hayashi T, Sack DA. Rotavirus infections and climate variability in Dhaka, Bangladesh: a time-series analysis. Epidemiol Infect 2008b; 136:1281–1289.

Hati AK. Studies on dengue and dengue haemorrhagic fever (DHF) in West Bengal State, India. J Commun Dis 2006; 38(2):124–129.

Hess JJ, Malilay JN, Parkinson AJ. Climate change: the importance of place. Am J Prev Med 2008; 35:468–78.

Hunter PR. Climate change and waterborne and vector-borne disease. J Applied Microbiol 2003; 94:37 S-46 S.
Huq S, Rahman A, Konate M, Sokona Y, Reid H. Mainstreaming adaptation to climate change in least developed countries (LDCs). London: International Institute for Environment and Development, 2003.
IPCC. Climate Change 2007: Synthesis Report. Contribution of Working Groups I, II and III to the Fourth Assessment Report of the Intergovernmental Panel on Climate Change [Core Writing Team, Pachauri, R.K and Reisinger, A. (eds.)]. IPCC, Geneva, Switzerland, 2007.
Islam MA, Ahmed MU, Begum N, Chowdhury NA, Khan AH, Parquet MC, Bipolo S, Inoue S, Hasebe F, Suzuki Y, Morita K. Molecular characterization and clinical evaluation of dengue outbreak in 2002 in Bangladesh. Jpn J Infect Dis 2006; 59:85–91.
Jabeen K, Hasan T. Re-emergence of Vibrio cholerae O139 in Pakistan: report from a tertiary care hospital. J Pak Med Assoc 2003; 53(8):335–8.
Jamil B, Hasan R, Zafar A, Bewley K, Chamberlain J, Mioulet V, Rowlands M, Hewson R. Dengue virus serotype 3, Karachi, Pakistan. Emerg Infect Dis 2007; 13(1):182–183.
Karande S, Ghandi D, Kulkarni M, Bharadwaj R, Pol S, Thakare J, De A. Concurrent outbreak of leptospirosis and dengue in Mumbai, India, 2002. J Trop Pediatr 2005; 51(3):174–181.
Kazmi JH, Pandit K. Disease and dislocation: the impact of refugee movements on the geography of malaria in NWFP, Pakistan. Soc Sci Med 2001; 52:1043–1055.
Khan MA, Mekan SF, Abbas Z, Smego RA Jr. Concurrent malaria and enteric fever in Pakistan. Singapore Med J 2005; 46(11):635–638.
Khan MA, Smego RA Jr., Razi ST, Beg MA. Emerging drug- resistance and guidelines for treatment of malaria. Med Today 2006; 4(3):81–87.
Khan E, Siddiqui J, Shakoor S, Mehraj V, Jamil B, Hasan R. Dengue outbreak in Karachi, Pakistan, 2006: experience at a tertiary care center. Trans R Soc Trop Med Hyg. 2007;101 (11):1114–9.
Khan MMH, Kraemer A. Socio-economic factors explain differences in public health-related variables among women in Bangladesh. BMC Public Health 2008: 8:254 doi:10.1186/1471-2458-8-254.
Khan TM, Quadir DA, Murty TS, Sarker MA. Seasonal and interannual sea surface temperature variability in the coastal cities of Arabian Sea and Bay of Bengal. Natural Hazards 2004; 31:549–560.
Khanna KK, Dhanvijay A, Riley LW, Sehgal S, Kumari S. Cholera outbreak in Delhi-1988. J Commun Dis 1990; 22(1):35–38.
Khasnis AA, Nettleman MD. Global warming and infectious disease. Arch Med Res 2005; 36 (6):689–96.
Kothari G. Diarrhoea in urban slums: Bombay. Dialogue Diarrhoea 1987; 31:4–5.
Kovats S and Akhtar R. Climate, climate change and human health in Asian cities. Environ Urban 2008; 20(1):165–175.
Kumat V. Resurgence of malaria in Bombay (Mumbai) in the 1990s: a historical perspective. Parassitologia 2000; 42(1–2):135–148.
Lipp EK, Huq A, Colwell RR. Effects of global climate on infectious disease: the cholera model. Clin Microbiol Rev 2002; 15(4):757–70.
Lobitz B, Beck L, Huq A, Wood B, Fuchs G, Faruque ASG, Colwell R. Climate and infectious disease: use of remote sensing for detection of Vibrio cholerae by indirect measurement. Science 2000; 97(4):1438–1443.
Mandal B, Mitra NK, Mukhopadhyay AK, Mukherjee H, Hati AK. Emerging plasmodium falciparum in an endemic area in Calcutta. J Indian Med Assoc 1998; 96(11):328–329.
McGeehin MA, Mirabelli M. The potential impacts of climate variability and change on temperature-related morbidity and mortality in the United States. Environ Health Perspect 2001; 109 (suppl 2): 185–189.
Mitchell T, Tanner T. Adapting to climate change: challenges and opportunities for the development community. Institute of Development Studies, Tearfund, UK, 2006 (www.ids.ac.uk/ids).

Mukhopadhyay AK, Karmakar P, Hati AK, Dey P. Recent epidemiological status of malaria in Calcutta Municipal Corporation area, West Bengal. Indian J Malariol 1997; 34(4):188–196.
Nagdeve DA. Environmental pollution and control: a case study of Delhi megacity. Popul Environ 2004; 25(5): 461–473.
Nalin DR, Mahood F, Rathor H, Muttalib A, Sakai R, Chowdhry MA, Safdar G, ul Huq I, Munir M, Suleiman M et al. A point survey of periurban and urban malaria in Karachi. J Trop Med Hyg 1985; 88(1):7–15.
Nandy A, Addy M, Maji AK, Bandyopadhyay AK. Monitoring the chloroquine sensitivity of Plasmodium vivax from Calcutta and Orissa, India. Ann Trop Med Parasitol 2003; 97(3): 215–220.
Nicholls, RJ. Coastal megacities and climate change. GeoJournal 1995; 37(3): 369–379.
Nishio O, Matsui K, Oka T, Ushijima H, Mubina A, Dure-Samin A, Isomura S. Rotavirus infection among infants with diarrhoea in Pakistan. Pediatr Int 2000; 42(4):425–7.
Pascual M, Rodo X, Ellner SP, Colwell R, Bouma MJ. Cholera dynamics and El Nino-Southern Oscillation. Science 2000; 289:1766–1769.
Patz JA, Kovats RS. Hotspots in climate change and human health. BMJ 2002; 325:1094–98.
Patz JA, Olson SH. Climate change and health: global to local influences on disease risk. Ann Trop Med Parasitol 2006; 100(5–6):535–49.
Paul RE, Patel AY, Mirza S, Fisher-Hoch SP, Luby SP. Expansion of epidemic dengue viral infections to Pakistan. Int J Infect Dis. 1998; 2(4):197–201.
Perveen A, Qaiser M, Sad-ul-Islam M. Airborne pollen survey of Karachi and adjacent areas in relation to allergy. World Applied Sci J 2007; 2(4):289–298.
Podder G, Breiman RF, Azim T, Thu HM, Velathanthiri N, Mai LQ, Lowry K, Aaskov JG. Origin of dengue type 3 viruses associated with the dengue outbreak in Dhaka, Bangladesh in 2000 and 2001. Am J Trop Med Hyg 2006; 74(2):263–265.
Pramanik MK, Aditya G, Raut SK. Seasonal prevalence of Aedes aegypti immatures in Kolkata, India. Southeast Asian J Trop Med Public Health 2007; 38(3):442–427.
Quadir DA, Shrestha ML, Khan TMA, Ferdousi N, Rahman M, Mannan A. Variations of surface air temperature over the land areas in and around the Bay of Bengal. Natural Hazards 2004; 31:561–584.
Qureshi JA, Notta NJ, Salahuddin N, Zaman V, Khan JA. An epidemic of Dengue fever in Karachi–associated clinical manifestations. J Pak Med Assoc 1997; 47(7):178–81.
Rahman M, Rahman K, Siddque AK, Shoma S, Kamal AHM, Ali KS, Nisaluk A, Breiman RF. First outbreak of dengue hemorrhagic fever, Bangladesh. Emerg Infect Dis 2002; 8(7): 738–740.
Ratho RK, Mishra B, Kaur J, Kakkar N, Sharma K. An outbreak of dengue fever in periurban slums of Chandigarh, India, with special reference to entomological and climatic factors. Indian J Med Sci 2005; 59(12):519–527.
Revi A. Climate change risk: an adaptation and mitigation agenda for Indian cities. Environ Urban 2008; 20(1):207–229.
Rodo X, Pascual M, Fuchs G, Faruque ASG. ENSO and cholera: a nonstationary link related to climate change? PNAS 2002; 99(2):12901–12906.
Sajjad SH, Hussain B, Khan MA, Raza A, Zaman B, Ahmed I. On rising temperature trends of Karachi in Pakistan. Climate Change 2009; 96:539–547.
Schwartz BS, Harris JB, Khan AI, Larocque RC, Sack DA, Malek MA, Faruque ASG, Qadri F, Calderwood SB, Luby SP, Ryan ET. Diarrhoeal epidemics in Dhaka, Bangladesh, during three consecutive floods: 1988, 1998, and 2004. Am J Trop Med Hyg 2006; 74(6):1067–1073.
Shah I, Deshpande GC, Tardeja PN. Outbreak of dengue in Mumbai and predictive markers for dengue shock syndrome. J Trop Pediatr 2004; 50(5):301–305.
Sharma NC, Mandal PK, Dhillon R, Jain M. Changing profile of Vibrio cholerae O1, O139 in Delhi and its periphery (2003–2005). Indian J Med Res 2007; 125: 633–640.
Sheikh A, Khan A, Malik T, Fisher-Hoch SP. Cholera in a developing megacity; Karachi, Pakistan. Epidemiol Infect 1997; 119(3):287–92.

Singh J, Bora D, Sharma RS, Khanna KK, Verghese T. Epidemiology of cholera in Delhi-1992. J Trop Pediatr 1995; 41(3):139–142.
Suleman M. Malaria in Afghan refugees in Pakistan. Trans R Soc Trop Med Hyg 1988; 82(1): 44–47.
Sur D, Deen JL, Manna B, et al. The burden of cholera in the slums of Kolkata, India: data from a prospective, community based study. Arch Dis Child 2005; 90:1175–1181.
Sur D, Sarkar BL, Manna B, Deen J, et al. Epidemiological, microbiological and electron microscopic study of a cholera outbreak in a slum community. Indian J Med Res 2006; 123:31–36.
Sutherst RW. Global change and human vulnerability to vector-borne diseases. Clin Microbiol Rev 2004; 17(1):136–173.
Tandon N, Raychoudhury S. Reinvasion of Calcutta city by Aedes albopictus: the proven vector of dengue in suburban areas. Indian J Public Health 1998; 42(1):24–25.
Tikar SN, Mendki MJ, Chandel K, Parashar BD, Prakash S. Susceptibility of immature stages of *Aedes (Stegomyia) aegypti*; vector of dengue and chikungunya to insecticides from India. Parasitol Res 2008; 102(5):907–913.
United Nations: World Urbanization Prospects. The 2007 Revision: Executive Summary. New York, 2008.
Wagatsuma Y, Breiman RF, Hossain A, Rahman M. Dengue fever outbreak in a recreation club, Dhaka, Bangladesh. Emerg Infect Dis 2004; 10(4):747–750.
WHO. Climate change and human health-risks and responses. Geneva: WHO, 2003.
Yunus EB, Bangali AM, Mahmood MAH, Rahman MM, Chowdhury AR, Talukder KR. Dengue outbreak 2000 in Bangladesh: from speculation to reality and exercises. Dengue Bulletin 2001; 25:15–20.
Zhang Y, Bi P, Hiller JE. Climate change and the transmission of vector-borne diseases: a review. Asia-Pacific Journal of Public Health 2008; 20(1):64–76.

Chapter 10
Human Bioclimate and Thermal Stress in the Megacity of Dhaka, Bangladesh: Application and Evaluation of Thermophysiological Indices

Katrin Burkart and Wilfried Endlicher

10.1 Introduction

Human bioclimate refers to the entirety of all climatological and meteorological parameters affecting the living organism. The relevance of climate and weather[1] for human health was already recognized by Hippocrates (Hippocrates Reprint). Later, Alexander von Humboldt defined climate as changes of the atmosphere affecting the human organism, thus putting human bioclimatological aspects in focus (von Humboldt 2004)[2]. Since then, numerous studies have been published focusing on the atmosphere-health relationship describing seasonal variations and non-linear relationships between multiple disease (e.g. cardio-respiratory, infectious) and temperature (Burkart and Endlicher 2009; Kunst et al. 1993; Braga et al. 2001; Braga et al. 2002; Basu and Samet 2002).

Apart from temperature, the thermal environment is influenced by several additional parameters such as humidity, radiation or air movement. The interplay of these parameters affects the human heat balance and triggers several physiological reactions to restore or maintain a constant core body-temperature (Parsons 2003; VDI 1998). Internal heat generated by metabolism is transferred through

[1]Commonly climate refers to the weather in some location averaged over some long period of time. Following this definition, climatological influences occur on a long-term scale and meteorological influences on a short-term scale. However, the direction and magnitude of short-term meteorological influences on human health depend on climate. Therefore, a strict distinction of the terms climate/climatological and weather/meteorological is often not possible or feasible. Particularly, in the realm of bioclimatic research this definition is not adhered to rigorously (e.g. climate definition given by Humboldt). In this article the terms climate and climatological comprise short-term and long-term influences.

[2]Energy released or absorbed by change of the aggregate state of water.

K. Burkart (✉) • W. Endlicher
Humboldt-Universität zu Berlin Department of Geography Climatological Section Unter den Linden 6 10066 Berlin, Germany
e-mail: katrin.burkart@geo.hu-berlin.de

A. Krämer et al. (eds.), *Health in Megacities and Urban Areas*, Contributions to Statistics, DOI 10.1007/978-3-7908-2733-0_10,

the skin to the surrounding atmosphere. If this heat exchange is impeded by the surrounding conditions, the core body temperature starts to rise with potential negative consequences for human health (Driscoll 1985; Robinson 2001). In contrast, if the body loses too much heat, the core body temperature drops which can result in cardiac irregularities or negatively affects the non-specific immune response (Cabanac and Brimmel 1987; Berk et al. 1987; Bull 1980). The magnitude and efficiency of heat exchange depends to a great extent on the temperature gradient between a body and its environment but is also influenced by other atmospheric parameters. Humidity, for instance, affects the latent energy flux, and short-wave radiation increases sensible heat, while air movement affects sensible and latent energy fluxes (Parsons 2003; VDI 1998). In view of the complex nature of these various interactions, many have pointed to the necessity of taking a modeling approach to this matter instead of considering the diverse parameters separately. A variety of models relating atmospheric-thermal conditions to human heat sensation have been developed (Büttner 1938; VDI 1998; Parsons 2003). In considering the overall heat balance of the human body, many of these models require meteorological information in addition to non-meteorological parameters concerning patient fitness and level of activity, clothing type and physiological adaptation to a particular environment (Parsons 2003; Staiger et al. 1997).

Apart from the general impact of thermal conditions, periods of extreme cold or heat can cause excess morbidity and mortality. These extreme events, usually referred to as cold or heat waves, can be assessed climatologically or epidemiologically. A climatological definition would imply the exceedance of a certain threshold value, while an epidemiological definition would imply adverse health outcomes, such as the occurrence of excess mortality or morbidity. Despite extensive research on this topic during recent years, a clear definition for heat or cold waves does not exist (Meehl and Tebaldi 2004; Robinson 2001). From a public health perspective, the focus of any such definition should lie on the impact on human health. Nevertheless, when assessing and forecasting the effect of weather or climate on public health, modeled or statistical values often constitute the only possible approaches. Representative parameters predicting the thermal impact are helpful for setting up early warning systems and preparing the population with adequate measures.

Thermal conditions vary not only with season and weather conditions but space. In addition to large scale differences resulting from geographical location, the modification of the meso- and microclimates are relevant. One widely observed mesoclimatic modification is the so-called urban climate, also referred to as urban heat island (UHI). Urban agglomerations generally exhibit higher temperatures than their surrounding areas, as the urban building structure profoundly affects short- and long-wave radiation fluxes, heat storage and the water balance (Oke 1973). Most studies on urban climate are conducted in mid-latitude regions and the knowledge derived is of only limited relevance to tropical urban areas due to differences in the prevailing climatological and hydrological conditions and the urban building structure. So far, the limited number of studies conducted in tropical climates generally allows us to state that the intensity of the urban heat island in tropical regions is lower and seasonal urban-rural differences are higher during the dry season (Roth 2007).

While the human bioclimate and thermal environment has been assessed on almost every scale for countries and regions in the mid-latitudes, little is known about tropical climates. However, understanding climatic conditions and their effect on human health in these latitudes can be a key factor in developing mitigating strategies. Climate adaptive architecture and urban planning, behavioral-adjustment or public health strategies represent just a few approaches to responding to atmospheric influences. Our study aims to describe the climate and human bioclimate in Bangladesh with especial focus on the urban anthropogenic modification of the mesoclimate in the megacity of Dhaka.

10.2 Data and Methods

10.2.1 Data

Meteorological data was collected from the Bangladesh Meteorological Department (BMD). This data comprises three hourly values of temperature, humidity, radiation, cloud coverage, wind speed and precipitation for three stations in Dhaka, Tangail and Mymensingh. The data was collected over a period of 10 years from 1998 to 2007. Measurements were recorded manually every 3 h at 0, 3, 6, 9, 12, 15, 18 and 21 GMT and sent to the BMD headquarters where they were organized in a database and subjected to several quality and plausibility controls. Daily values were calculated for complete daily data sets and monthly values were calculated if at least two thirds of the monthly data was available (approximately 10% of the data were missing). Thermophysiological indices (TPIs) were calculated on the basis of the three hourly values from which the mean, maximum and minimum values were determined. We acknowledge that in the case of minimum and maximum TPIs, the value thus produced do not necessarily comply with the highest or lowest values occurring on that day. Data analysis was conducted using *R* (Version 2.10.1).

10.2.2 Thermophysiological Models and Indices

TPIs are output parameters of thermophysiological models. The complexity of these models and number of parameters considered varies. The following section provides a short introduction to the models and indices used in this study. The Heat Index (HI) developed by Steadman (named apparent temperature) and modified by the US National Weather Service combines air temperature and humidity (Robinson 2001; Steadman 1979). The HI is a parameter assessing heat sensation and is defined for temperatures and humidity above 26.7°C and 40%. For cold stress assessment, the Windchill Index (WCI), also based on a model developed by Steadman, is usually applied and is defined for temperatures below 10°C and wind speeds above 4.8 km/h (Steadman 1971). These two indices were used as under hot conditions, humidity increases heat sensation whereas under cold conditions air movement increases cold sensation (Steadman 1971). In the case of both indices a reference environment with constant humidity (50% relative humidity) or wind speed (1.34 m/s) is defined which would result in the same energy gain as the actual environment. For the purposes of this study, we calculated HI whenever the thresholds were surpassed; WCI was calculated whenever temperatures fell below and wind speed exceeded defined thresholds. In-between the measured air temperature remained.

The physiological equivalent temperature (PET) (Höppe 1999) is based on the Munich Energy-balance Model for Individuals (MEMI). PET is defined as the air temperature at which, in a typical indoor setting (without wind and solar radiation), the heat budget of the human body is balanced with the same core and skin temperature as under the complex outdoor conditions to be assessed. In this way, PET allows us to compare the integral effects of complex thermal conditions outside with the experience indoors (Höppe 1999). PET requires the input parameters temperature, humidity, radiation temperature and wind speed, whereby the radiation temperature is modeled as a function of cloud coverage and temperature.

The universal thermal climate index (UTCI) was developed within the frame of the COST action 730 (www.utci.org) established by the International Society of Biometeorology (ISB). The index is based on the Fiala model, a thermophysiological model which has been extensively validated by experimental data from numerous groups (Jendritzky et al. 2007). The index claims to be applicable for all environments, conditions and regions. The model incorporates two interacting systems of thermoregulation: the controlling, active system and the controlled passive system. The passive system is a multi-segmental, multi-layered representation of the human body with spatial subdivisions including a detailed representation of the anatomic, thermophysical and thermophysiological properties of the human body. The model accounts for the heat transfers occurring inside the human body (blood circulation, metabolic heat generation, -conduction and -accumulation) and at its surface (free and forced surface convection, long- and short-wave radiation, skin moisture evaporation, diffusion and accumulation) (Fiala et al. 1999). The active system simulates the different responses of the human thermoregulatory system to thermal conditions, i.e. the suppression (vasoconstriction) and elevation (vasodilation) of the cutaneous blood flow, sweat moisture excretion and changes in metabolic heat production by shivering and thermogenesis (Fiala et al. 1999; Fiala et al. 2001). Like other indices, UTCI follows the concept of an equivalent temperature. A reference environment with 50% relative humidity, still air and radiant temperature equaling air temperature is defined.

10.2.3 Extreme Heat and Cold Stress Assessment

In order to assess heat and cold waves, we adopted a statistical approach which defines an extreme event as the exceedance of a statistically derived threshold. Maximum temperatures provide a good measure of extremely hot or cold days, whereas the use of minimum temperatures seems to be important in assessing conditions under which there is little relief for persons during night-time (Medina-Ramón et al. 2006). For our analysis, days with heat stress were defined as those days on which the maximum temperature surpassed the 95th percentile, whereas nights with heat stress were defined as nights during which the minimum temperature exceeded the 95th percentile. Reciprocal maximum and minimum temperatures falling below the 5th percentile were defined as days with cold stress or nights with cold stress respectively. As there is evidence that mortality is more likely during or after a period of several days, when the interior of a building is more likely to reflect the outdoor apparent temperature (Kalkstein and Smoyer 1993) and when there is no intermittent relief, a duration criterion was integrated in the definition of a heat wave. We determined the frequency of heat and cold stress during day- and night-times for a particular day of the year. Additionally we determined the probability of heat and cold waves in a particular month. In order to account for the different lengths of heat and cold waves, the concept of heat and cold wave days was introduced. A heat or cold wave day refers to a 24-h period (gliding intervals) which is part of a 48-h period of ongoing heat or cold over which the threshold values are permanently exceeded. In order to determine the probability of a 24-h period being a heat or cold wave day, we divided the number of days that were part of a heat or cold wave by the number of possible days[3].

10.2.4 Urban Heat Island Assessment

The UHI was assessed by calculating the differences in monthly average values of the mean, maximum and minimum (equivalent) temperatures between Dhaka and two reference stations located in Tangail and Mymensingh. Tangail and Mymensingh are two small towns in close proximity to the megacity area, which differ considerably in their building density and structure compared to Dhaka. Dhaka constitutes a classical urban site while the stations in Tangail and Mymensigh serve as reference stations with rural characteristics. The site in Mymensingh is situated in an agricultural environment surrounded by fields and

[3]For example: Three heat waves were observed in May over the 10-year data period with the following duration time: (a) 2 days (48 h), (b) 4 days (96 h) and (c) 3½ days (60 h). The number of occurring heat wave days was divided by the number of possible heat wave days: (2 + 4 + 3½)/310.

water. The Tangail site constitutes a more built up environment than Mymensingh and might be considered as semi-rural. The difference in temperature or indices served as an indicator for the UHI and is displayed in its seasonal distribution. Before determining the differences between stations, we matched the data sets in such a way that measurement values for both sites were concurrent. Monthly differences were displayed for the mean, maximum and minimum values. Monthly differences in 3-houly values were displayed as isopleths.

10.3 Results

10.3.1 Seasonal Bioclimate of Bangladesh

Generally, three seasons can be distinguished in Bangladesh. The cold season, from November to February, the hot and humid pre-monsoon season (summer), from March to May, and the monsoon season with heavy rainfall from June to October (also referred to as rainy season). About 90% of precipitation fell in the period May to October, while the rest of the year was relatively dry.

The lowest (equivalent) temperature were recorded in December and January. Average mean temperatures were almost equally high from April to September. The HIWCI peaked in August, while PET and UTCI reached maximum values from June to August. During the warm season, TPIs surpassed the temperature values (Fig. 10.1). The HIWCI and the UTCI run almost parallel for all three measuring sites. According to the assessment scale of UTCI, no thermal stress occurs between 9°C and 26°C. The average mean temperatures of UTCI exceeded this value in March and did not fall below 26°C (UTCI) before October. Considering average maximum temperatures, the threshold is surpassed from February to November. On the contrary, cold stress never occurred concerning average values on the UTCI assessment scale.

Figure 10.2 displays temperature and TPIs as isopleths. Dhaka exhibited typical characteristics of a diurnal climate from May to September. Monthly changes were minor, while diurnal differences were pronounced. Between October and March, the isopleths followed the pattern of a seasonal climate (usually observed in the mid-latitudes). Diurnal differences were diminished and a strong gradient between months was observed. Seasonal difference in monthly average mean temperatures amounted to 10 K. A strongly pronounced diurnal gradient with quickly changing values from noon to early evening was observed for PET and UTCI.

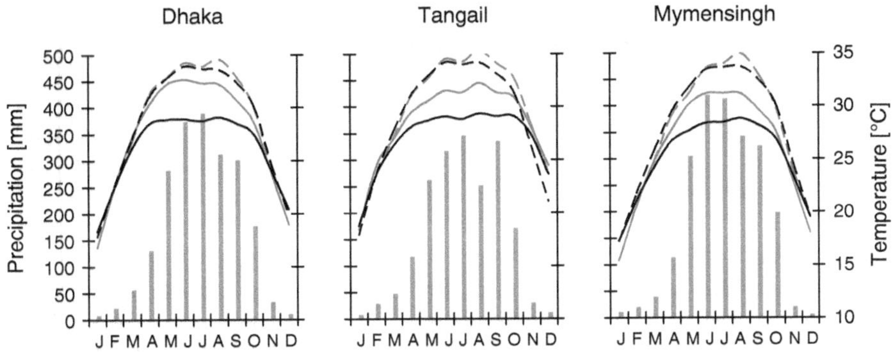

Fig. 10.1 Annual variations of monthly mean temperature (*black solid line*), mean HIWCI (*grey dashed line*), mean PET (*grey solid line*) and mean UTCI (*black dashed line*), and precipitation (*grey bars*) in Dhaka, Tangail, and Mymensingh

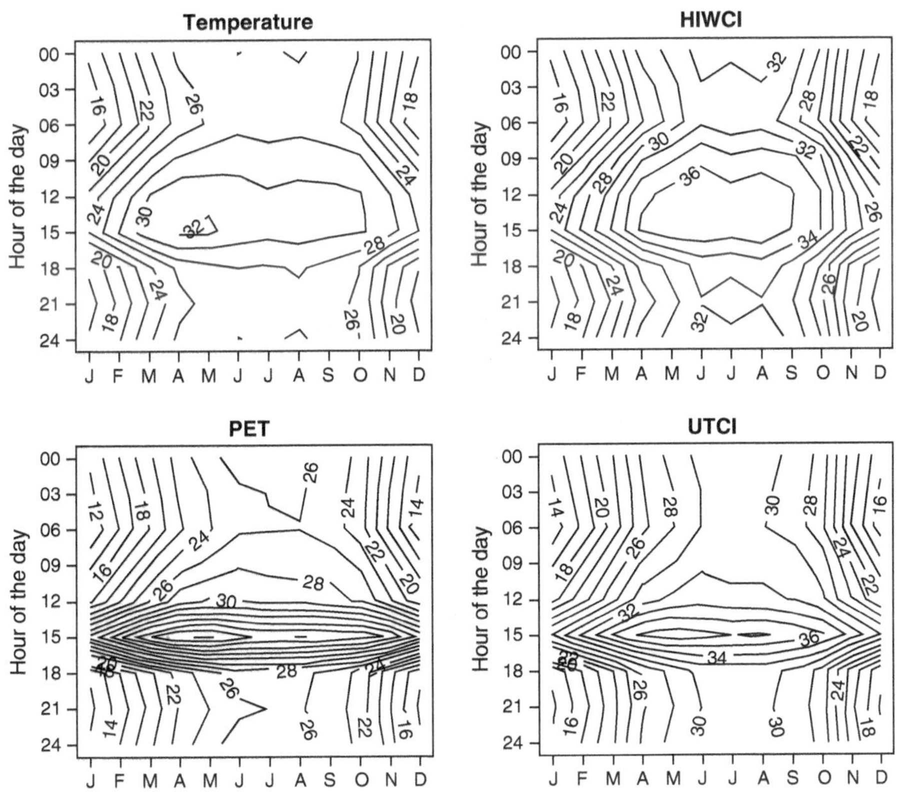

Fig. 10.2 Isopleth diagrams displaying seasonal and diurnal temperature, HIWCI, PET, and UTCI distribution in Dhaka

10.3.2 Temporal Occurrence and Frequency of Heat and Cold Stress in Dhaka

Figure 10.3 displays the frequency of heat and cold stress during day- and night-times at a particular day of the year. Threshold values given by the 5th and 95th percentile of minimum and maximum values are displayed in Table 10.1.

Considering heat stress, most daytime temperature extremes occurred from March to July, while night-time extremes occurred from May to September, with a peak in July. The highest frequency of day- and night-time temperature extremes occurring together was observed from the mid April to the beginning of June. Extremes of HIWCI occurred from mid April to mid October, with the highest frequency being measured around June. Considering PET, daytime extremes occurred between mid-May and August, whereas night-time extremes were broadly distributed between March and October. In the case of UTCI, extremes of highest frequency during daytime can be observed from April to June, while the highest frequency during night-time can be observed between June and September. Concerning all the considered indices and temperature, cold stress is mostly limited to December and January; the highest frequency was observed in January.

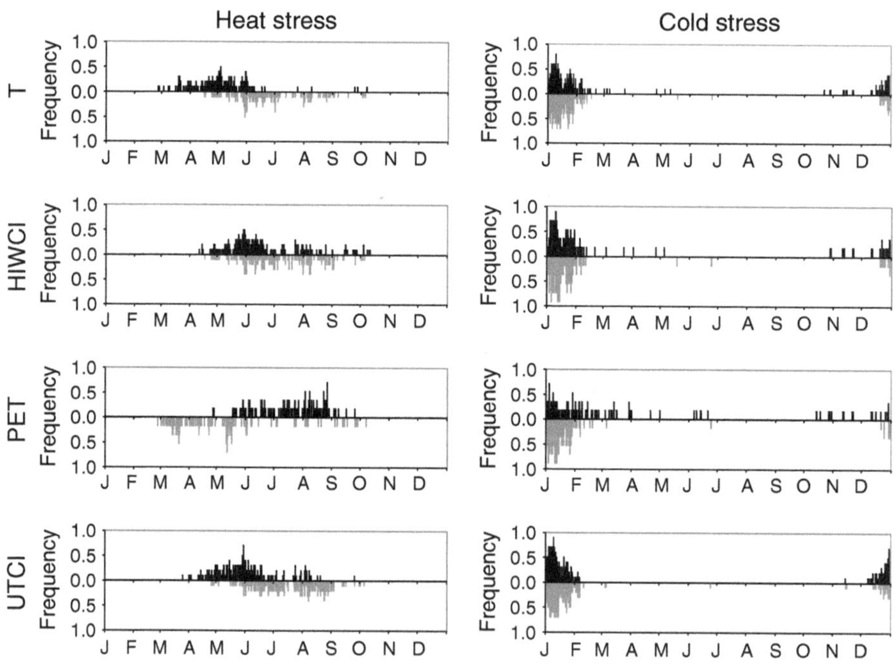

Fig. 10.3 Frequency of days and nights with heat stress (*left-hand column*) and cold stress (*right-hand columns*) defined by the exceedance and undercutting of the 95th and 5th percentile of maximum and minimum temperature, HIWCI, PET, and UTCI. (Daytime frequency is displayed in the upper half of the figure and night-time frequency is displayed in the lower half of the figure)

Table 10.1 5th and 95th percentile of minimum and maximum temperatures, HIWCI, PET, and UTCI

	T_{min}	T_{max}	$HIWCI_{min}$	$HIWCI_{max}$	PET_{min}	PET_{max}	UTC_{min}	$UTCI_{max}$
5th percentile	13.0	24.0	13.0	24.0	9	34.6	11.5	25.2
95th percentile	28.0	35.0	33.5	42.9	26.4	52.9	30.5	41.7

Table 10.2 Probability of the occurrence of heat and cold wave days (periods of 24 h) in a particular month

	Heat waves				Cold waves			
	T	HIWCI	PET	UTCI	*T*	HIWCI	PET	UTCI
Jan	–	–	–	–	$3.3 \cdot 10^{-1}$	$3.3 \cdot 10^{-1}$	$0.9 \cdot 10^{-1}$	$3.0 \cdot 10^{-1}$
Feb	–	–	–	–	$0.3 \cdot 10^{-1}$	$0.3 \cdot 10^{-1}$	$0.1 \cdot 10^{-1}$	$0.3 \cdot 10^{-1}$
Mar	–	–	–	–	–	–	–	–
Apr	$0.2 \cdot 10^{-1}$	$0.2 \cdot 10^{-1}$	–	–	–	–	–	–
May	$1.0 \cdot 10^{-1}$	$0.5 \cdot 10^{-1}$	–	$0.3 \cdot 10^{-1}$	–	–	–	–
Jun	$0.3 \cdot 10^{-1}$	$1.5 \cdot 10^{-1}$	–	$0.6 \cdot 10^{-1}$	–	–	–	–
Jul	–	$0.2 \cdot 10^{-1}$	–	$0.03 \cdot 10^{-1}$	–	–	–	–
Aug	–	$0.1 \cdot 10^{-1}$	$0.1 \cdot 10^{-1}$	–	–	–	–	–
Sep	$0.1 \cdot 10^{-1}$	$0.1 \cdot 10^{-1}$	–	–	–	–	–	–
Oct	–	–	–	–	–	–	–	–
Nov	–	–	–	–	–	–	–	–
Dec	–	–	–	–	$0.5 \cdot 10^{-1}$	$0.5 \cdot 10^{-1}$	$0.2 \cdot 10^{-1}$	$0.5 \cdot 10^{-1}$

Table 10.2 depicts the probability of a day (24-h period) being embedded in a heat or cold wave. Considering temperature, the highest probability was observed in May, while the adjacent months April and June also showed an increased probability. On the contrary, regarding HIWCI, heat waves occurred between April and September with the highest probability in June. The probability of a heat wave day occurring in June is 15%. Considering PET, heat wave probability is rather low. A somewhat higher probability was observed for UTCI with the highest probability registered in June. As already observed for the frequency of cold stress days, the occurrence of cold waves is restricted between December and February. No major differences were observed between different indices, but a reduced probability was observed in terms of PET. The probability of the occurrence of a cold wave is many times higher than the probability of a heat wave.

10.3.3 Urban Heat Island

Figures 10.4 and 10.5 display the seasonal distribution of differences in temperature, HIWCI, PET, and UTCI between Dhaka and the two reference sites. In both cases it can be seen that urban-rural differences are reduced during the rainy season. During the dry season, differences between Dhaka and the reference stations ranged between 1 and 3 K.

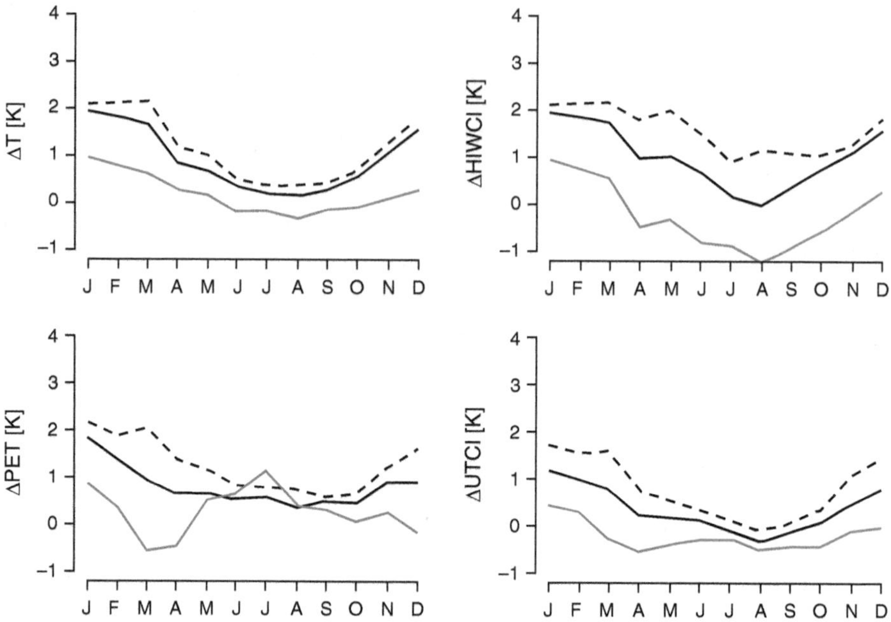

Fig. 10.4 Differences in monthly mean average values (*black solid line*), mean maximum values (*grey solid line*) and mean minimum values (*black dashed line*) of temperature, HIWCI, PET, and UTCI between Dhaka and Tangail

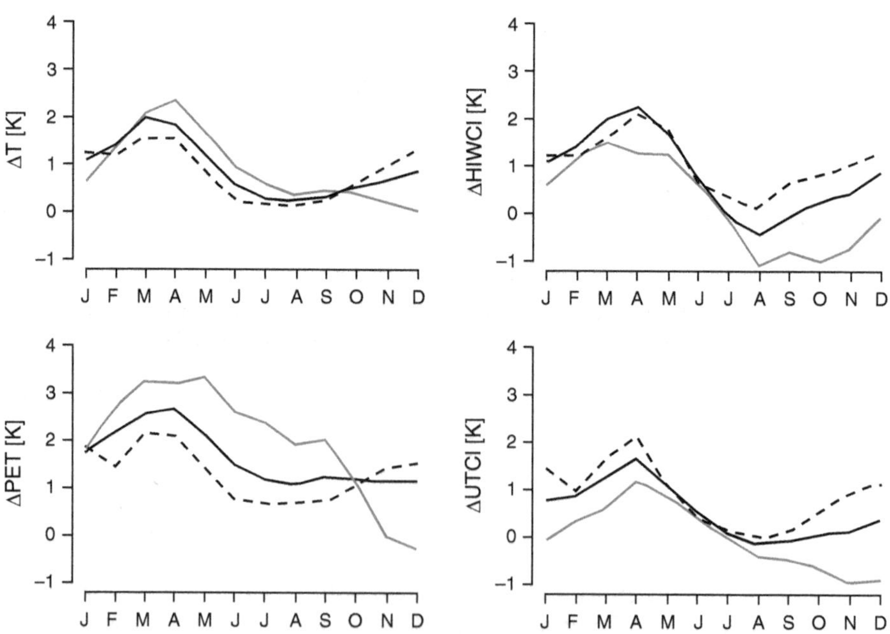

Fig. 10.5 Differences in monthly mean average values (*black solid line*), mean maximum values (*grey solid line*) and mean minimum values of temperature (*black dashed line*), HIWCI, PET, and UTCI between Dhaka and Mymensingh

Urban-rural differences between Dhaka and Mymensingh were most pronounced during the summer season in March and April. Temporarily, monthly values in Dhaka fell below those of the reference stations. Temperature and TPI differences follow a similar seasonal distribution. The magnitude of the UHI was most strongly pronounced for minimum temperatures using Tangail as references station. Using Mymensingh as reference station, highest differences regarding HIWCI and UTCI were observed for mean and minimum values. Concerning temperature and PET, however, differences in maximum values were highest.

The seasonal and temporal distribution of the UHI magnitude is reflected in the isopleth diagrams. Differences in 3-hourly values between Dhaka and Tangail are most pronounced during evening and night-times throughout the year, but particularly from October to March (Fig. 10.6). Daytime Dhaka-Mymensingh differences reach their maxima around March and April. During the rainy season, differences are equally high in their diurnal distribution (Fig. 10.7).

In addition to differences in temperature and TPIs, urban-rural differences were also observed for humidity, cloud coverage, mean radiation temperature and wind

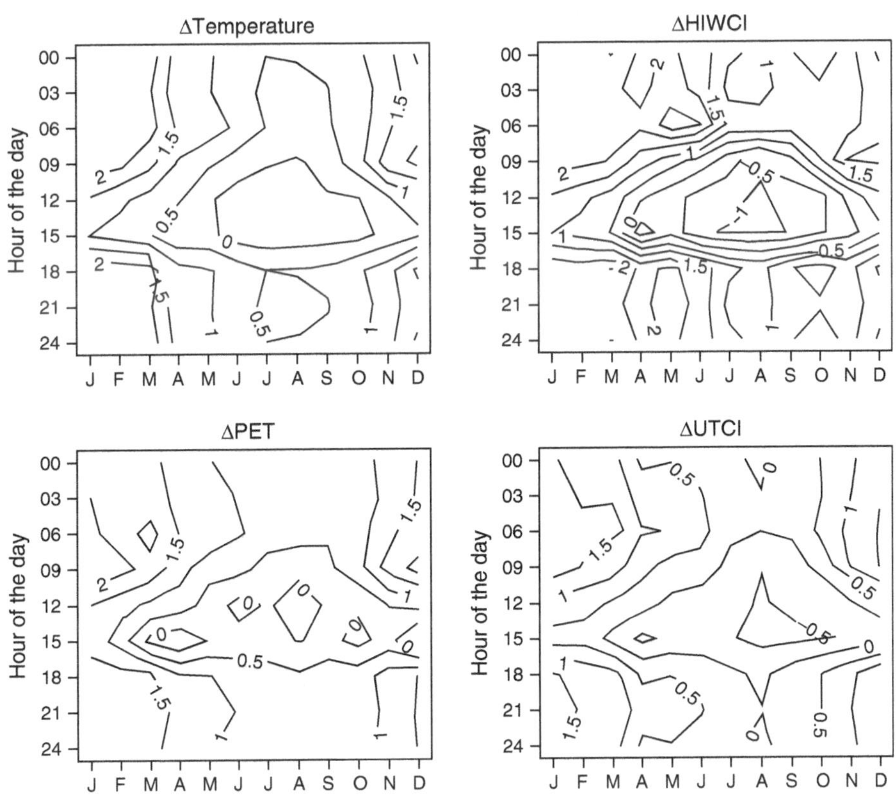

Fig. 10.6 Differences in temperature, HIWCI, PET, and UTCI between Dhaka and Tangail displayed as isopleths

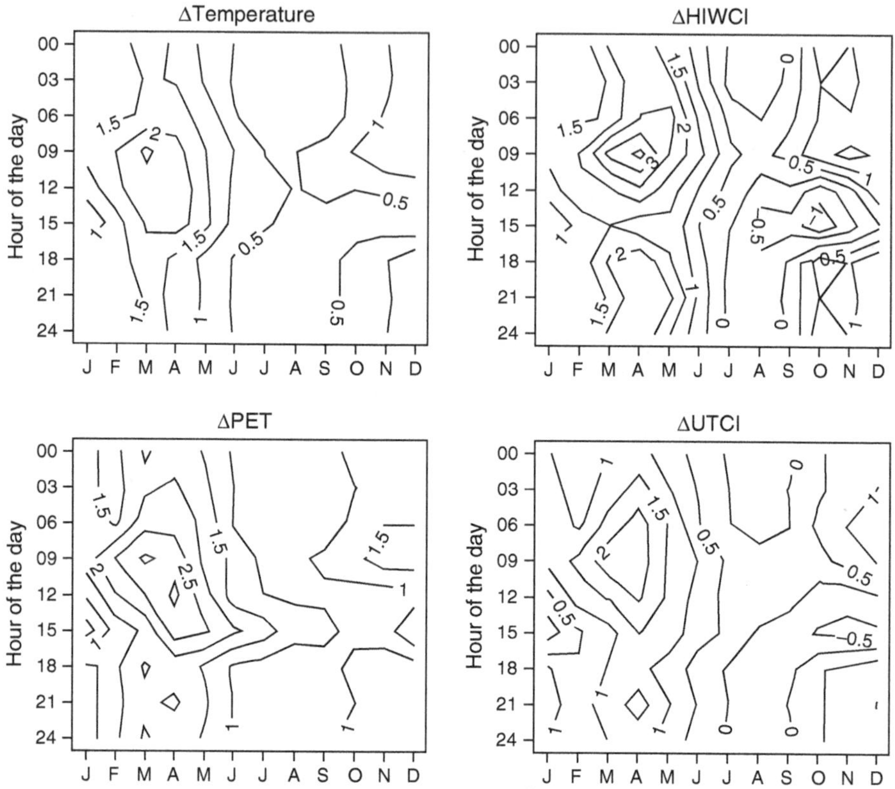

Fig. 10.7 Differences in temperature, HIWCI, PET, and UTCI between Dhaka and Mymensingh displayed as isopleths

speed (data not shown). Humidity was higher in rural areas, particularly in Mymensingh (approximately 10% relative humidity). Mean radiation temperature was higher in Dhaka, as was cloud coverage, particularly during winter. Wind speed was higher in Dhaka compared to Tangail but lower compared to Mymensingh.

10.4 Discussion

Tropical regions are usually associated with high temperatures and humidity as well as small seasonal fluctuations. According to the Köppen-Geiger classification, Bangladesh's climate is categorized as tropical winter dry (A_w) (Kottek et al. 2006). Our analysis demonstrated that climatic conditions in Bangladesh are typically tropical during the monsoon season but show characteristics of a seasonal climate during winter. Cold air masses from the Asian continent cause an abrupt fall in temperatures during the Northeast monsoon. Nevertheless, the thresholds indicating extreme cold, should rather be considered as moderate values in comparison to mid-latitude countries (or according to the UTCI assessment scale). The thresholds, indicating heat stress according to the UTCI assessment however, are surpassed most of the year.

In this context, the suitability of an absolute assessment and the information value of TPIs require further discussion. One question of importance is whether TPIs should be regarded as indicators of well-being and thermal perception rather than predictors for human morbidity or mortality. While the winter season in Bangladesh is commonly perceived as preferable compared to the hot and humid season, the winter mortality rate is characteristically higher and there is evidence of cold-related mortality (Becker and Weng 1998; Burkart et al. 2011). The crucial research question is the extent to which the human heat balance is connected to human health outcomes. Apart from human thermophysiological regulation, external parameters such as the prevalence of certain pathogens (themselves dependent on meteorological parameters) are relevant to the atmosphere-health relationship. Furthermore, biochemical reactions influenced by temperatures could be of importance. Bull (1980) argued that excess winter mortality is due to physiological changes in cellular and humoral immunity. In addition to changes in blood pressure and vasoconstriction, exposure to cold can lead to increases in blood viscosity, higher red blood cell counts, and increased levels of plasma, cholesterol, C-reactive protein, Interleukin-6 and fibrinogen, which can result in arterial thrombosis and other cold-induced cardiovascular reflexes (Keatinge et al. 1984; Keatinge and Donaldson 1995; Neild et al. 1994). There is further evidence to suggest that the adverse effects of cold on the immune system can be ascribed to stress hormones, or to the direct effects of cold on the respiratory tract, for example bronchoconstriction (Millqvist et al. 1987; Ophir and Elad 1987; Berk et al. 1987). Such mechanisms are not considered in current thermophysiological models. Unless TPIs are checked against measurable health outcomes no meaningful conclusions can be drawn. Considering complexity, the UTCI clearly outclasses other indices. However, a simpler index such as the HIWCI might be beneficial for application. Further research is needed on this matter in order to provide conclusive indicators for health impacts.

Although thermal levels are comparably high or moderate throughout the year, there is evidence that cold does matter in (sub)tropical regions. High mortality during the cold season as well as cold related-mortality was observed in studies conducted in Kuwait (Douglas et al. 1991) or Bangladesh (Becker 1981; Becker and Weng 1998; Burkart et al. 2011). Douglas et al. (1991) argued that the adverse effects of cold are not a consequence of low absolute temperatures, but of a seasonal fall below annual mean temperatures. In addition to physiological mechanisms, social, cultural and behavioural adaptation strategies determine the impact of cold (or heat). Research conducted by the *Eurowinter Group* demonstrated that regions with harsh winter climates exhibit a lower level of excess winter mortality than those with moderate winter climates (Eurowinter Group 1997). Housing and clothing in Bangladesh are adapted to the hot weather conditions prevailing for most of the year, while adaptation to the limited time-frame of relative cold is probably insufficient. Cold stress and increased cold wave probability occur over a relatively short time-frame. In the case of heat stress, the time-frame of days and nights with heat stress is broader. Thus the probability for the occurrence of a cold wave is considerably higher than for heat waves. The highest probabilities of heat waves were determined during the summer season regarding temperature, but shifted toward the monsoon season regarding TPIs. The combination of a high prevailing humidity, low diurnal amplitudes, persisting elevated daytime thermal conditions and little night-time cooling (this is due to reduced net long-wave emission) results in persisting thermal stress during the monsoon season.

Excess (equivalent) temperatures, marking the UHI were equally high for the different parameters considered. The UHI was most intense during the cold season but excess (equivalent) temperatures were still recorded throughout the summer and rainy seasons. While the urban heat island phenomenon might mitigate cold stress during the cold season, urban excess temperatures increase the thermal load during the hot and humid (pre-)monsoon season. In a climate of persistently high thermal levels even small excess temperatures might serve to cause excess morbidity and mortality if a certain breakpoint is passed. Indeed, there is evidence which suggests that in rural regions, cardiovascular mortality exhibit no heat effects, while urban areas show a strong heat-related increase in mortality above a specific threshold (Burkart et al. 2011). This could either be caused by urban excess temperatures or by the higher susceptibility of urban populations to heat effects. It most likely represents an interaction of both causes.

In mid-latitude regions, the UHI has often been described as a night-time phenomenon. Urban areas heat up more slowly than rural areas and show lower temperature maxima, as building materials divert and store heat into the building structure. At night, the cooling rate of urban areas is lower as the structure emits heat only gradually. These mechanisms could be responsible for the UHI differences observed between Dhaka and Tangail. However, building structures and materials in developing countries differ strongly from those used in industrialized countries. While the construction materials used in western countries usually have a high heat conductivity and specific heat capacity, this is not the case for the corrugated metals and brick types often used as building material in developing countries. In addition to the modifying effects of the building materials used, the association between sensible and latent heat could also be of particular significance in explaining characteristics of the tropical UHI. Water vapour capacity increases exponentially with temperature. Tropical air is able to contain exponentially more humidity than the air found in mid-latitude climates. Due to the high water availability in rural regions and the high atmospheric intake capacity, sensible heat flux is reduced and temperatures rise more slowly and not to the same extent as in urban areas. This could represent the cause of daytime urban excess temperatures[4]. During night-time, energy is released as water vapour condensates

[4]The energy amount needed to evaporate 1 g of water, increasing relative humidity of one cubic meter air about 2–3% is up to about 7 kJ. The same amount of energy would increase the sensible heat of one cubic meter air about 2 K. (Evaporation enthalpy, specific heat capacity and air mass per cubic meter are temperature dependent. The calculations are based on average values for approximately 30°C.)

leading to reduced cooling (in rural areas). Mymensingh can be considered a more rural environment compared to Tangail in terms of structure of buildings. The Mymensingh measurement site is located in an agricultural area in proximity to water bodies. The relative humidity level is approximately 10% higher than in Dhaka. Tangail which is more built up than Mymensingh only showed 2–3% increased levels. We conclude that in reference to areas with high water availability, the UHI is a daytime phenomenon (excess temperatures higher during daytimes) as latent energy fluxes reduce daytime heating as well as night-time cooling. With decreasing water availability due to increased building structure and sealing the tropical UHI is more and more shifted towards a night-time phenomenon as observed in mid-latitudes.

The pattern of increased mean radiant temperature follows the distribution of temperature and TPIs and can probably be explained by the same mechanisms. The high cloud coverage in Dhaka, particularly during winter, is most likely to be caused by urban aerosols serving as condensation nuclei. Concerning wind speed the mechanisms seem to be more complex. Surface roughness in the urban area may reduce wind speed, but the canalization of air movement (Bernoulli effect) or thermally induced wind could serve to increase wind speeds. The open field environment with little surface roughness gives a good explanation for high wind speeds in Mymensingh. In Tangail, a more built up environment, winds might already have been slowed down. Although, surface roughness is higher in Dhaka, canalization and thermal effect might cause increased wind speeds in comparison to Tangail.

Data availability constitutes a general problem in tropical developing regions. While numerous measurement campaigns have been launched in western countries designed to assess the urban heat island, this study had to rely on secondary data from the Bangladesh Meteorological Department. This brought the advantage of a long study period in the time series (10 years). However, the measurement sites were chosen to serve synoptic purposes, meaning that they are more likely to be representative of the macro- rather than the mesoclimate.

10.5 Conclusion

Until today, only few studies have been conducted on bioclimate and the health-atmosphere relationship in tropical regions. Heat stress is commonly believed to be a major issue in the tropics and the premonsoon season is supposed to be a period of high thermal stress. In this study we discussed several climatological approaches to health relevance assessment of weather conditions. We pointed out that in addition to the summer/pre-monsoon season, other seasons require attention concerning their health risk. During the monsoon season little relief is offered during night-time and the probability of a heat wave is increased. Furthermore, we argued that low temperatures and cold stress need to be considered. Although temperatures and modeled temperatures (TPIs) are constantly high (according to the absolute assessment scale provided with TPIs), there is evidence that periods of relative cold constitute health threats due to inadequate adaptation in (sub)tropical counties. We followed a statistical percentile-based approach for assessing cold stress, and found that extremes are restricted to the months of December and January and the probability of a cold wave is thusly increased. The megacity of Dhaka exhibited considerable excess temperatures, particularly during winter but also during the pre-monsoon season. Although the temperature differences remain below those observed in mid-latitude regions, the UHI might be epidemiologically relevant for tropical regions due to the persisting high levels of temperature and thermophysiological temperatures. Nevertheless, we point out the necessity of checking thermophysiological models and statistical approaches against measurable health outcomes in order to reach reliable conclusions about their explanatory power.

Acknowledgements The authors are very grateful to the Bangladesh Meteorological Department for providing meteorological data. Furthermore, we would like to thank the German Research Foundation (DFG) for funding the Dhaka INNOVATE project within the priority programme 1233 "Megacities-Megachallenge – Informal Dynamics of Global Change".

References

Basu R, Samet JM (2002) Relation between Elevated Ambient Temperature and Mortality: A Review of the Epidemiologic Evidence. Epidemiologic Reviews; 24 (2):190-202. doi:10.1093/epirev/mxf007

Becker S (1981) Seasonality of Deaths in Matlab, Bangladesh. International Journal of Epidemiology; 10 (3):271-280

Becker S, Weng S (1998) Seasonal Patterns of Deaths in Matlab, Bangladesh. International Journal of Epidemiology; 27 (5):814-823

Braga AL, Zanobetti A, Schwartz J (2001) The Time Course of Weather Related Deaths. Epidemiology; 12 (6):662-667

Braga AL, Zanobetti A, Schwartz J (2002) The Effect of Weather on Respiratory and Cardiovascular Deaths in 12 U.S. Cities. Environmental Health Perspectives; 110 (9):859-863

Bull G (1980) The weather and deaths from pneumonia. The Lancet; 315 (8183):1405-1408

Burkart K, Endlicher W (2009) Assessing the Atmospheric Impact on Public Health in the Megacity of Dhaka, Bangladesh. Die Erde; 140 (1):93-109

Burkart K, Breitner S, Schneider A, Khan M, Krämer A, Endlicher W (2011): The effect of atmospheric thermal conditions and urban thermal pollution on all-cause and cardiovascular mortality in Bangladesh. Environmental Pollution, *in press*

Büttner K (1938) Physikalische Bioklimatologie. Akademische Verlagsgesellschaft Leipzig

Cabanac M, Brimmel H (1987) The pathology of human temperature regulation: Thermiatrics. Experientia; 43:19-27

Douglas AS, Rawles JM, Al-Sayer H, Allan TM (1991) Seasonality of disease in Kuwait. The Lancet; 337 (8754):1393-1397

Driscoll DM (1985) Human health. Handbook of Applied Meteorology. John Wiley and Sons,

Eurowinter Group (1997) Cold exposure and winter mortality from ischaemic heart disease, cerebrovascular disease, respiratory disease, and all causes in warm and cold regions of Europe. Lancet; 349:1341-1346

Fiala D, Lomas KJ, Stohrer M (1999) A computer model of human thermoregulation for a wide range of environmental conditions: The passive system. Journal of Applied Physiology 87 (5): 1957-1972

Fiala D, Lomas KJ, Stohrer M (2001) Computer prediction of human thermoregulatory and temperature responses to a wide range of environmental conditions. International Journal of Biometeorology 45 (2):143-159

Hippocrates. (Written 400 B.C.E) On Airs, Waters, and Places (Reprint). Kessinger Publishing's Rare Reprints,

Höppe P (1999) The physiological equivalent temperature - a universal index for the biometeorological assessment of the thermal environment. International Journal of Biometeorology; 43: 71-75

Huynen MMTE, Martens P, Schram D, Weijenberg MP, Kunst AE (2001) The Impact of Heat Waves and Cold Spells on Mortality Rates in the Dutch Population. Environmental Health Perspectives; 109 (5):463–470

Jendritzky G, Havenith G, Weihs P, Batchvarova E, DeDear R The Universal Thermal Climate Index UTCI. In: NCUB London, London, September 20 2007.

Kalkstein LS, Smoyer KE (1993) The impact of climate change on human health: Some international implications. Experimentia; 49:44-64

Keatinge W, Donaldson G (1995) Cardiovascular mortality in winter. Arctic Medical Research; 54 (suppl 2):16-18

Keatinge W, Coleshaw S, Cotter F, Mattock M, Murphy M, Chelliah R (1984) Increases in platelet and red cell counts, blood viscosity, and arterial pressure during mild surface cooling: factors in mortality from coronary and cerebral thrombosis in winter. British Medical Journal (Clin Res Ed); 289 (6456):1405-1408

Kottek M, Grieser J, Beck C, Rudolf B, Rubel F (2006) World Map of the Köppen-Geiger climate classification updated. Meteorologische Zeitschrift; 15:259-263

Kunst AE, Looman CWN, Mackenbach JP (1993) Outdoor Air Temperature and Mortality in the Netherlands: A Time-Series Analysis. American Journal of Epidemiology; 137 (3):331-341

Medina-Ramón M, Zanobetti A, Cavanagh DP, Schwartz J (2006) Extreme Temperatures and Mortality: Assessing Effect Modification by Personal Characteristics and Specific Cause of Death in a Multi-City Case-Only Analysis. Environmental Health Perspectives; 114 (9): 1331-1336

Meehl GA, Tebaldi C (2004) More Intense, More Frequent, and Longer Lasting Heat Waves in the 21st Century. Science; 305 (5686):994-997

Millqvist E, Bengtsson U, Bake B (1987) Occurrence of breathing problems induced by cold climate in asthmatics–a questionnaire survey. European Journal of Respiratory Diseases; 71 (5):444-449

Neild P, Syndercombe-Court D, Keatinge W, Donaldson G, Mattock M, Caunce M (1994) Cold-induced increases in erythrocyte count, plasma cholesterol and plasma fibrinogen of elderly people without a comparable rise in protein C or factor X. Clinical Science; 86:43-48

Oke T (1973) City size and the urban heat island. Atmospheric Environment; 7 (8):769-779

Ophir D, Elad Y (1987) Effects of steam inhalation on nasal patency and nasal symptoms in patients with the common cold. American Journal of Otolaryngology 8(3):149-153

Parsons K (2003) Human Thermal Environments: The Effects of Hot, Moderate, and Cold Environments on Human Health, Comfort and Performance, Second Edition. Taylor & Francis, London

Robinson PJ (2001) On the Definition of a Heat Wave. Journal of Applied Meteorology; 40 (4):762-775

Roth M (2007) Review of urban climate research in (sub)tropical regions. International Journal of Climatology; 27 (14):1859-1873

Staiger H, Bucher K, Jendritzky G (1997) Gefühlte Temperatur. Die physiologisch gerechte Bewertung von Wärmebelastung und Kältestress beim Aufenthalt im Freien mit der Maßzahl Grad Celsius. Annalen der Meteorologie; 33:100 - 107

Steadman R (1971) Indices of Windchill of Clothed Persons. Journal of Applied Meteorology; 10:674-683

Steadman R (1979) The Assessment of Sultriness. Part I: A Temperature-Humidity Index Based on Human Physiology and Clothing Science. Journal of Applied Meteorology 18:861-873

Verein Deutscher Ingenieure (VDI) (1998) VDI-Richtlinie 3787 Blatt 2 (Technische Regel) Ausgabe 1998-01 Umweltmeteorologie - Methoden zur human-biometeorologischen Bewertung von Klima- und Lufthygiene für die Stadt- und Regionalplanung - Teil 1: Klima.

von Humboldt A (2004) Kosmos. Entwurf einer physischen Weltbeschreibung, vol Auflage: 1. Eichborn,

Part IV
Informality and Health

Chapter 11
Marketization and Informalization of Health Care Services in Mega-Urban China

Tabea Bork, Bettina Gransow, Frauke Kraas, and Yuan Yuan

11.1 Marketization of Health Care in China Under Transition Conditions

Introduction of the market, privatization and decentralization have been the dominant corner stones throughout the first two decades of China's reform line after the introduction of the open door policy in 1978. Many China researchers (e.g. Wang 2008; Wu 2008) thereby judge, that China's development path was not merely a transition from planned economy to market-oriented economy, but that a "market society" emerged, in which market principles permeate also noneconomic arenas and "threatened to become the dominant mechanism integrating all of society (and even political life)" (Wang 2008: 18). The marketization of the health sector thereby entailed that social security schemes and therewith financing of public health care collapsed almost completely and out-of-pocket payment became the dominant factor defining people's access to health care. China's health care sector became one of the most commercialized in the world, social polarization between high- and low-income groups accelerated, increasing inequities in access to health care and increasing gaps in health status between population groups of different socio-economic levels emerged. Consequentially, appraisals of the impact of China's transition path on public health is harsh: e.g. Liu and Mills (2002), Sun (2006), Yip and Mahal (2008) and even the Chinese authority itself (Ge et al. 2005)

T. Bork (✉) • F. Kraas
Institute of Geography, Cologne University, Cologne, Germany
e-mail: t.bork@uni-koeln.de

B. Gransow
Seminar of East Asian Studies, Free University Berlin, Berlin, Germany
and
School of Sociology and Anthropology, Sun Yat-sen University, Guangzhou, China

Y. Yuan
Seminar of East Asian Studies, Free University Berlin, Berlin, Germany

A. Krämer et al. (eds.), *Health in Megacities and Urban Areas*,
Contributions to Statistics, DOI 10.1007/978-3-7908-2733-0_11,

consent that all up-to-date reforms of the health system have either failed or are not profound enough.

However, in the last years an academic discourse started, which discusses if the Chinese state reform line lately made a major turnaround since which increased efforts to care for equity and for establishing social welfare are taken. It grounds, firstly, on a changed government rhetoric, which started at the beginning of the new millennium, which admits mistakes in reform policies and the disastrous effects it had on social development, equity and access of people to social security and which emphasizes need for action. Secondly, indeed, since the end of the 1990s various and numerous social security schemes have been experimented with and policies released and partly been transferred to the whole country (e.g. the new cooperative medical insurance schemes in rural areas in 2003 [NCMS] and the basic urban medical insurance [BIS] introduced for employees in 1998 and for non-employee urban residents in 2007). Doubters, though, point at actual effects of promulgations and new policies and conclude that these are rather humble. Wu (2008: 1094, 1096), for example, claims that

> Although it is possible to provide some 'empirical' evidence of enlarged state capacity and increasing social expenditure... We can also find evidence of increasing marketization [... while] the political economy has not shifted.

This article's aim is to judge the focal point of the academic dispute against the background of the so far restricted access of rural-urban migrants to health care and the increasing informalization of health care services – both of which have been negative by-products of the reform era and against structural reality in China pose special challenges for adequate intervention. It will be evaluated in how far a change in government's efforts in increasing migrant's access to health care and in reacting to increasingly economically-oriented informal strategies of health care providers is certifiable in terms of their actual impact and scope.

The argumentation is based on fieldwork in Guangzhou, China and was conducted between January 2007 and December 2008 and consisted of a variety of methods. It firstly included 29 expert interviews with representatives of urban and health administration in Guangzhou, health care personnel from public and private and informal facilities and NGO representatives. These are the source for judgements on the development of informal strategies of health care providers and partly for the evaluation of migrant's access to health care. Secondly, 68 in-depth interviews as well as a quantitative survey with 450 rural-urban migrants were conducted out of which conclusions for the development of migrant's actual access to health care were drawn.

Guangzhou, which was base for the conduction of fieldwork, is part of the Pearl River Delta, a mega-urban area that experienced accelerated economic development and urbanization in the course of economic opening. Changes have been especially profound and fast here and pose special challenges for adequate response of local administration and planning. Therefore a section on the role of mega-urbanization in the transition process is included hereafter.

11.2 The Role of China's Mega-Urban Development in Transition and Challenges for the Health Care Sector

The UN World Urbanization Prospects forecast that in 2010 7 megacities with more than 5 million inhabitants will have developed in China and that their number will grow to 12 megacities in 2025 (United Nations 2008: 169–171). In the course of China's transition megacities took a special role in several respects. Firstly, they were centres of economic opening, industrialization and development. Consequentially, they also attracted high numbers of China's, up-to-date, 225.4 million rural-urban migrants (National Bureau of Statistics 2009), who search for jobs outside their hometowns, and many megacities only emerged due to this huge in-migration (e.g. Dongguan and Shenzhen). Therefore they were and are hubs of China's post-opening industrialization, of urbanization in general and exogenous urbanization (i.e. urbanization triggered considerably by FDI) in particular. Hence, they experience not only up to now unknown dimensions of expansion, highest concentrations of population, infrastructure, economic power, capital, and decisions as well as highest dynamics, but above all also a simultaneity and overlapping of different processes with mutual feedback, an increase in informality, in disparities and in numerous situations of urban stress – and therefore pose highest challenges for governance on different levels (Kraas 2007).

With regard to the changing health sector, China's economically booming megacities were, firstly, especially burdened with the inflationary increase in prices for medical services and pharmaceuticals, which resulted from price reforms. Likewise, due to – in average – much higher income levels and higher demand, negative effects of the double-tracked price system accelerated here. Its intention was to ensure access to primary health care and pharmaceuticals through price-regulation, while non-basic services became subject to market-driven prices. Quite the converse it led to a mushrooming of high end diagnostic services and a massive over-prescription of expensive drugs (Yang and Shi 2006), while provision of primary health care became neglected. Reasons are the increasing pressure on public providers to obtain own financing due to heavily shortened state subsidies, but beyond that also efforts of the facilities' managers and staff to increase their own income. Secondly, intra-urban disparities in access to health care and in health status are especially high here due to huge and widening socio-economic gaps in megacities. Eventually, in some cases (emerging) megacities were bases for experiments of reforms, e.g. experiments with migrant insurances that were launched successively in Shanghai, Chengdu, Beijing, Shenzhen, Guangzhou since 2002 (Gov.cn 2006; Wang 2008: 34).

11.3 Findings: Informal Answers to Migrants' Lacking Access to Health Care

The reform of health care services as part of China's transition process was carried out along the lines of the dual (urban/rural) structure of the *hukou*-system. In the urban medical sector, China established a Public Fund Medical Care System after 1949 for government staff and the employees of state-owned enterprises that covered both of these groups via their places of work. Since 1994, China has begun to reform its urban medical system. Now every employee has an individual account at a medical fund managed by the local health safety management bureau (BIS for employees). All employees of urban enterprises are required to join this program, while the funding is managed by the state and not confined to the enterprises. Thus employees can accumulate money in their accounts and change jobs without changing their accounts (Xiang 2005). A second basic urban medical insurance system was introduced to non-employed urban residents in 2007 (cf. Fig. 11.1).

In the countryside, China used to have a rural cooperative medical system whereby production brigades contributed part of their annual resources to a collective fund covering their members' medical expenses. Yet, the system collapsed in

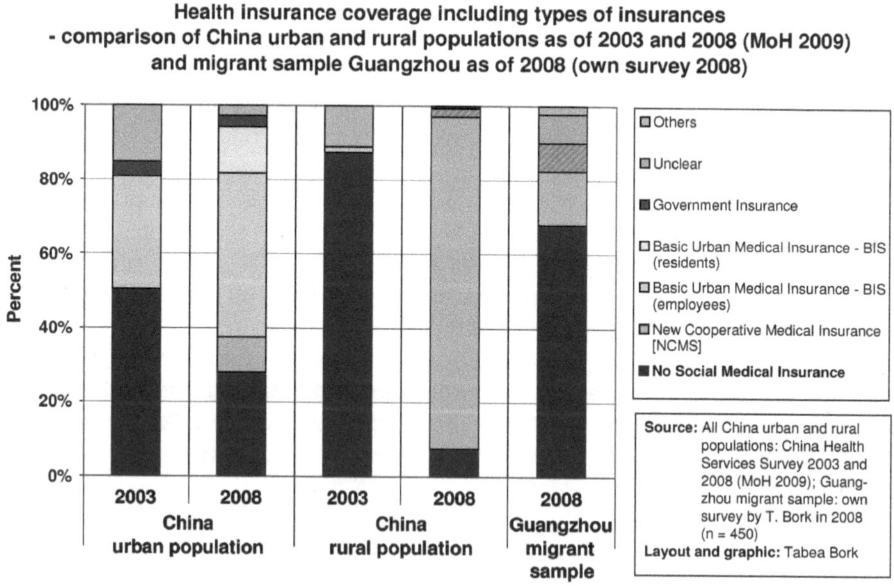

Fig. 11.1 Comparison of health insurance coverage among China rural and urban populations and the Guangzhou migrant sample. Only 5 years after its introduction in 2003 the NCMS achieved amazing coverage rates for the rural population, while the BIS (employees) is spreading comparatively slow. The BIS (residents) was introduced in 2007 to allow an inclusion of the large group of non-employees in urban areas. Almost all migrants among the sample where still uncovered – considering that the small percentage carrying NCMS insurance cannot make use of these in the city

the 1980s when the commune system was abolished and the new individual household responsibility system was adopted: coverage rates declined substantially so already in the mid-1980s the peasants had to shoulder around 90% of healthcare costs themselves. Only in 2003 a new cooperative medical scheme (NCMS) was introduced. Although it is a voluntary scheme it achieved amazing coverage rates since then as Fig. 11.1 shows. Nevertheless, while it had a positive effect on the utilization rate of health care, it did not have a significant impact on per capita out-of-pocket spending and catastrophic expenditure risk. Reasons are that firstly its budget is too small – it amounts to only around 20% of the average per capita total health expenditure and it is confined to the reimbursement of inpatient services and catastrophic outpatient service, neglecting preventive and basic medical care (You and Kobayashi 2009: 7).

As part of the rural-hukou population, migrants are supposed to claim medical benefits at the rural locations where they are registered, but due to the above-explained limited nature of the NCMS migrants cannot expect much if anything from it. In addition costs to return to their places of origin to obtain care would use up an important share of their salaries. Among the generation of political leaders under Hu Jintao, attitudes towards migration have changed from observation and tolerance to policies that more actively promote migration as part of an overall development strategy (cf. Holdaway 2008). This includes reforming the healthcare system with the aim of including all citizens and guaranteeing basic health services at reasonable prices. The reform plan has been released at the beginning of 2009 and call for the new system to be tested in a few provinces (cf. Chen 2009). However, this type of program is still more theory than practice, and the question arises of how the far-reaching sociopolitical goals and prescriptions of the central government can be implemented on the various different local levels. Furthermore, with BIS and NCMS only recently new schemes have been introduced, which again cement the rural-urban divide in the medical system: Another contemporary change to an integrating system might appear as a self-defeating action on behalf of the government.

The Pearl River Delta has been an attractive destination for rural-urban migration within China since the 1980s. Hence problems connected with mass migration such as insufficient health care and insurance for migrants attracted attention relatively early on. However, initial attempts to include migrants in the social security system are still quite far from being fully implemented. In 2005, only 4.87 million migrants (out of 22 million) in Guangdong Province had an old-age pension (Guangdong Research Team 2006: 432), 3.14 million had a health insurance, 3.29 million had an unemployment insurance, 5.94 million had an accident insurance, and 360,000 were eligible for maternity benefits (Guangdong Research Team 2006: 437). Why were such comparatively small numbers of migrants integrated into the social security system? The exclusion of migrants from the healthcare system has been explained as a result of the government's intention to reform the urban healthcare system so as to relieve state-owned enterprises from the financial burden of nearly unlimited medical care for their employees. It therefore became common practice for enterprises to replace older employees who held generous welfare coverage with migrants who were not entitled to these benefits. In addition, it is feared that the inclusion of migrants in urban healthcare systems would encourage rural inhabitants to rush to the city when sick and falsely claim to be migrant workers (Xiang 2005). Additional factors hindering success are corporate interests, and fears by migrants themselves that they would bear additional costs that would ultimately not pay off due to their non-localized working patterns.

To prevent the spread of contagious diseases, Guangdong province has begun to install five types of health services free of charge on the local level:

- vaccination of migrant children according to the same guidelines valid for local residents,
- diagnosis, treatment and isolation of migrants with contagious diseases,
- basic health care provision and treatment in cases of tuberculosis for migrants residing in Guangdong for more than 6 months,
- HIV/AIDS-counselling for migrants and
- health education for migrants (Labour and Social Security Office of Guangdong Province 2007).

This minimum of health care measures is far from covering the needs of the migrant population. Medical supply is insufficient, costs are prohibitive to migrants and can ruin whole families and smaller enterprises in the informal sector lack safety measures against occupational accidents and diseases.

As a result the migrants themselves have to bear almost all the cost out-of-pocket; to the majority of them medical treatment is unaffordable. Among the sample of migrants surveyed in Guangzhou 32% carried an insurance of which half were members of the NCMS of which, as explained above, they cannot make much use (cf. Fig. 11.1). While a migrant's budget has an average limit of 100 Yuan a month for health expenditure, the treatment of a minor illness (such as a cold) in a large hospital amounts to 500 Yuan, which would consume almost a complete monthly wage of a migrant worker. Financial hardship has ever and anon led to migrants' aborting necessary subsequent treatment following emergency treatment in a hospital. The ward for external injuries of the People's Hospital in Guangdong, for instance, treats an average of 200 migrants a year of whom a third is unable to pay their bill (Xiang 2005: 162). Some hospitals have thereupon stopped to accept migrants as patients. In turn, many migrants do not seek treatment when they feel indications of illness; they rather wait and see – which might lead to more severe diseases. At the same time living and working conditions of migrants are characterised by heightened health risks. Reports on occupational accidents and diseases identify pollution, noise, dust and poisonous substances as the main sources of migrants' health problems. The high health risks of so-called 3-D-jobs (dirty, dangerous and demanding) have created an army of incapacitated migrant workers that will keep on growing if adequate occupational safety measures and health care provision are not established. Among the migrant sample, which reflected the cross-section through typical migrant employments and did not particularly focus on 3-D-jobs, still 17.3% reported that they had suffered from a medical problem before, which was caused at their workplace and almost 6% stated they had

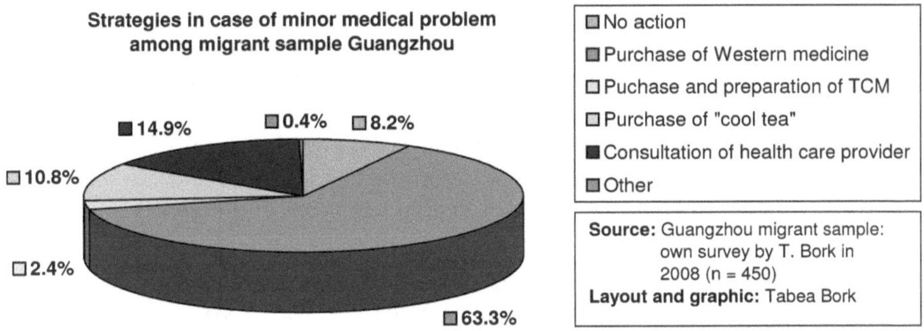

Fig. 11.2 Strategies in case of minor medical problem among migrant sample Guangzhou. Most interviewees would choose self-treatment strategies (purchase of medicine or the preparation of traditional Chinese medicine, including the Guangdong speciality "cool tea" [liang cha]) without consulting a health professional

lost a job before due to sickness during an average length of stay in Guangzhou of slightly less than 6 years.

Against this background of insufficient access to the formal health care system in the cities, areas with a high concentration of migrant populations – such as the so-called "villages-in-the-city" (cf. Gransow 2007) – have developed a multiplicity of small informal clinics and medical practices that attempt to fill the market niche of migrant health care – addressing their needs but also giving rise to new risks (cf. Part IV). However, the survey showed that migrants only hesitantly make use of this new supply. It seems that the dominant strategy remains to wait and see how a disease develops instead of taking on the almost non-verifiable offerings of informal health services. In the migrant survey 76.5% said they would chose self-treatment strategies when they come across a minor medical problem, while only 15% would consult a health professional (cf. Fig. 11.2).

11.4 Findings: Informalization of Health Care Services

Given the insufficient access to healthcare in cities and areas with large migrant populations a large number of small unregistered, informal health stations have arisen in an attempt to fill emerging market niches for migrant's health care. In 1998 a new regulation was introduced to increase quality in health care and ensure tax payment, which determined that all health care facilities need to obtain three licenses – a license for the health care facility, a license for every practicing doctor and a business license (Lim et al. 2002; Meng 2005). In addition, in the course of privatization in the health sector, a wide variety of different types of cooperation between public and private units has arisen. Not all of these newly established facilities are legal or formally acknowledged and possess necessary licenses. However, informal strategies today can be found in all health facilities – ranging from public facilities of all hierarchical levels to private and informal ones, and, as must be noted, are also applied by other actors in the health care system. An overview on informal actions lacking good governance performance and their possible health impacts is given in Table 11.1. With regard to public health, a distinction is made between those actions that have severe directly linked consequences for health and those actions without directly linked severe consequences for people's health. Results reflect that stakeholders quite well manage to evade or go round existing policies and regulation.

Attempts to contain profit-oriented behavior of all stakeholders are only being undertaken in the last years. They include campaigns against corruption of public servants and for food and drug safety (Yong and Ran 2006), but embracing campaigns on improving health care quality are still lacking. The 2009 health care reform plan includes the aims to ensure the non-profit character of public facilities, stop the sale of medications by public hospitals and clinics and call for the government to take over more responsibility (cf. Chen 2009). All of these points are important in containing profit-oriented action in the health sector. However, the plan neglects private and informal facilities from the beginning. Its success cannot be evaluated at this stage.

Hereafter findings from fieldwork in Guangzhou amending to the current state of the art will be elaborated on in more detail. A special focus was put on the observation of unregistered health care providers, which have been neglected by research and policy makers. Field observations and expert interviews indicated that these types of facilities are mushrooming. Reasons not to register are not only lack of education: Due to the flourishing of health care providers, the Guangzhou administration started to set limits to new registration of providers. In some cases then, practitioners open a facility without acquiring a registration. In other cases practitioners do not register to obviate having to pay taxes. Still, in many cases practitioners do lack any professional training. A doctor interviewed in a public hospital reported that in several cases patients, who received mistreatments in unregistered clinics ended up in his hospital:

> So after several days passed the patients get worse. So they transfer to this hospital, but … the disease is getting very serious then. … They just see the symptoms and give normal treatment, so if there are special cases they will make their disease worse.

On their business signs informal practitioners canvass with treatment of respiratory diseases, diarrhoea and many claim to be specialized in gynaecology, in the treatment of sexually transmitted diseases, conduction of sterilisation, ultrasonic testing during pregnancy and abortions. Hence, next to underbidding prices of the formal health sector, unregistered providers try to fill market gaps through offering services, which are illegal in China, as e.g. X-ray gender determination and gender-selective abortions. Informal health care facilities may provide cheap services needed at the grass-root level, but they may also include a variety of informal actions which might cause severe health consequences for the patients.

Table 11.1 Informal actions in China's health care sector

	Informal actions	
Actors	With knowingly, directly linked possible severe health consequences	Without directly linked health consequences
Administration, regulation, control units	• Accepting bribes from health care facilities or personnel, e.g. in order to overlook certain irregularities • Attempts to cover up the epidemic or incompetence in the SARS outbreak 2003 or in the milk scandal in 2008 • Condoning informal health care providers and activities (e.g. due to understaffing or bribing)	• Embezzlement and appropriation of public funds found in 41 out of 55 central government departments • Accepting bribes associated with the purchase of pharmaceuticals • Personal gains of public servants • Selling of promotions for cash
Pharmaceutical companies	• Production of counterfeit products and frauds • Export of fake drugs • Export of expired drugs • Exaggeration of benefits of medical products • Advertisements for fake products	• Offering and/or paying bribes to increase sales • Tax evasion through not registering company
Public and private providers (registered)	• Distribution of expired or counterfeit drugs • Misuse of pharmaceuticals due to lack of educational training (especially antibiotics) • Overprescription of pharmaceuticals • Cooperation with employers to deceive employees from receiving compensation for occupational diseases • Performing medical treatment beyond knowledge and educational training • Hiring practitioners without licenses	• Demanding different types of illegal payments or commissions ("red packages") from patients and thereby giving preferred treatment to certain groups of patients • Acceptance of kickback s from suppliers of pharmaceuticals • Linking of doctor's rewards with their prescription and sale of drugs • Requiring and/or accepting bribes • Overuse of high-technology, expensive diagnostic services • Public hospital managers largely employ for-profit means and partly use it to increase their and the hospital's staff income • Strong internal cohesion among health facility personnel in covering up and mutually conducting for-profit activities

(continued)

Table 11.1 (continued)

Actors	Informal actions: With knowingly, directly linked possible severe health consequences	Informal actions: Without directly linked health consequences
Private providers (unregistered)	• Performing medical treatment beyond knowledge and educational training	• Threatening of competing doctors/facilities
	• Distribution of fake medical products	• Tax evasion through not obtaining business license
	• Hiring practitioners without licenses	• Bribing of executive personnel to be tolerated or to make the executive personnel drive out other informal competitors
		• Underbidding of prices in formal health care facilities

Sources: Akunyili (2006: 99), Bloom et al. (2000: 29–30), Chai (1997: 1045), Choi et al. (1999: 314), Cohen (2006: 83), Lewis (2006: 2), Li (2006: 90, 91, 94), Savedoff and Hussmann (2006: 12), Yang (2006: 71), Yang and Shi (2006: 125), Yip and Mahal (2008: 928), Yong and Ran (2006: 142–144), China Daily and People's Daily as well as expert interviews and field observation in Guangzhou, conducted between January 2007 and May 2008

All interviewed unregistered practitioners report that competition has been increasing lately, as a result of which some use different strategies to drive out rivals, among which are threatening of rivals or bribing executive personnel to inspect their rivals. An unregistered practitioner reports about a competitor:

> He is really angry for I opened another clinic here. And he hired few people to come here, that it is better if I close the door, because if I open this clinic it will effect his economic situation. Or if you insist on opening this clinic you have to pay me 2000 Yuan.

Among the findings are that control agencies and executive personnel responsible for checking registering of medical institutions and medical personnel in Guangzhou were reported to accept bribes from health care facilities or personnel to oversee certain irregularities. For example, one unregistered practitioner interviewed reports about a competing unregistered practitioner:

> . . ., it is also hard to survive here. That man is familiar with the police here. Sometimes they even invite the police for dinner. And we have just been here for half a year and we are not familiar with the police or other people, as they are.

As another example, clinic owners, simply close their facilities in case they suspect that controls are being made and open again after the controls are over. Hence, effects of campaigns are temporary. Next to bribing and evasion strategies mere understaffing and lack of financial resources, though, is a reason for condoning deteriorating provider activities. According to an expert from the Guangzhou health administration, responsible departments and the police would need at least twice up to treble staff to be able to supervise health care providers adequately. In addition, interest in controlling private facilities is not government's priority according to the same interviewee:

> ...the private clinics... do not use national health resources, so the government may not pay too much attention to this. Same about drugstores. There are too many drugstores now in China, but many of them belong to private companies, so the country also does not pay too much attention to this.

All in all, the 1998 licensing regulation proved irrelevant in increasing service quality in case of informal facilities as controls are either lacking or ineffective.

With regard to formal public health care providers, NGO representatives interviewed reported that in case of occupational diseases health professionals have been detected to cooperate with employers instead of the employees by refusing to diagnose that disease or injuries were caused at the workplace. Another strategy is to purposely underrate the severity of disease or injuries in order to allow the employers to avoid or to reduce the compensation they have to pay to the concerned employees. Modalities by the employers are to send their employees to facilities in which they cooperate with the practicing physicians. The physicians are then asked to degrade the injury level or to send the medical report to the employer, who then refuses to hand it out to the employees. Furthermore, employers urge their employees to go to certain facilities to which they refer all their employees. Benefiting from this procedure, the medical personnel are willing to cooperate with the employers in covering the reasons for injuries or diseases. Health care providers, moreover, cooperate with employers through changing names of employees in case of occupational injuries: Companies insure only a certain number of their employees and in case accidents happen to other employees they are being registered under the name of an insured. Later on, however, they will not be able to claim compensation for occupational injury as the medical data refers to a different person.

The quantitative survey with rural-urban migrants in villages-in-the-city in Guangzhou showed that among the interviewees almost one third believes that doctors would treat them better if they paid them under-the-table-fees, and of those patients, who had visited a health facility in Guangzhou before (72.9% of which were public facilities), two thirds believe that the health personnel had carried out more examinations or prescribed more medicine than necessary.

11.5 Discussion: The Role of China's Transition Path in the Development of Increasing Informalities

Above-explained examples show that indeed in terms of reaching its targeted goals of integrating migrants into the social security scheme and in containing deteriorating health care service delivery in the health care sector China is far behind. It is argued here that there are certain similar structural and institutional reasons grounded on the national level, which will be discussed below.

1. "Gradual" but still rapid reform process with weak legislative foundation and executive power
   It is argued here that despite the general gradual transition path the implementation and effects of reforms in many cases were by no means gradual. Impacts of

many new reforms were fundamental for the society and economy and furthermore the pace of change was rapid, while the real extent and long-term consequences were traceable at the earliest years after their implementation. Legislative foundation thereby was not developed coevally but only as a response to newly soaring problems. Decentralization of management, financing and regulation of health care, for example, did not go in hand with the provision of directives to local administrations. Thus, national guidelines were implemented quite differently locally, and were influenced by local interests, which often deviate from national aims. According to Gong (2006: 80) it also explains the mushrooming of informal strategies among providers: "*Given the absence of macro-management [...] and inappropriate micro-management [...] possibilities for the suppliers' abuse of their position become a reality*".
The almost two decades that the government waited to introduce effective legislation in the health system formed a period in which stakeholders had almost no limits and could freely make use of gaps in the system. Furthermore in spite of the – lately – large number of newly released laws, regulations and rules in different levels, they are often not going along with the securing of implementation and enforcement. Financial and personnel shortages, lack of experiences and corruption among government personnel – which resulted from lack of power control and of accountability mechanisms – aggravate enforcement.

2. Lack of a health system governance framework and target-oriented sound reform line
From the beginning of the reform policy China lacked an all-embracing health system governance framework guiding its reform line with clear cut targets. Hence, reforms were guided by short- and medium-term goals, which aimed at dealing with symptoms, but lacked long-term vision of how to create structures that are able to sustainably improve public health. This explains why actual effects of, e.g., the new insurance schemes, is limited, as their design was inappropriate even if their implementation (in terms of coverage rates) was successful. Even more government repeatedly bucked against long-term vision, which became especially obvious in its long persistency in labeling and treating rural-urban migration as a temporary phenomenon, which was but an avoidance strategy of having to deal with including this group in social security schemes in the cities. Reforms for a long time have been very selective and even the 2009 reform plan of the health care system, which is more embracing, again focuses on public and formal providers. In addition, the restricted nature of certain elements of the plan, as e.g. of the above-explained NCMS, limit its scope from the start.
3. Institutional fragmentation
China's administrative system is highly fragmented with a variety of government ministries, bureaus and departments being responsible for different health system- and health care-related fields. As a result, policies are being developed, which are often uncoordinated and effective coordination of responsibilities and control through the agencies is aggravated (Peng 2004; Sun 2006; Xu and Zhang 2006).

4. Financing problems and decentralization
   After the economic opening and the consequential breakdown of financing schemes, financial constraints throughout the whole reform process have been the decisive problem. As an answer, decentralization of financing, organization and management of health care were among the dominant health system reform lines of the Chinese government since the mid-1980s, all of which were targeted at cost recovery (Bogg et al. 1996). Between 1997 and 2003, for example, more than 90% of the health care costs were shouldered by local governments, while health care expenditure by central government simultaneously never exceeded 10% (Ministry of Finance 2006a; Ministry of Finance 2006b). Lack of financial capital as well as skilled personnel as explained above for Guangzhou aggravated effective planning and control of health care providers.
5. Cultural tradition of informal relations and negotiation
   Competition with regard to power distribution between central and local government has a long tradition in Chinese history. This tradition is still taking effects in the present Chinese political regime. The decentralization process since the 1980s, to a large degree has not been accompanied by an institutionalization of the new central-local power relation due to the lack of a clear governance framework. Therefore decentralization has locally been applied very differently (Orban et al. 2003). Furthermore "informal rules and institutions . . . play a[n] . . . important role in the interaction between the center and the provinces" (Zheng 2000: 221). This fact gave space for informal actions among personnel employed in government units.
6. "Experimental informality"
   Rooting in the rapidness and profoundness of economic and social changes as well as in a long tradition of local experimentation as a way of preparing and carrying out reform measures in China (cf. Heilmann 2008), public administration, had no chance but to become subject to practical learning. Such local experiments are often embedded in informal settings, officially neither allowed nor forbidden, with the administration "keeping one eye shut and one eye open". As a peculiarity of China's transition process, therefore, in many instances local reform experiments precede their national implementation. However, time spans between local experiments and national layout are in most cases much too short to really await and evaluate the success of experiments. Furthermore, here again, experiments, do not go in hand with legislative adaptation.
7. Lack of social responsibility
   Next to financial hardships faced, lack of social responsibility among stakeholders in China must be taken as explanation for increasing commercial orientation and application of informal strategies to improve revenues and personal gain.

11.6 Conclusions

The article discusses the impact of changes in China's reform line during the last 10 years on migrant's access to health care and advanced informalization of health care services. Among the main findings are reasons of substantial deficiencies in the health sector which point out that due to the reform processes, involving an economy and society of incomparable dimension, changes de facto have not been gradual but rapid and profound in many instances. Most important problems thereby have been and are the absence of a clearly defined health system governance framework, sound reform line and constitution and adjustment of legislative foundation – which has proven elsewhere to be most crucial for the entire well-functioning of the health system. Due to this lack numerous informal practices and strategies have developed among administration, health care providers and pharmaceutical companies. "Experimental informality" through local reform experiments – serves as a field of trial and error of adaptive transition in the Chinese context as well as an "creative floor for negotiation processes" between stakeholders with all negative (e.g. issues of marginalization or social justice) and positive implications (e.g. essential stabilization and compensation of governmental deficits).

Findings indicate China's development an at least temporary, if not lasting degradation of the health system as economic profit-making behavior is dominating. Thereby governmental and public duties as well as responsibilities to provide and sustain basic and advanced health care for all are neglected. Despite some regulations adopted during the last few years the findings in respect to the health sector support Wu's (2008: 1096) notion that China has not yet reached the tipping point, which marks "the 'historical transition' from having only economic policy to the development of social policies" – which would also imply that its results would be perceivable in terms of stakeholder behavior and in integration of China's huge bulk of rural-urban migrants in the social security scheme.

Steadying of transition processes in China cannot be expected soon: As China's urban population is prospected to count only for 48.5% of the total population in 2010 but is projected to rise to 59.2% in 2025 or – if one regards these figures to be projectable – to 74.1% in 2050 (United Nations 2008: 77), according to the connected dynamics it will be unrealistic to achieve consistent, coherent and stable framework conditions throughout the country in a near future. Accompanying increasing urbanization, great changes embracing the whole Chinese society are still to be awaited. Many changes are and will be especially profound in megacities, which can be regarded as laboratories of urban future, as they are the basis of experiments and precursors for reforms in experiencing highest development pace and dynamics in multi-stakeholder environments under the circumstances of global change.

References

Akunyili D (2006) The fight against counterfeit drugs in Nigeria. In: Transparency International (ed(s).) Global corruption report. Health and corruption. Pluto Press, London, Ann Arbor, pp 96–100

Bloom G, Han L and Li X (2000) How health workers earn a living in China. In: Human Resources for Health Development Journal 5: 25–35

Bogg L, Dong H, Wang K, Cai W and Diwan V (1996) The cost of coverage: rural health insurance in China. In: Health Policy and Planning 11: 238–252

Chai BK (1997) China's floating population and its implications. In: International Journal of Social Economics 24: 1038–1051

Chen Z (2009) Launch of the health-care reform plan in China. In: Lancet 373: 1322–1324

Choi KH, Zheng X, Zhou H, Chen W and Mandel J (1999) Treatment delay and reliance on private physicians among patients with sexually transmitted diseases in China. In: International Journal of STD & AIDS 10: 309–315

Cohen JC (2006) Pharmaceuticals and corruption: a risk assessment. In: Transparency International (ed(s).) Global corruption report. Health and corruption. Pluto Press, London, Ann Arbor, pp 77–85

Ge Y, Ding N, Gong S, Wang L, Wang X, Shi G and Lei H (2005) "Dui zhongguo yiliao weisheng tizhi gaige de pingjia yu jianyi 对中国医疗卫生体制改革的评价与建议" (China's health care system reform, the evaluation and recommendations). In: Zhongguo fazhan pinglun 中国发展评论 (China Development Review) Special Edition 1: 1–14

Gong S (2006) Supply and demand in China's health resources and health care reform. In: Social Sciences in China Special Issue: Reform of the Public Health System: 76–84

Gov.Cn (2006) Medical insurance regulation benefits migrant workers in Shenzhen. Accessible at: http://www.gov.cn/english/2006-05/16/content_281392.htm (01.10.2009)

Gransow B (2007) „Dörfer in Städten" - Typen chinesischer Marginalsiedlungen am Beispiel Beijing und Guangzhou. In: Bronger D. (ed(s).) Marginalsiedlungen in Megastädten Asiens. Lit. Verlag, Münster, pp 343–377

Guangdong Research Team (2006) „Guangdong sheng jiaqiang nongmingong guanli he fuwu zhuyao zuofa " (广东省加强农民工管理和服务主要做法. (Important practises to encourage the administration and services for rural migrants in Guangdong province). In: Research Team of the State Council (ed(s).) Zhongguo nongmingong diaoyan baogao (中国农民工调研报告 (Research report on rural migrants in China). Zhongguo yanshi Verlag, Beijing, pp 432–446

Heilmann S (2008) Policy experimentation in China's economic rise. In: Studies in Comparative International Development 43: 1–26

Holdaway J (2008) Migration and health in China: An introduction to problems, policy and research. In: The Yale-China Health Journal 5: 7–23

Kraas F (2007) Megacities in Asia. In: Asien 103: 7–8

Labour and Social Security Office of Guangdong Province (2007) Guangdong sheng laodong he shehui baozhang ting 广东省劳动和社会保障厅2006/2007. Unpublished report on the situation of migrants in Guangdong.

Lewis M (2006) Tackling health care corruption and governance woes in developing countries. CGD brief of the Center for Global Development Accessible at: http://www.google.de/search?hl=de&q=Tackling+health+care+corruption+and+governance+woes+in+developing+countries&btnG=Google-Suche&meta=&aq=f&oq= (17.08.2009)

Li W (2006) Institutional reform and governance of public hospitals. In: Social Sciences in China Special Issue: Reform of the Public Health System: 85–97

Lim MK, Yang H, Zhang T, Zhou Z, Feng W and Chen Y (2002) The role and scope of private medical practice in China. Commissioned by UNDP and WHO. Accessible at: http://www.worldbank.org.cn/english/content/Private_Medical_in_China.pdf (14.04.2007)

Liu X and Mills A (2002) Financing reforms of public health services in China: lessons for other nations. In: Social Science & Medicine 54: 1691–1698

Meng Q (2005) Review of health care provider payment reforms in China. Worldbank Background Paper. Accessible at: http://siteresources.worldbank.org/INTEAPREGTOPHEANUT/Resources/502734-1129734318233/Reviewofproviderorganization-0730-Acceptanceofchanges.pdf

Ministry of Finance (2006a) Central budgetary expenditure on operation of culture, education, science & health care. Accessible at: http://www.mof.gov.cn/news/uploadfile/zhongyang002.xls (11.12.2006)

Ministry of Finance (2006b) Local budgetary expenditure on operation of culture, education, science & health care. Accessible at: http://www.mof.gov.cn/news/uploadfile/difang000.xls (11.12.2006)

National Bureau of Statistics (2009) Guojia tongjiju 2009: 2008 nianmo quanguo nongmingong zongliang wei 22542 wan ren (A total of 225,42 million rural migrants in China at the end of 2008). Accessible at: http://www.stats.gov.cn/tjfx/fxbg/t20090325_402547406.htm (12.10.2009)

Orban E, Chen X and Koehn PH (2003) Great power decentralization and the management of global/local economic policy and relations: lessons in fluidity from the People's Republic of China. In: International Review of Administrative Sciences 69: 235–258

Peng X (2004) Establishing and perfecting a social policy system for society on an ultra-large scale. In: Social Sciences in China 25: 146–153

Savedoff WD and Hussmann K (2006) Why are health systems prone to corruption? In: Transparency International (ed(s).) Global corruption report. Health and corruption. Pluto Press, London, Ann Arbor, pp 4–16

Sun B (2006) New breakthrough required for China's public health system reform. In: Social Sciences in China Special Issue: Reform of the Public Health System: 98–106

United Nations (2008) World Urbanization Prospects. The 2007 Revision. New York

Wang SG (2008) The 'great transformation': The double movement in China (Karl Polanyi). In: Boundary 2 35: 15–47

Wu FL (2008) China's great transformation: Neoliberalization as establishing a market society. In: Geoforum 39: 1093–1096

Xiang B (2005) An institutional approach towards migration and health in China. In: Jatrana S., Toyota M. and Yeoh BSA. (ed(s).) Migration and health in Asia. Routledge, London, pp 161–176

Xu Y and Zhang X (2006) Reform of health care system seen in the light of institution-building. In: Social Sciences in China Special Issue: Reform of the Public Health System: 134–145

Yang T (2006) Introduction: reform of the public health system requires that the public share responsibilities. In: Social Sciences in China Special Issue: Reform of the Public Health System: 69–75

Yang T and Shi Y (2006) Governance and regulation: an alternative to the stalemate in health reform program in China. In: Social Sciences in China Special Issue: Reform of the Public Health System: 117–133

Yip W and Mahal A (2008) The health care systems of China and India: performance and future challenges. In: Health Affairs 27: 921–932

Yong G and Ran L (2006) China country report. In: Transparency International (ed(s).) Global corruption report. Health and corruption. Pluto Press, London, Ann Arbor, pp 142–145

You X and Kobayashi Y (2009) The new cooperative medical scheme in China. In: Health Policy 91: 1–9

Zheng Y (2000) Institutionalizing de facto federalism in Post-Deng China. In: Tien HM. and Chu YH. (ed(s).) China under Jiang Zemin. Lynne Rienner, Boulder, Colorado, pp 215–231

Chapter 12
Migration and Health in Megacities: A Chinese Example from Guangzhou, China

Heiko J. Jahn, Li Ling, Lu Han, Yinghua Xia, and Alexander Krämer

12.1 Introduction

Migration has influence on health in various aspects. It affects public health in home and host countries and can cause severe health consequences for the migrants. Within this paper, general migration patterns and processes will be introduced and the various associations to health will be discussed. We describe the situation of internal migration in China and emphasise the importance of the Chinese household registration (hukou) system. Using the example of first results of a public health field study, we describe different urban life-world dimensions and their influences on health of working migrants in the megacity of Guangzhou, South China.

12.2 Migration and Health

Among other things, there are two major aspects that should be taken into account while studying the effects of migration: the influence of migration on both the individual health status of migrants as well as the public health effects of migration in the place of origin and destination. Carballo and colleagues (1998) even stated that migration "... has probably become one of the most important determinants of global health and social development" (Carballo et al. 1998:936).

H.J. Jahn (✉) • A. Krämer
(Europ.) MSc Public Health Department of Public Health Medicine School of Public Health, Bielefeld University, P.O. Box: 100131, D-33501 Bielefeld, Germany

L. Ling • Y. Xia
Department of Medical Statistics and Epidemiology, School of Public Health, Sun Yat-sen University, Guangzhou, China

L. Han
Department of Social Medicine and Health Management, School of Public Health, Sun Yat-sen University, Guangzhou, China

A. Krämer et al. (eds.), *Health in Megacities and Urban Areas*,
Contributions to Statistics, DOI 10.1007/978-3-7908-2733-0_12,

Emigration can have negative effects on public health in the countries of origin. It is of international concern, that the "brain drain" of health care professionals may lead to deterioration of health care in the affected, particularly in developing countries (Diallo 2004:601). Specifically countries with a low health care worker density, like some countries in Africa, are strongly affected, which can result in poor health care provision (Brush and Sochalski 2007:42).

In the scientific literature but also in public and political discussions, migration and health consequences are often discussed as problems which occur in the host countries in terms of infectious disease burden. Migrants are considered as populations, who may carry their disease burden like acute or chronic communicable infections into the countries of destination (Gushulak and MacPherson 2000:778).

In one way or the other, any type of migration has to some extent influence on an individual's social, biological or psychological well-being and health. Whereas high-skilled and well-paid professionals' health is generally less strongly affected, other types of migrants undergo stronger changes. People, who do not voluntarily migrate e.g. those, who are displaced due to war or social unrest, people, who are forced to move because of environmental changes (e.g. water shortage and desertification) or natural disasters (e.g. tsunamis or earthquakes) or people, who are affected by human trafficking are forced to cope with higher burden of migration-related health consequences (Carballo 2007:1).

There are various explanations for the different health statuses in migrants as compared to non-migrants. *Cultural and social differences* are responsible for the critical adaptation process to the new conditions in the host countries (Schenk 2007:90).

Lack of access to health care is a frequently experienced problem for migrants. Cultural differences between migrants and health care professionals and language barrier cause lack of health-related information resulting in limited health care access (Carballo 2007:3; Schenk 2007:91). Additionally, the legal status of migrants and legal regulations in the host countries can have an important effect on migrants' access to health care facilities because their status may not entitle them to benefit from public health care systems (Schenk 2007:91).

Education and socioeconomic status are interdependent and their impact on health among migrants is frequently discussed (Nguyen and White 2007:108–109). The *socioeconomic status* determines migrants' health in several ways. Underprivileged migrants often suffer from poor *housing conditions* (Carballo 2007:2) including overcrowding and poor sanitation (Carballo et al. 1998:937). The same holds for their *working conditions*, which are often coined by limited work safety. Migrants are also more likely to experience accidental injuries compared to their non-migrant counterparts, who do not want to engage in unsafe jobs (Carballo et al. 1998:939).

Many migrants who adopt the *life-styles* of the host countries may be under additional health risks. It is known, e.g., that Asian Americans with Indian, Bangladeshi or Pakistani origin are under higher risk of developing diabetes type II after life-style acculturation including a western high calorie nutrition as compared to ethnic groups of European descent (Abate and Chandalia 2001:320).

The development of diabetes is not only associated with life-style changes but also with *genetic predispositions*. It was reported that Asian Indian people, e.g., have a special predisposition and are therefore under higher risk for diabetes type II after migration-induced nutritional changes (Mohan 2004:468–469).

Mental health is also threatened by migration due to complex interactions. Migrants often move without spouses and children, leave other family members and friends behind and have difficulties to stay in touch with them. They therefore lack *social networks and social support*. They may be disenchanted after realising the difficulties in their new environment, including cultural and religious differences, language barriers and they are forced to cope with these new conditions on their own. These factors can lead to affected mental health (Carballo 2007:1–2) and it was observed that migration and concomitant circumstances can foster the onset of schizophrenia (Bhugra 2004:247–248; Carballo et al. 1998:941), depressive moods and suicides (Carballo et al. 1998:941) among migrants.

12.3 Migration in China

After introducing the open door policy in 1979, several reforms of the hukou system and the relaxation of rural-urban migration policies, tremendous internal migration occurred in China since the early 1980s (Chai and Chai 1997:1049; Zhang and Song 2003:391). There are different numbers of internal migrants reported. It is estimated that 100–200 million people left their hometowns to move elsewhere in China; mainly from rural areas to the prosperous coastal cities (Chan and Zhang 1999:8; Wen 2006:22).

High numbers of migrants floated into Guangdong province in South China (Fig. 12.1), particularly into the Pearl River Delta (PRD). The megacity of Guangzhou is the largest city in PRD and the capital of the Guangdong province. It has a population reaching 10 (People's Government of Guangzhou Municipality 2007) to 12 million (China Daily 2007) inhabitants.

Guangzhou is home to a high proportion of migrant workers, but it is difficult to determine their exact figure. There are numbers ranging from 1.6 million "migrant workers" reported by the newspaper *China Daily* (Liang 2009) to approximately 3.91 million "migrants" by the end of 2006 (People's Government of Guangzhou Municipality 2007).

Typical reasons for migration are earning more money (Ping and Pieke 2003:8; Seeborg et al. 2000:46), improving the living standard (Shen and Huang 2003:58) or to be able to support the left-behind rural relatives (Huang and Zhan 2005; Ping and Pieke 2003:6). There are also life-style-related reasons for rural-urban migration. New life experiences related to an urban life-style pull rural people beyond the aim of gaining more money (Chai and Chai 1997:1038; Huang and Zhan 2005:6; Li 2007–87; Ping and Pieke 2003:13).

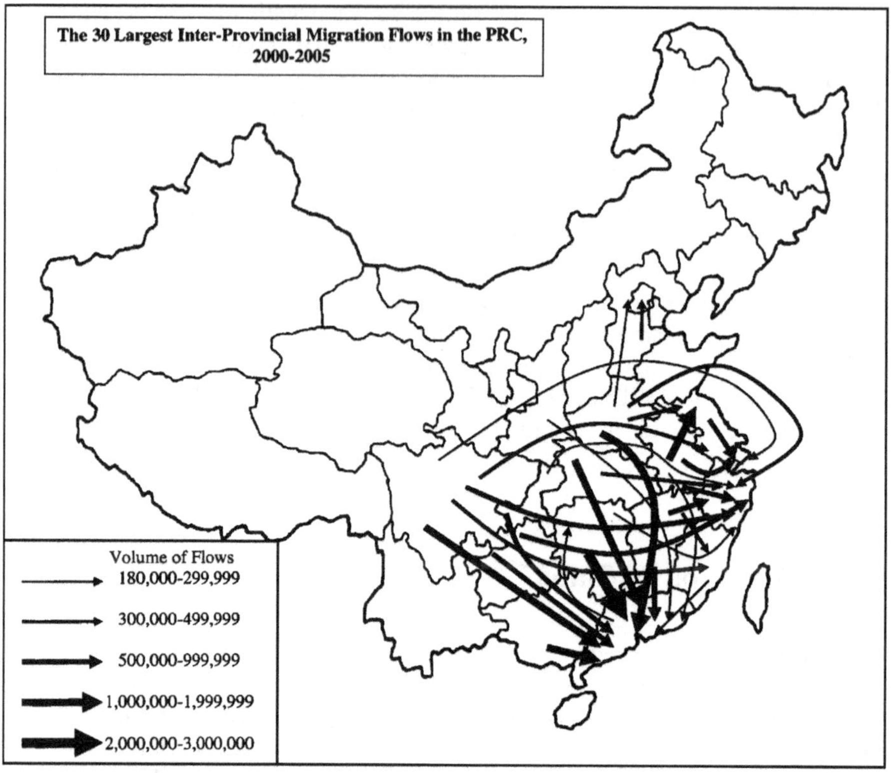

Fig. 12.1 Directions and amount of rural-urban migration in the PRC (People's Republic of China) in 2000–2005
Source: State Council, Population Census Office and Department of Population Statistics, State Statistical Bureau (2007), quoted in Chan (2008:16)

12.4 Internal Migration in China and the Chinese Household Registration System

The Chinese internal migration processes are inextricably linked with the hukou system in China. In China each household and its members have to register with local household registration authorities at the place of residence (Chan and Zhang 1999:821; Wu and Treiman 2004:3). All Chinese citizens are obliged to provide personal information including their residential address, religion and employment details. The hukou status is based on a dual classification according to (1) the locality of residence (hukou suozaidi) and (2) the socioeconomic eligibility (hukou leibie) expressed by the so-called agricultural (rural) and non-agricultural (urban) hukou status (Chan and Zhang 1999:821–822).

The hukou entitles the holder in his/her place of residence to benefit from state-provided social services. He/she is eligible to participate in state-provided health insurance and pension schemes and has unrestricted access to educational institutions for himself/herself and/or his/her children. Urban residents not holding a regular urban hukou, however, have only restricted or no access to these benefits.

Chinese authorities distinguish basically between two types of migration: first *hukou migration*, meaning that people have the official permission to migrate internally from one to another place and obtain a local hukou or a preliminary hukou which can be transferred to a regular local hukou. The other type is migration without obtaining a local (urban) hukou. This *non-hukou migration* is considered to be *informal migration* because these persons (mainly working migrants) do not or cannot change their status of local residence and socioeconomic eligibility (transfer from rural to urban hukou).

12.5 Health Determinants of Chinese Rural-Urban Working Migrants

This chapter particularly focuses on the health determinants of Chinese rural-to-urban working migrants. We consider a working migrant as a person, who migrates from his/her rural birthplace and place of upbringing in order to stay for a certain time or permanently in a city without having a local hukou.

12.5.1 Demographic Characteristics

The majority of the working migrants is relatively young and commonly in the age group of 15–39 years (Liang and Chen 2004:429; Zheng and Lian 2006:197). They are better educated than their counterparts, who are staying in their hometowns but less educated than the urban residents (He 2007:74). Often men account for a larger proportion of migrant groups (Ping and Pieke 2003:8; Shen 2002:365; Zhan 2005:21) but depending e.g. on the kind of work the migrants do, females can also represent half or even a higher share of migrant populations (He 2007:74; Hesketh et al. 2008:192). The overall income of migrants is difficult to estimate and figures in the literature vary. Whereas Fan reported that by the late 1990s, jobs in industrial or services sectors could offer monthly wages up to 1,000 Yuan including overtime (Fan 2002:121), other scientists stated that the majority of working migrants earn 300–600 Yuan per month (Wong et al. 2007:35; Zhan 2005:14). In comparison to these figures the "... Wages of Staff and Workers in Urban State-owned Units (2007)" were on average over 4,600 Yuan per month in Guangzhou (Statistics Bureau of Guangzhou Municipality 2007).

12.5.2 Environment

Working migrants often live in informal or marginal settlements (Chai and Chai 1997:1038) and they are more frequently exposed to low-standard living and working conditions (Ping and Pieke 2003:17 ff.; Zheng and Lian 2006:197). These adverse living conditions are often coined by poor hygiene and crowded living space increasing the risk for (infectious) diseases (Fan 2006:13; Zheng and Lian 2006:203). Furthermore, many migrant workers suffer from unhealthy or dangerous working conditions causing serious injuries (Human Rights in China 2002:93; Wen 2006:23).

12.5.3 Mental Health

Migrants are also threatened by psychological problems, e.g., due to stress or discrimination in the cities. They are often not considered to belong to the urban society what was found to be associated with poor mental health (Li et al. 2006:24). Wong et al. (2008) reported that in their study 25% of the male migrant workers suffered from poor mental health due to stress because of financial and employment difficulties (Wong et al. 2008:486). In a previous study of Wong and Lee, the authors found that 63% of the migrants were at risk for mental health problems.

12.5.4 Health Care Access

Since migrant workers do not have a local hukou and do generally not have well paid jobs in a company which supports health insurance, they often suffer from restricted access to health care due to financial problems. They mainly have to pay out-of-pocket, which can be a high financial burden. Migrants also lack health-related information, e.g. about sexual and reproductive health (Amnesty International 2007:16) and about health care facilities in their neighbourhoods (Amnesty International 2007:21).

12.5.5 Social Exclusion, Segregation, Discrimination

Working migrants, particular the ones coming from distant regions who do not speak the local dialect, tend to live together with migrants from the same hometowns or from the same ethnic group (Chai and Chai 1997:1045). Mostly the social segregation is driven by the urban population. The local urbanites often look down on the migrants and attribute a lower social status to them (Ping and Pieke 2003; Wong et al. 2007:36; Wong et al. 2008:484). This kind of social exclusion can vice versa cause the migrants to reject integration into the urban society as a means of self-protection (Li et al. 2006:26).

In summary, Chinese rural-urban migrants are on average less educated, have a lower income, are exposed to low-standard working and living conditions, are less integrated in the local urban society and have only restricted access to urban social services like education and health care as compared to the local urban hukou holders.

12.6 Background and Framework of Own Research

The following findings result from the first part of a quantitative public health field study conducted in spring 2008. It deals in the first point with informal living conditions in the megacity of Guangzhou and their influence on human health. It was performed within the framework of the Priority Programme 1233: "Megacities – Megachallenge: Informal Dynamics of Global Change" funded by the German Research Foundation (DFG). The overall aims of our project are quantitative assessments of major disease burdens for selected subpopulations and the associations between risk determinants and these disease burdens. In this chapter we provide data with respect to the first subpopulation studied, namely production workers and workers in the service sector with a high proportion of migrant workers.

12.7 Migration in China, Informality and Health in Working Migrants

The DFG-Priority Programme 1233 primarily focuses on the process dynamics of global change, mega-urbanisation and informal phenomena and their relationships and interactions.

Informal processes are mainly a result of lacking structural resources to cope with the high influx of migrating people, insufficient urban planning capacities and limited urban governability. Formal structures like regulated real estate or labour markets are often incapable to cope with the increasing number of new inhabitants (Kraas 2007:81). In Chinese megacities one can identify several dimensions of informality (see also Gransow 2008:2) and mainly working migrants without local hukou live in such informal conditions. Therefore, this population and its health influencing factors are of interest. We consider the following dimensions of informality:

12.7.1 Hukou Status

As addressed earlier, the hukou status has important implications in terms of many aspects in Chinese livelihood and migration. Within the context of this research non-hukou migration (migrating to Guangzhou not holding a local urban hukou) is considered an informal migration status (see also Fan 2002:108; Wu and Treiman 2004:363). This status is tied to the above described (health-related) dimensions like housing, working, access to healthcare and social exclusion. Consequently, we examine these different dimensions in association with the interviewees' hukou status.

12.7.2 Informal Housing Conditions

Due to the rapid influx of rural-urban migrants to China's cities, affordable housing became increasingly needed. This contributed to the emergence of the so-called "villages-in-the-cities" (Gransow 2007:347–348). These settlements resulted from former villages which were increasingly surrounded by the strongly expanding cities like Guangzhou due to rapid urbanisation. Farm land was confiscated and used for non-agricultural purposes (Gransow 2007:365–366). The villagers for their part restructured their settlements from rural houses to densely built multi-storied buildings and took advantage by renting out the additional living space to working migrants. Migrants also live in other dwellings like employer-provided dormitories and in private households depending on the kind of jobs (Gransow 2008:11–12).

12.7.3 Informal Working Conditions

By law employees should have a working contract. Therefore the legal working status is one criterion of informality in this study. We also examine the kind of work the migrants do, their income and their employment status. Also the workload and job satisfaction are of significance. So we link these work- and health-related dimensions to the hukou status to obtain a deeper insight whether and in how far informality has an influence on working conditions and thus on health.

12.7.4 Informal Health Service Utilization

After reforms of the Chinese health care system from a state-sponsored health care to a rather market-oriented financing system, health care costs rose for patients and created high barriers for the poorer segments of society to access health care services. Due to a lack of a local hukou and their low socioeconomic status, working migrants are forced to seek health care beyond the formal sector. In the cities various informal service providers such as small more or less illegal clinics and pharmacies exist. We examine in which cases (severe/minor diseases) the study participants consult formal versus informal health care providers and how satisfied they are with the service.

12.7.5 Informal Networks and Support

Aside from the protective effect of social support against mental health problems, social networks and support provide important information particularly for new in-migrating people. Such networks are often hallmarked by a high degree of informality. They can provide information to find a dwelling, a job or to accomplish the necessary administrative paperwork like applying for a temporary hukou. We therefore study in how far the migrants receive social support from family, friends and others.

12.8 Methods

This first part of the study was carried out from May to July 2008 in Guangzhou in three inner city districts (Huangpu, Yuexiu, Tianhe) by means of a standardised questionnaire. For study purposes, we considered people who had not been born in Guangzhou as persons with migration background. In the analysis of informality and social disparities we stratified for the hukou status (holding local urban Guangzhou hukou yes versus no). The questionnaire covers four broad health-related dimensions, which are interacting with informal living conditions (Fig. 12.2). To obtain information about the self-perceived health status we used a question from the "SF 36 Health Survey": "In general, would you say your health is 'excellent', 'good', 'so-so', 'fair', or 'poor'[1]"?

Mental health was measured by means of the WHO-5 Well-Being Index (1998 version). This index uses five items to examine how the respondents felt over the last 2 weeks. The raw score of the scale ranges from 0 to 25. 0 represent the worst possible and 25 the best possible well-being. An additional question with respect to self-perceived health was used. It asks for satisfaction with the health condition using a 5-point Likert-type scale ranging from "highly unsatisfied" to "highly satisfied". Social support was assessed by the Multidimensional Scale of Perceived

[1]We slightly changed the answer options of the original SF-36 scale from "very good" to "good" and from "good" to "so-so" because in the Chinese context the differentiation between "excellent" and "very good" seemed to be difficult.

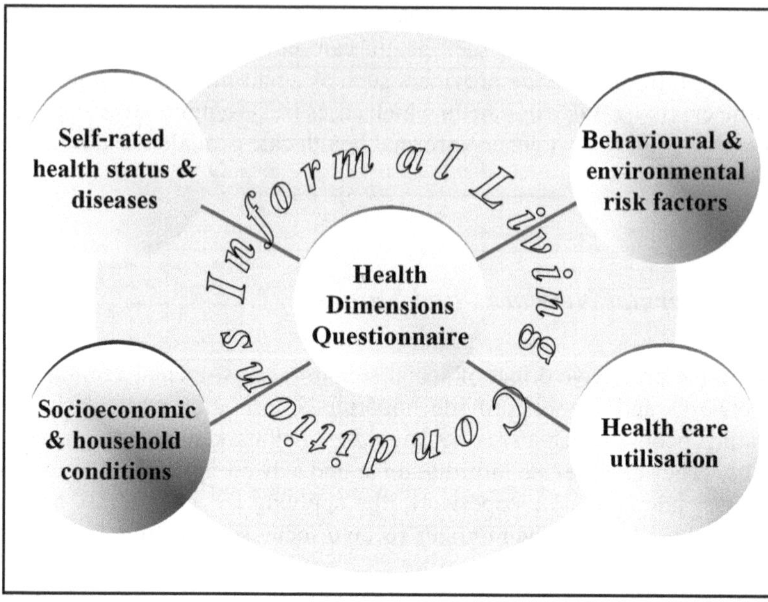

Fig. 12.2 Dimensions examined during part I of the public health field study among different subpopulations in Guangzhou

Social Support (MSPSS) developed by Gregory D. Zimet and colleagues (Zimet et al. 1990). During the statistical analysis we used different significance tests according to the type of data like the Mann-Whitney test, Kruskal-Wallis test, Chi-square test and a Spearman's rho. A 5% level of significance was determined.

12.9 Results

We aimed to reach mainly migrant workers. We therefore chose particularly employees of occupations that are typically chosen by working migrants. As a result, the proportion of people with migrational background is high. We aimed to understand the influence of the hukou status on health hypothesising that non-hukou holders generally face more health problems than local urban hukou holders.

12.9.1 Demographic Characteristics

Data were obtained from 302 employees (mainly from industrial production and service sector). It was a relatively young population with a mean age of 29.4 years (women: 28.8; men: 30.2). The gender distribution was nearly balanced with 158 women (53%) and 141 men (47%). The majority ($n = 213$, 70.5%) did not hold an urban hukou and 90.5% ($n = 143$) of the women and 84.4% ($n = 119$) of men reported to have a migration background. Only 5 women and 5 men were born in Guangzhou. The sociodemographic characteristics by hukou status are shown in Table. 12.1.

Non-hukou holders were younger ($p < 0.001$), were more likely to be born in rural areas, single ($p < 0.001$ each) and to be male ($p = 0.008$) than local residents. They were less educated ($p < 0.001$) and earned less money ($p = 0.036$) as compared to the local residents. The most frequent reasons for rural-urban migration to Guangzhou were related to working purposes reported by 165 (77.5%) of the non-hukou respondents. Other statements were not conclusive because of the wide variability of answers.

Table 12.1 Sociodemographic characteristics by hukou status

		Local GZ hukou *n* (valid %)	
	Total *n* (valid %)[a]	No	Yes
Age			
10–19	29 (11.0)	27 (13.9)	2 (2.9)
20–29	128 (48.7)	106 (54.6)	22 (31.9)
30–39	59 (22.4)	41 (21.1)	18 (26.1)
40–49	33 (12,5)	18 (9.3)	15 (21.7)
50–60	14 (5.3)	2 (1.0)	12 (17.4)
Sex			
Male	141 (47.2)	110 (52.1)	31 (35.2)
Female	158 (52.8)	101 (47.9)	57 (64.8)
Education			
Not attended school	–	–	–
Elementary (grade 1–6)	15 (5.0)	14 (6.6)	1 (1.1)
Junior middle (grade 7–9)	146 (48.3)	132 (62.0)	14 (15.7)
Senior middle (grade 10–12)	110 (36.4)	63 (29.6)	47 (52.8)
University	31 (10.3)	4 (1.9)	27 (30.3)
Marital status			
Single	149 (50.7)	127 (61.7)	22 (25)
Married/living partner	138 (46.9)	75 (36.4)	63 (71.6)
Separated/divorced/widowed	7 (2.4)	4 (1.9)	3 (3.4)
Income (Yuan per month)			
≤1,000	71 (44.9)	61 (49.6)	10 (28.6)
1,001–1,500	54 (34.2)	41 (33.3)	13 (37.1)
>1,500	33 (20.9)	21 (17.1)	12 (34.3)
Place of birth			
Urban	125 (42.8)	56 (27.5)	69 (78.2)
Rural	167 (57.2)	148 (72.5)	19 (21.6)

[a]In this text only valid percentages are reported. In case missing values, inconclusive data or refusals exceed 5% it is mentioned in the text.

12.9.2 Health Status and Social Support

12.9.2.1 Diseases and Symptoms

We asked for diseases and symptoms during the last 3 months. As expected, this young group reported a low level of disease burden (Fig. 12.3).

Striking was that 128 (42%) of the respondents reported to have experienced "cold/cough" in the past 3 month. We also examined the participants' smoking habits but smoking was not associated with "could/cough". The individuals without local hukou were stronger affected than the local residents (45.1% vs. 36%, $p = 0.144$). Further analysis regarding diseases and symptoms was inconclusive due to the small number of reported cases.

12.9.2.2 Self-rated General Health

Overall, the self-rated health status was moderate. Slightly more than one-third (37.3%) rated their health status as good or excellent, whereas 28% stated "so-so". A substantial proportion (34.7%) perceived their status as fair/poor. Whereas 27.9% of the males perceived their health status as fair/poor, 40.8% of the women rated their health as fair/poor ($p = 0.054$). Consistent over all categories, non-hukou

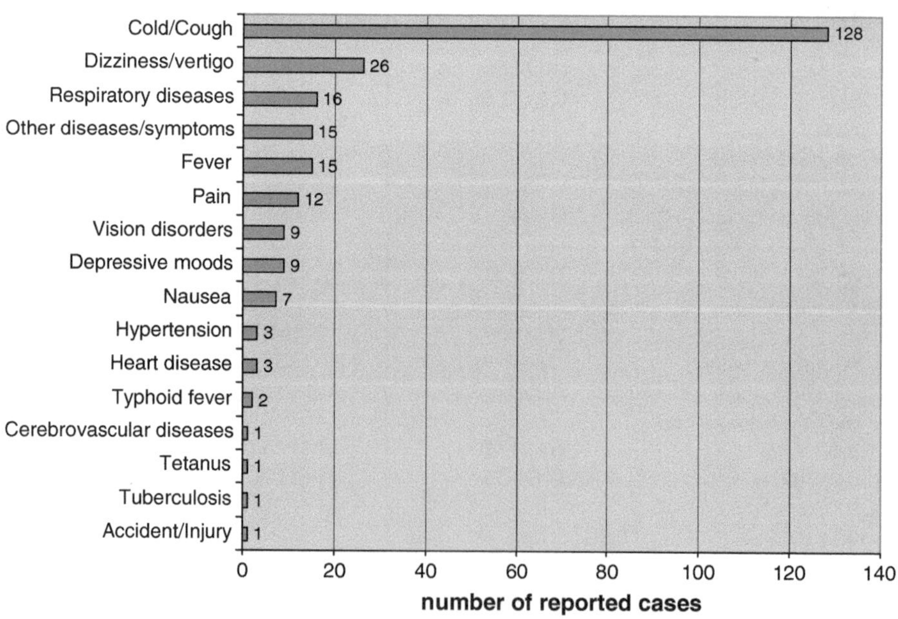

Fig. 12.3 Reported diseases and symptoms

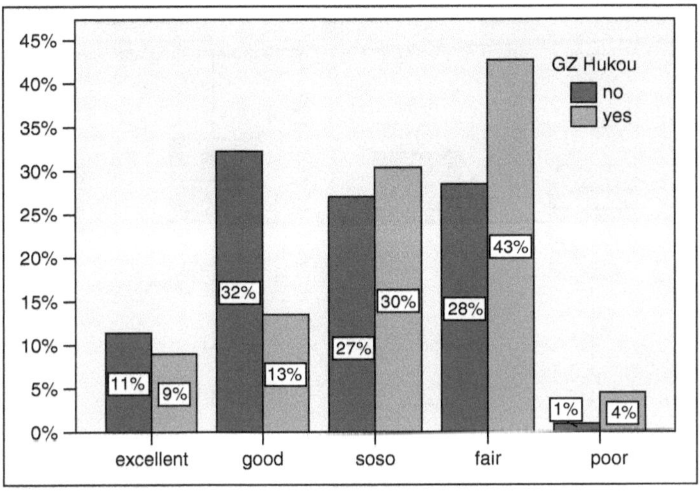

Fig. 12.4 Self-rated general health stratified for hukou status

holders seemed to have a better self-perceived health status. The percentage of non-hukou holders who rated their health as good/excellent was nearly twice as much as the percentage of the hukou holders (43.6% vs. 22.5%, $p = 0.001$) (Fig. 12.4).

12.9.2.3 Satisfaction with Health Status

More than half of the interviewees were satisfied/highly satisfied with their health status (53.9%), 109 (37.5%) participants were moderately and 25 (8.6%) were unsatisfied/highly unsatisfied. Males were more satisfied (60.5%) than females (49.3%) ($p = 0.049$). Whereas 61.7% of the non-hukou interviewees reported to be satisfied/highly satisfied, a smaller proportion (35.3%) of the hukou holders was satisfied/highly satisfied ($p < 0.001$).

12.9.2.4 Mental Health

The mean value of the WHO-5 Well-Being Index for the whole sample was 14.1 (standard deviation, SD = 5.1). Women reached slightly higher values (14.6, SD = 4.8) than men (13.5, SD = 5.3) ($p = 0.08$). Local hukou holders reported a mean well-being of 13.6 (SD = 4.9) compared to 14.3 (SD = 5.2) in non-locals with no significant differences ($p = 0.28$), but a relatively low level overall (Fig. 12.5).

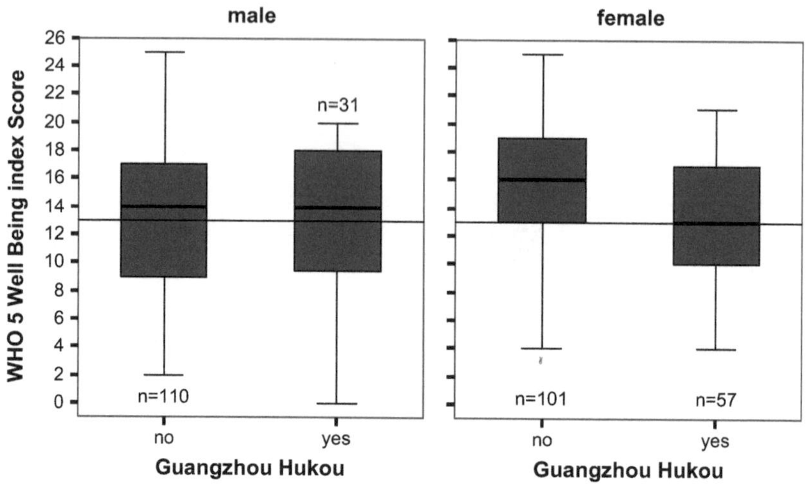

Fig. 12.5 WHO Well-Being Index score by hukou status and sex. The score ranges from 0 (worst well-being) to 25 (best well-being). The reference line marks the level under which the WHO recommends to test for depression

12.9.2.5 Social Support

The overall level of social support was 5.06 (SD = 0.85). Women were likely to receive a slightly higher social support (5.14, SD = 0.82) as compared to men (4.97, SD = 0.87) ($p = 0.055$). Interviewees who reported a high level of social support were more likely to be satisfied with their health status than persons with lower social support (Spearman's rho = 0.18, $p < 0.001$) and reported a better well-being (Spearman's rho = 0.25, $p < 0.001$).

12.9.3 Living Conditions

Living conditions can vary strongly between formal settlements like rented or bought apartments in the urban environment and typical settlements inhabited by working migrants of low socioeconomic status (e.g. villages-in-the-cities, company-provided dormitories). The latter frequently show a lack of hygiene, space and privacy, what can cause adverse health consequences. We therefore aimed to have a closer look on the relation between the interviewees' hukou status and the kind of settlements they live in.

12.9.4 Housing Conditions

A minority of the respondents reported to live in villages-in-the-city ($n = 83$, 27.8%) and only a small stratum of the hukou-holders lived in these settlements

($n = 12$, 14.5% vs. non-hukou holders: $n = 71$, 33.6%, $p < 0.001$). Some of the respondents ($n = 17$, 5.6%) refused to report their type of housing (apartment/dormitory/others). The overall majority ($n = 155$, 54.6%) of the responding participants lived in dormitories and significant differences were found between the proportion of non-hukou holders (69%) and hukou holders (15.5%, $p < 0.001$).

Hukou holders were more likely to live with their families ($n = 71$, 79.8%) compared to non-hukou holders ($n = 50$, 23.5%, $p < 0.001$).

Among the people, who shared their rooms with friends or colleagues, 142 reported how many persons shared their room. On average, 6.4 persons shared one room. Whereas the few hukou-holders shared their rooms with 4.6 persons, the non-hukou holders shared their rooms with 6.6 persons on average ($p = 0.034$).

A substantial proportion used group/collective toilets ($n = 120$, 39.7%). Only 2% used mainly public toilets ($n = 6$). Hukou holders generally used their own toilet in the apartment or house ($n = 78$, 87.6%; non-hukou holders: $n = 98$, 46.0%). Many of the latter used also collective/group toilets ($n = 110$, 51.6%, $p < 0.001$). People without local hukou shared their toilets with more people (6.9 persons) than the small number ($n = 13$) of local urbanites, who lived with friends/colleagues (3.9 persons, $p = 0.007$). We also assessed further aspects of health-related living conditions like used energy source for cooking or ways and frequency of garbage disposal but no significant results were found. Anyhow, 40 respondents (13.2%) thought that their housing conditions may have a negative influence on their health but there was no difference between non-hukou and hukou holders.

12.9.5 Working Conditions

Almost all participants were employed (99.3%) and 217 (71.9%) had a working contract ($n = 22$ refused). They worked 51 h per week on average. Non-hukou holders had a higher workload (55.6 vs. 40 h per week) than local hukou holders ($p < 0.001$).

Nearly two-thirds were satisfied or neutral with respect to their income ($n = 188$, 62.3%) and about one third was dissatisfied or highly dissatisfied ($n = 100$, 33.1%), hukou holders being less satisfied with their salary compared to non-hukou holders ($p < 0.001$).

Overall, 27.4% ($n = 78$) of the interviewees thought that their job had negative effects on their health (non-hukou holders 20.8%, $n = 42$ vs. hukou holders 43.4%, $n = 36$). The most frequently stated reasons by both non-hukou and hukou holders for a possible negative influence on health were related to air pollution at the workplace.

12.9.6 Health Care Utilization

One question referred to the health seeking behaviour: "Have you ever or do you currently suffer from a medical problem without visiting a doctor?" More than one-third ($n = 92$, 34.1%) stated "yes", hukou holders being more likely to not seeking medical service ($n = 37$, 46.3%) compared to the non-hukou holders ($n = 55$, 28.9%, $p = 0.006$).

We asked also for the actually used health care. The non-hukou holders preferred to approach pharmacies ($n = 12$, 33.3%) and smaller health care facilities compared to hukou holders, who approached pharmacies ($n = 4$, 23.5%) but also preferred the larger governmental providers ($p = 0.075$).

12.10 Discussion

In this article we sought to have a closer look at the living conditions of employees with migration background in Guangzhou and aimed to better understand the role of a hukou and non-hukou status the latter representing a certain level of informality in China among this population.

One limitation of this article is the small sample out of three inner-city districts of Guangzhou with relatively similar occupations. Inferences about migrant workers across the city and across other occupational domains therefore cannot be made.

Another limitation is the use of self-reported data. Participants may tend to provide socially desired answers.

A further frequently discussed methodological problem in migrant health studies is the so-called "healthy migrant effect". It is assumed, that particularly healthy people with a lower health risk profile decide to migrate – a self selection towards healthier migrants as compared to the people who stay at home. Furthermore, it is possible that migrants, who contract health problems, may travel home to their families, which is another selection process leading to an underestimation of health problems in migrant populations (Krämer and Prüfer-Krämer 2004:15).

The demographic characteristics of the non-hukou holders compared to local hukou holders were similar to other studies examining rural-urban migration. They were younger and more likely to be singles. They were less educated and earned less money as compared to the local hukou holders (Liang and Chen 2004:429; Wong et al. 2007:34).

Overall, only few symptoms/diseases were reported. Solely "cold/cough" was often stated and more frequently so by the non-hukou holders. These symptoms may be related to poor living conditions but also to overall air pollution. Air pollution at work was the most frequently stated reason for negative health effects at the work place. Smoking status had no influence on "cold/cough".

Besides the high prevalence of cold/cough, this group seemed to be relatively healthy, what is plausible since young adults do generally not suffer from high burden of disease, especially with respect to chronic diseases.

A considerable proportion of people did not seem to be satisfied with their health status. First, a low mean score of well-being of about 14 was reported. Taking into account the WHO's statement that a Well-Being Index score below 13 indicates poor well-being and is an indication for testing for depression (Psychiatric Research Unit 2003), our findings suggest a low level of mental health in this group. Poor mental health among migrants was also reported in other international studies on migrant health (Li et al. 2006:24; Wong et al. 2008:486). Second, more than one-third of the sample and even 47% of the hukou holders reported fair/poor health and third, a high proportion (46.0%) of the interviewees did not seem to be satisfied with their health status. These findings suggest that this group suffered substantially from impaired well-being and mental health problems.

Therefore, action should be taken in order to reduce mental health problems. These are of multifactorial pathogenesis but one approach could be to improve social support from family, friends, colleagues and institutions (e.g. advisory services concerning workers' rights and social services, self-help groups, etc.) depending on the kind of social support needed. Social support was moderate with a mean score of 5.06 of reachable 7 underlining the proposed intervention approach.

Our study identified a number of significant differences between migrants with informal status and local urban hukou holders. On the one hand the objectively disadvantaged non-hukou holders reported more "could/cough", lived in poorer housing conditions, suffered more from worse working conditions, were less likely to live with their families and earned less money than the local urban hukou holders. On the other hand, they reported on average better general health, were more likely to report a higher level of satisfaction with their health status and complained less about their salary as compared to the local urbanites. At a first glance it seems somehow inconsistent that the disadvantaged seem to be more satisfied with their health and salary compared to the group of local urban hukou holders, who live and work on average in better circumstances (higher income, less workload, more likely to live in apartments with their families) but there are maybe plausible explanations: It is possible that the reported "cold" or "cough", which was more frequent reported among the non-hukou holders, was not considered as illness and did therefore not strongly influence the self-reported health status results.

Additionally, the non-hukou holders were on average about 10 years younger and stayed a shorter period of time in Guangzhou compared to the locals. Young people are generally healthier than older ones and the adverse living and working conditions may not have affected them so much during the (on average) short living period in Guangzhou.

That the working migrants were more satisfied with their lower salary compared to the local hukou holders with higher earnings may be explainable by a recent increasing income after migrating to Guangzhou. They were probably confident to further improve their financial conditions, whereas the hukou holders, who stayed longer in Guangzhou, maybe were disenchanted to a certain extent.

Further results from the second part of this study will provide a deeper understanding concerning these aspects and will allow group comparisons between different social subgroups. These comparisons will lead to a more comprehensive and detailed picture of migration, informal living conditions and related health consequences in Guangzhou.

Acknowledgements We thank the German Research Foundation (DFG) for funding this research conducted in the framework of the subproject "Satellite-based aerosol mapping over megacities: Development of methodology and application in health and climate related studies" under DFG Priority Programme 1233.

References

Abate N, Chandalia M (2001) Ethnicity and type 2 diabetes: focus on Asian Indians. Journal of diabetes and its complications 15:320–327

Administrative Regions and Population. The People‘s Government of Guangzhou Municipality, 2007. (Accessed May 21, 2009, at http://www.gz.gov.cn/vfs/subsite/JGIN7QPB-AZE4-2MTO-EA6G-R281E8V2SFJH/category/category07.jsp?catId=5713&PageNo=5.)

Amnesty International (2007) People's Republic of China. Internal migrants: Discrimination and abuse. The human cost of an economic 'miracle'. London: Amnesty International

Bhugra D (2004) Migration and mental health. Acta psychiatrica Scandinavica 109:243–258

Brush BL, Sochalski J (2007) International nurse migration: lessons from the Philippines. Policy, politics & nursing practice 8:37–46

Carballo M (2007) The Challenge of Migration and Health. In. Vernier: International Centre for Migration and Health.

Carballo M, Divino JJ, Zeric D (1998) Migration and health in the European Union. Tropical Medicine and International Health 3:936–944

Chai JCH, Chai BK (1997) Chinas's floating population and its implications. International Journal of Social Economics 24:1038–1051

Chan KW, Zhang L (1999) The Hukou System and Rural-Urban Migration in China: Processes and Changes. The China Quarterly 160:818–855

Chan KW (2008) Internal labour migration in China: Trends, geographical distribution and policies. New York, United Nations

Diallo K (2004) Data on the migration of health-care workers: sources, uses, and challenges. Bull World Health Organ 82:601–607

Fan CC (2002) The Elite, the Natives, and the Outsiders: Migration and Labor Market Segmentation in Urban China. Annals of the Association of American Geographers 92:103–124

Fan Y (2006) Achievements and challenges of the rural migrant workers in access to healthcare service in urban China - A case study of Nanjing [Master thesis]. Lund: Lund University, Sweden

Gransow B (2007) "Dörfer in Städten" - Typen chinesischer Marginalsiedlungen am Beispiel Beijing und Guangzhou. In: Bronger D, ed. Marginalsiedlungen in Megastädten Asiens. Münster, Westf.: LIT Verlag:343–378.

Gransow B (2008) Zwischen Informalisierung und Formalisierung – Migration, Stadtentwicklung und Transformation im Perlflussdelta In: Megastädte in Asien. Japanisch-Deutsches Zentrum Berlin.

Gushulak BD, MacPherson DW (2000) Population mobility and infectious diseases: the diminishing impact of classical infectious diseases and new approaches for the 21st century. Clin Infect Dis 31:776–780

He N (2007) Sociodemographic characteristics, sexual behavior, and HIV risks of rural-to-urban migrants in China. Biosci Trends 1:72–80

Hesketh T, Ye XJ, Li L, Wang HM (2008) Health status and access to health care of migrant workers in China. Public Health Rep 123:189–197

Huang P, Zhan S (2005) Internal migration in China: linking it to development. In: Regional conference on migration and development in Asia. Lanzhou.

Human Rights in China (2002) Institutionalized Exclusion: The tenuous legal status of internal migrants in China's major cities Human Rights in China.

Kraas F (2007) Megacities and global change: key priorities. Geographical Journal 173:79–82

Krämer A, Prüfer-Krämer L, eds (2004) Gesundheit von Migranten - Eine internationale Bestandsaufnahme und Perspektiven. Weinheim: Juventa Verlag

Li X, Stanton B, Fang X, Lin D (2006) Social Stigma and Mental Health among Rural-to-Urban Migrants in China: A Conceptual Framework and Future Research Needs. World Health Popul 8:14–31

Li Y (2007) Migration and Spatial Development. Cases from the Coastal and Interior Regions in Contemporary China: Shantou University Press

Liang Q (China Daily 2009) Guangzhou to provide cheaper medical insurance for migrant workers. March 19, 2009.

Liang Z, Chen YP (2004) Migration and Gender in China: An Origin-Destination Linked Approach. In. Chicago: University of Chicago:423–443.

Mohan V (2004) Why are Indians more prone to diabetes? The Journal of the Association of Physicians of India 52:468–474

Number and Wages of Staff and Workers in Urban State-owned Units (2007). Statistics Bureau of Guangzhou Municipality, 2007. (Accessed May 26, 2009, at http://www.gzstats.gov.cn/tjsj/tjnj/2008nj/2008/3-13.htm.)

Nguyen LT, White MJ (2007) Health Status of Temporary Migrants in Urban Areas in Vietnam. International Migration 45:101–134

Ping H, Pieke FN (2003) China Migration Country Study. In: Regional Conference on Migration, Development and Pro-Poor Policy Choices in Asia June 22–24; Dhaka, Bangladesh: Institute of Development Studies; 2003.

Population of Guangzhou approaches 12 million. China Daily, 2007. (Accessed May 24, 2009, at http://www2.chinadaily.com.cn/china/2007-02/12/content_807703.htm.)

Schenk L (2007) Migration and health - developing an explanatory and analytical model for epidemiological studies. International journal of public health 52:87–96

Seeborg MC, Jin Z, Zhu Y (2000) The new rural-urban labor mobility in China: Causes and implications. Journal of Socio-Economics 29:39–56

Shen J (2002) A study of the temporary population in Chinese cities. Habitat International 26:363–377

Shen J, Huang Y (2003) The working and the living space of the "floating population" in China. Asia Pacific Viewpoint 44:51–62

Wen D (2006) China copes with globalization. A mixed review. San Francisco: The International Forum on Globalization (IFG) December.

WHO (Five) Well-Being Index (1998 version). Psychiatric Research Unit, Frederiksborg General Hospital, 2003. (Accessed December, 16, 2007, at http://www.cure4you.dk/354/WHO-5_English.pdf.)

Wong DFK, Chang Y, He X (2007) Rural migrant workers in urban China: living a marginalised life. International Journal of Social Welfare 16:32–40

Wong DFK, He X, Leung G, Lau Y, Chang Y (2008) Mental health of migrant workers in China: prevalence and correlates. Social Psychiatry and Psychiatric Epidemiology 43: 483–489

Wu X, Treiman DJ (2004) The Household Registration System and Social Stratification in China: 1955–1996. Demography 41:363–384

Zhan S (2005) Rural labour migration in China: Challenges for policies. Paris: UNESCO

Zhang KH, Song S (2003) Rural–urban migration and urbanization in China: Evidence from time-series and cross-section analyses. China Economic Review 14:386–400

Zheng Z, Lian P (2006) Health vulnerability among temporary migrants in urban China. In: Wuyi W, Krafft T, Kraas F, eds. Global change, urbanization and Health. Beijing: China Meteorological Press:197–207

Zimet GD, Powell SS, Farley GK, Werkman S, Berkoff KA (1990) Psychometric characteristics of the Multidimensional Scale of Perceived Social Support. Journal of personality assessment 55:610–617

Chapter 13
Informal Employment and Health Conditions in Dhaka's Plastic Recycling and Processing Industry

Ronny Staffeld and Elmar Kulke

13.1 Introduction

The urban economy of mega-cities located in developing countries is often characterized by the dominance of informal activities. In some urban agglomerations more than two-thirds of the workforce is engaged in this labour segment (ILO 2002a). However, the phenomenon of informal employment is not restricted to a specific economic branch or industry but encompasses a broad spectrum of diverse groups of workers and enterprises. It includes self-employed survival activities, such as street vendors, shoe shiners or garbage collectors (c.f. Hansen 2004; Rouse 2006; Wilson et al. 2006), as well as paid domestic workers employed by middle or high income group families or informal production-oriented activities taking place in small and medium backyard factories (c.f. Kamete 2004; Kulke and Staffeld 2009). Furthermore, millions of employees in formal enterprises located in special economic or export processing zones work under conditions of informal employment (Kabeer and Mahmud 2004; Staffeld 2007; Kilian et al. 2010). It is important to note here that informal employment is not only a phenomenon of developing countries, but also exists in the industrialized world, e.g. in the form of the employment of illegal, unprotected migrants on plantations or as domestic workers (c.f. ILO 2002a: 26ff; Cyrus 2008).

Due to the more and more heterogeneous and complex characteristics of informal employment the use of the term "informal sector" to describe all these groups of workers and enterprises is now regarded as being inadequate and misleading. Rather, the term "informal economy" is widely used (e.g. Castells and Portes 1989; ILO 2002b; Chen 2005) to convey a substantially different understanding (see Table 13.1). According to the old point of view the informal sector was considered to consist basically of marginal small scale and mostly self-employed survival activities that were clearly separate from the modern and formal economy. This

R. Staffeld (✉) • E. Kulke
Department of Geography, Humboldt-Universität zu Berlin, Berlin, Germany
e-mail: ronny.staffeld@geo.hu-berlin.de

A. Krämer et al. (eds.), *Health in Megacities and Urban Areas*, Contributions to Statistics, DOI 10.1007/978-3-7908-2733-0_13,

Table 13.1 Old and new view of the informal economy

The old view	The new view
The informal sector is comprised mostly of small scale, self-employed survival activities characterized by low market entrance barriers, low level of necessary qualifications and low productivity.	The informal economy includes not only survival activities but also stable enterprises and dynamic growing businesses.
It is only marginally productive.	It is a major provider of employment, goods and services for lower-income groups and contributes significantly to GDP.
It will wither away with the industrial growth of the country.	It is 'here to stay' and expanding.
It exists separately from the formal economy.	It is linked to the formal economy – it produces for, trades with, distributes for and provides services to the formal economy.
Most of the actors in this sector run illegal and unregistered enterprises in order to avoid regulation and taxation.	Most entrepreneurs and self-employed persons would welcome efforts to reduce barriers to registration and related transaction costs and to increase benefits from regulation.

Source: Chen 2005

notion has shifted towards an understanding that dynamic growing businesses are also part of the informal economy. Moreover, the idea of a clear-cut duality consisting of a "formal" and an "informal" economic segment has been rejected. As recent studies point out, informal activities are often one element of an informal-formal continuum (e.g. Chen 2005; Etzold et al. 2009; Kulke and Staffeld 2009). This continuum may even reach global dimensions as the informal economy is increasingly included in global economic structures, for instance in the form of low cost manufacturing units integrated into global flexible production networks or as the lower part of international commodity and value chains (Carr and Chen 2001; ILO 2002b; Revilla-Diez et al. 2008).

In the same way that the understanding of the role and impact of the informal economy has been transformed, the definition of informality itself has changed over recent decades. For a long time informality was seen as "alien to modernity and capitalism" (Misztal 2000: 9). In contrast to this notion, recent academic discussions focus on the expansion of the "informal sphere" into numerous aspects of the modern world (Altvater and Mahnkopf 2002; Roy and AlSayyad 2004; Kraas 2007). Generally authors emphasis the unregulated character of informality (Castells and Portes 1989; Daniels 2004; Chen 2005). Etzold et al. (2009), in contrast, argue that informality is highly regulated. However, the scope and authority of rules largely depend on the position of the involved actors and their agency (c.f. Giddens 1984). In today's megacities, effective rules and regulations are negotiated from day to day and are dominated by the most "powerful players of the game".

As mentioned above, informal employment encompasses a broad variety of activities. One distinctive characteristic of working under informal conditions is, however, that these activities are "not recognized or protected under the legal and regulatory frameworks" (ILO 2002b: 3). Due to the lack of social protection informal workers are often confronted with a high degree of vulnerability (ibid.). Usually employment is unstable and the incomes are low and irregular. There is a strong relationship between informal work and poverty, as Chen and Vanek (2005) emphasise. Moreover, informal employment is associated with massive deficits in work security (ILO 2002b). Excessive working hours, lack of protection against accidents and exposure to harmful materials at work are among the most common problems. Usually, health and safety regulations do not exist. Additionally, most employees are unaware of the risks they face, and are in any case in no position to change them. Low levels of technology as well as inadequate technical skills increase the exposure of workers to occupational accidents and diseases.

13.2 Methods

In order to focus on the working and occupational health conditions of the employees of the plastic recycling industry in Dhaka, a full-standardized quantitative survey was conducted in two steps. During the first phase, between November and December 2007, 83 workers were interviewed. About half a year later, between April and May 2008, a further 135 employees were polled. Before conducting the survey, we attempted to assess the overall situation of the recycling industry itself, with a special focus on analysing the functional relations between different steps of the production process. Based on a number of indicators such as size, capital investment and kind of legal registration, we were then able to divide the different enterprises involved in the recycling industry into three different groups of business types: (a) informal enterprises, (b) semi-formal enterprises and (c) formal enterprises.

For the employee survey enterprises from all three groups were selected randomly. Before starting the fieldwork a questionnaire was developed according to our research interests. The questionnaire included both various socio-economic aspects (like age, sex, income, housing situation and kind of contract) and questions regarding various health dimensions (such as perception of health status and recent history of diseases or health problems). Students from the Bangladesh University of Engineering and Technology (BUET) were trained to conduct the survey as it was evident that it would be necessary to collaborate with native speakers in order to perform the survey successfully. At the randomly chosen enterprise interviews included all current employees involved in the various production processes, from operating the machinery to the finishing and packaging of the final product. Due to the heavy work load of the employees the interview time was limited to between 5 and 15 min.

It is important to note here several other limitations regarding the interpretation of our findings. Since the scope of our research did not permit a proper medical examination of the employees, the health outcomes of the employee survey are based on self-reported diseases and symptoms. Estimations of the severity of the reported diseases or symptoms thus lack professional medical expertise.

Furthermore, we have no information about employees whose occupationally induced illness became so severe that they were no longer able to work but had to stay at home.

In addition to the employee survey, standardised surveys with different actors involved in the collection and intermediate trade processes were conducted. As described in the following section these actors play a crucial part in the working steps between the production of recyclable plastic waste and its processing. It was necessary to include these actors in the study in order to gain a full understanding of the recycling process. We therefore interviewed in a quantitative manner 33 waste pickers, 20 door-to-door collectors, 28 small waste dealer shops, 21 wholesalers and 28 granulate retailers during the first fieldwork phase between November and December 2007.

13.3 The Recycling Process of Plastic Waste in Dhaka

With the overall economic development the use of plastic materials in Dhaka has increased drastically over recent decades. So has the amount of plastic waste. Today, 124 t of plastic waste are generated per day in the Dhaka City area (PCI 2005). An astonishing volume of 103 t per day (83%) is collected from the streets and waste bins and eventually processed into new plastic items due to an efficient recycling system which is based on the informal economy. In this section the general structure of Dhaka's plastic recycling process is described, tracing the recycling and processing chain of the recovered plastic waste and highlighting the specific characteristics of the actors involved.

In the Megacity of Dhaka plastic waste, like other materials such as glass, paper or metal, has an economic value. It is therefore gathered from the streets and waste bins by thousands of waste pickers or collected directly from the households by ambulant door-to-door collectors. Especially the waste pickers are highly vulnerable to lack of income and inadequate living conditions. With an average daily income of 1.02 Euro[1] (about 27 Euro a month; authors' survey), waste pickers live in extreme poverty. 17 of the 33 waste pickers surveyed during fieldwork are slum-dwellers living in houses built of non-permanent materials (e.g. bamboo). Even worse, another 15 of the 33 waste pickers interviewed live on the street without any shelter at all. In contrast, ambulant door-to-door collectors, locally known as ferrywallas, earn about 50% more with an average daily income of 1.49 Euro, resulting in somewhat better living conditions. For instance, seven of the 20 ferrywallas interviewed reported living in permanent structures such as brick or stone houses (authors' survey).

It is apparently difficult to estimate the number of waste pickers and door-to-door collectors in Dhaka. While a study conducted by Waste Concern Consults

[1]The average income was found to be 107.42 Tk which equals with 1.02 Euro, as per 30.04.2008.

(2006) calculate that there are approximately 2,500 waste pickers and 1,600 ambulant door-to-door collectors, Sinha and Amin (1995) estimate the number as being much higher: 12,000 and 10,000 respectively.

Waste pickers and door-to-door collectors sell their recovered goods to small waste dealer shops, locally called vangari dokans. These shops are located in the neighbourhood of residential and commercial areas where the collectors can easily access them. Here, the materials are roughly sorted, cleaned and stored until a sufficient quantity has been accumulated to be sold to wholesalers. While the majority of small waste dealers operate their businesses informally without adhering to any formal regulations and without paying taxes, there are also some bigger shops which possess formal documents and licenses. According to the calculations of Waste Concern Consult (2006) approximately 650 small waste dealers are active in Dhaka.

Wholesalers, the next group of actors in the plastic recycling and processing chain, buy the recovered plastic waste from the vangari dokans. They operate on a large scale: on average half a ton of plastic materials are obtained per day (authors' survey results). One third of this amount arrives from outside Dhaka. The wholesalers' stores are located in the southwestern part of Old Dhaka along the river Buriganga and in close proximity to the plastic processing industry. Here the plastic materials are sorted into categories according to type, solidity and colour, etc. Usually three to five people are employed by one wholesaler. The majority of the wholesale shops belong to the informal economy as they do not have any licenses or official documents for their business.

After being sorted into different categories the plastic materials end up in Dhaka's plastic processing industry which is also located in the southwestern part of Old Dhaka. This area, named Lalbagh, is densely populated and the immense lack of space results in a mixture of residential and industrial land use. Houses often accommodate some plastic processing activities on the ground floor while the floors above are used for residential purposes. More than 2,500 small and medium, informal and formal plastic pre-processing and processing enterprises operate in the Lalbagh area. Spatial proximity and intense interlinkages, not only vertically along the processing process but also through various types of cooperation, make this area to what Marshall (1927) has described as an industrial district (see also Kulke 2008: 127f.).

Before being moulded into new items the sorted recycled plastic waste is cut into small "flakes". This takes place in so-called shredder enterprises, small and usually informal plants. After being shredded, the plastic flakes are transformed into granulate at pelletizing enterprises. These firms are also usually small and informal businesses. Finally, the granulate is used by the moulding enterprises. It is possible to distinguish between two different groups among the moulding enterprises depending on the machinery they use: (a) enterprises with simple compressing moulding machines and (b) enterprises using injection moulding machines. The later are much more sophisticated and require a high capital investment. Owners of injection moulding enterprises usually have all obligatory legal documentation. This seems logical as in this way they can protect their investment from any kind of harassment. Furthermore, it is often necessary to have these documents to gain access to bank credits. In contrast, simple compressing moulding enterprises operate with simple equipment and thus require much lower capital investment. While some of these enterprises possess the necessary licences and documentation, others do not and run their business informally.

Working conditions in all these small and medium, informal, semi-formal and formal factories have the characteristics of informal employment. However, substantial differences can be found, corresponding with the different types of enterprises described above.

13.4 Working Conditions

Approximately 20,000 people (authors' estimation based on survey and expert interviews) are employed by the plastic recycling enterprises in Old Dhaka (Lalbagh) converting plastic waste into new goods such as household items (buckets, jars, mugs etc.), irrigation pipes, toys, foil and shoes. The industry, as may be seen from Table 13.2, is dominated by young male workers. Women are basically deployed as machine helpers or for finishing or packaging the final products. At visits to injection moulding enterprises no female workers were found at all. As in other economic branches in Bangladesh, the employees face harsh working conditions. Usually they work 12 h a day, 6 days a week. Despite these physical efforts their income remains extremely low. The average earnings of an employee were found to be 3,181 Tk per month (about 30.28 Euro).

However, the income varies depending on the type of enterprise (Table 13.2). At formal injection moulding factories workers earn about 50% more than their colleagues in informal pre-processing enterprises, with 4,388 Tk. per month versus 2,931 Tk. per month. In the simple moulding companies (semi-formal) the average income of the polled workers was found to be 3,057 Tk. per month. Generally women earn about 35% less than their male counterparts.

Due to the low income most workers live below the poverty line. Many of them have their homes in slum areas. Illiteracy is common among the labour force (Table 13.2). In total, 39% of the polled employees indicated that they were not able to either read or write. This high rate indicates that special skills or training are not needed for most of the tasks executed in the recycling factories. In enterprises with injection moulding machines, however, better skilled employees are deployed since the handling of this type of machinery requires special knowledge. This is reflected by the relatively low illiteracy rate of 18% (see Table 13.2). It may also be seen from Table 13.2 that employment is based on oral contracts. In all polled enterprises, no matter whether informal or formal, workers have only this informal working contract, making them entirely dependent on the good-will of their employers. Unions or workers' associations do not exist.

Table 13.2 Socio-economic characteristics of employees in the plastic recycling industry

Variables	Type of enterprises: Pre- processing (informal) ($n = 115$)	%	Simple moulding (semi-formal) ($n = 59$)	%	Injection moulding (formal) ($n = 33$)	%
Sex						
Male	92	80.0	49	83.1	33	100.0
Female	23	20.0	10	16.9	0	0.0
Age						
12–16	2	1.7	4	6.8	3	9.1
17–26	66	57.4	34	57.6	23	69.7
27–36	41	35.7	14	23.7	7	21.1
>36	6	5.2	7	11.9	0	0.0
Income[a]						
<999 Tk	0	0.0	1	1.7	0	0.0
1,000–2,999 Tk.	63	54.8	25	42.4	7	21.2
3,000–4,999 Tk.	52	45.2	32	54.2	14	42.4
>4,999 Tk.	0	0.0	1	1.7	12	36.4
Illiteracy status						
Illiterate	47	40.9	30	50.8	6	18.2
Literate	68	59.1	29	49.2	27	81.8
Type of working contract						
Oral	115	100.0	59	100.0	33	100.0
Written	0	0.0	0	0.0	0	0.0
Job satisfaction						
Like my job	9	7.8	6	10.2	6	18.2
Job is ok	53	46.1	25	42.4	24	72.7
Don't like my job	53	46.1	28	47.4	3	9.1

[a]Income in Tk. per month (1,000 Tk equals with 9.49 Euro; as per 30.04.2008)

13.5 Occupational Health and Safety

In addition to the generally demanding workload and exhausting working hours, occupational health risks are very common in the plastic recycling industry in Lalbagh. Workers at shredder enterprises are exposed to a high level of dust and noise resulting from the crushing of the plastic waste, thus creating an unhealthy occupational environment. Furthermore, the handling of the shredder machine can cause serious injuries such as amputation of fingers or parts of the arm. In pelletizing factories dust and fumes are emitted and workers are exposed to hazardous chemicals without the benefit of protective gear. At compression moulding enterprises employees often complain about the extreme heat caused by the moulding machines. Usually these small factories have no ventilation facilities. In contrast, working conditions at injection moulding enterprises were found to be substantially better than those in other types of enterprises. As Fig. 13.1 indicates,

Fig. 13.1 Workers (in %) perceiving their occupation as harmful to their health

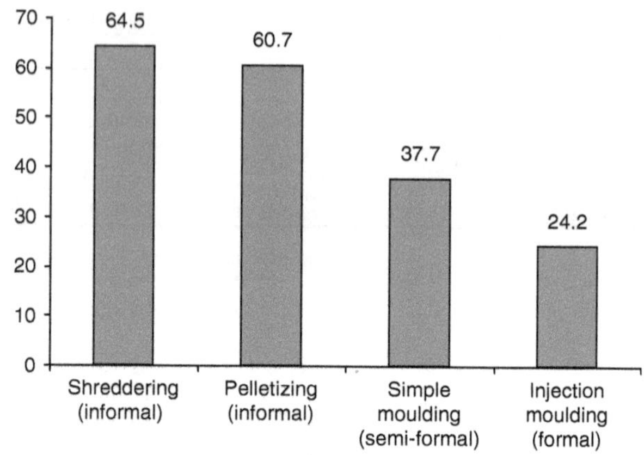

Fig. 13.2 Workers (in %) who have suffered from at least one disease or symptom of ill-health in the last 3 months

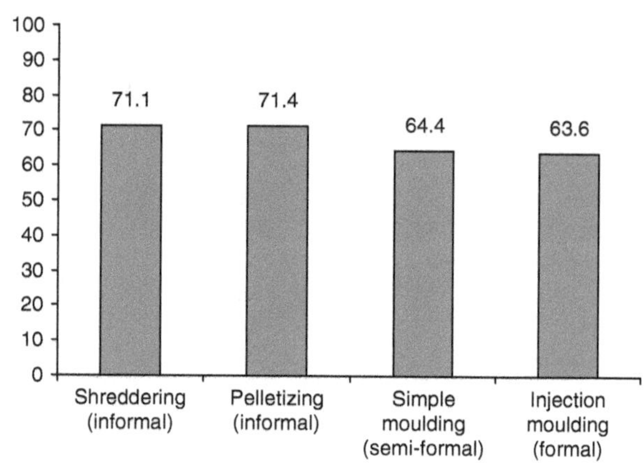

in formal companies (injection moulding) every fourth worker polled perceived his work as being harmful to his health. While this might be an alarmingly high proportion in an industrialized country, it has to be considered as comparatively low if the picture is completed by the figures obtained from the informal enterprises (shredder or pelletizing firms). Here the share of workers perceiving their occupation as harmful was found to be 65% and 61% respectively.

Furthermore, illness is a common problem among the workers. The majority of employees reported having suffered from at least one illness during the three months preceding the interview (Fig. 13.2). While 71.1% of workers employed at informal shredder enterprises and 71.4% of informal pelletizing firms reported having suffered from an illness, the number at semi-formal and formal moulding enterprises was not considerably lower, standing at 64.4% and 63.6% respectively.

As Table 13.3 indicates, fever, colds/coughs and pain are the main problems among the workers, followed by hepatitis, gastric problems, diarrhoea and respiratory diseases. For example, 33.8% of all interviewed employees reported having suffered from fever in the last three months. However, it is important to note here

Table 13.3 Reported health problems and symptoms by type of enterprise

	Pre-processing – informal ($n = 115$) (%)	Simple moulding – semi-formal ($n = 59$) (%)	Injection moulding – formal ($n = 33$) (%)	Total ($n = 218$) (%)	Khan et al. Slum health outcomes ($n = 1{,}444$)
Fever	32.2	39.0	30.9	33.8	33.9
Cold/cough	20.4	27.1	20.0	22.2	17.5
Pain	10.4	16.9	9.1	12.1	15.2
Hepatitis	11.3	8.5	6.1	9.7	Not incl.
Gastric problems	7.8	8.5	12.1	8.7	6.9
Diarrhoea	4.3	5.1	6.1	4.8	5.6
Respiratory disease	3.5	3.4	3.0	3.4	2.4

that the prevalence of fever is subject to substantial seasonable variation. During the winter period (fieldwork carried out between November and December 2007) twice as many workers suffered from fever than in the spring season (fieldwork carried out between April and May 2008). The prevalence of the symptom cold/cough was also subject to a very high seasonal variation.

With regard to the different types of enterprises (informal, semi-formal and formal) the results give a complex picture. Generally, in formal enterprises the number of workers who reported having suffered from a disease or symptom of ill-health was lower than in informal or semi-formal enterprises. This is true for the symptoms fever, cold/cough and pain as well as for hepatitis and respiratory disease. For example, the number of workers who reported having suffered from hepatitis was twice as high in informal enterprises as in formal companies. In contrast, for gastric problems and diarrhoea the number was found to be higher in formal firms than in informal or semi-formal. Furthermore, in semi-formal enterprises the share of workers who had suffered from fever, colds/coughs or pain was found to be significantly higher than in formal and even than in informal companies.

13.6 Conclusion

This study documents various aspects related to working conditions and occupational health status in both informal and formal enterprises of the recycling industry located in Lalbagh, Dhaka. As demonstrated above, the workforce in this industry generally lack legal employment contracts and are employed on the basis of oral arrangements agreed with the owner or manager of the company. Furthermore, there is no nationwide legal or regulatory framework thus leaving these workers without any social protection. Unions or workers' associations that could advocate the interests of the employees do not exist. Employment in the plastic recycling industry can therefore be generally considered as informal. However, substantial differences exist depending on the type of enterprises. Workers in informal or semi-formal companies not only earn less than their colleagues in formal injection moulding companies, but they are also less satisfied with their jobs. In terms of the occupational health situation the number of workers perceiving their occupation as being harmful to their health was alarmingly high, especially in informal companies. Here, more than 60% considered their job to be harmful. Although in formal enterprises the number was found to be much lower, at 24%, this is still an unacceptable situation indicating the necessity of new safety regulations and protective measures for the entire industry. The high proportion of workers who have suffered from an illness provides another argument. Over 70% of the work-force engaged in informal recycling plants reported suffering from at least one disease or symptom in the preceding 3 months. For semi-formal and even for formal enterprises the number was not significantly lower.

Interestingly, the overall prevalence of different diseases and symptoms (not sub-divided by type of enterprise) corresponds with the findings of Khan et al. (2009) who conducted a study of health outcomes in several slums in Dhaka. This leads to the conclusion that the health status of workers in the recycling industry is not so much dependent on the type of enterprise in which they work, but is far more a result of being poor and living below the poverty line. New safety regulations, campaigns to increase occupational risk awareness and the introduction of legal working contracts are instruments that can improve the working conditions of the employees. This may not, however, suffice for a substantial change in health status. Decent work, which includes a healthy working environment, is built upon a decent income. But this seems far away for the workers in the plastic recycling factories in Lalbagh, Dhaka.

References

Altvater E, Mahnkopf B (2002) Globalisierung der Unsicherheit - Arbeit im Schatten, schmutziges Geld und informelle Politik. Münster

Carr M, Chen M (2001) Globalization and the Informal Economy: How Global Trade and Investment impact on the Working Poor. Sussex

Castells M, Portes A (1989) World Underneath: The Origins, Dynamics and Effects of the Informal Economy. In: Portes A, Castells M, Benton A (eds) The Informal Economy: Studies in Advanced and Less Developed Countries. Baltimore, pp 11–37

Chen M (2005) Rethinking the informal economy: Linkages with the Formal Economy and Formal Regulatory Environment. Paper presented at the EGDI and UNU-WIDER Conference 17–18 September 2004, Helsinski

Chen M, Vanek J (2005) Informal employment: rethinking workforce development. In: Avirgan T, Bivens LJ, Gammage S (eds) Good Jobs, Bad Jobs, No Jobs: Labour Markets and Informal Work in Egypt, El Salvador, India, Russia, and South Africa. , Washington, pp 491–502

Cyrus N (2008) Being Illegal in Europe: Strategies and Policies for Fairer Treatment of Migrant Domestic Workers. In: Lutz H (ed): Migration and Domestic Work - A European Perspective on a Global Theme. Studies in Migration and Diaspora, Padstow/ Cornwall, pp 177–194

Daniels PW (2004) Urban Challenges: The Formal and Informal Economies in Mega-Cities. In: Cities vol. 21 (6), pp 501–511

Etzold B, Keck M, Bohle HG, Zingel WP (2009) Informality as Agency – Negotiating Food Security in Dhaka. In: Die Erde vol. 140 (1), pp 3–24

Giddens A (1984) The Constitution of Society: Outline of the Theory of Structuration. Cambridge

Hansen KT (2004) Who Rules the Streets? The Politics of Vending Space in Lusaka. In: Hansen KT, Vaa M (eds) Reconsidering Informality - Perspectives from Urban Africa. Uppsala, pp 120–138

ILO (2002a) Women and Men in the Informal Economy: A statistical picture. International Labour Office (ILO), Employment Sector, Geneva

ILO (2002b) Decent Work and the Informal Economy. Report VI, International Labour Conference, 90th Session, Geneva

Kabeer N, Mahmud S (2004) Rags, Riches and Women Workers: Export-oriented Garment Manufacturing in Bangladesh. In: Carr M. (ed): Chains of Fortune: Linking Women Producers and Workers with Global Markets. London, pp 133–162

Kamete AY (2004) Home Industries and the Formal City in Harare/ Zimbabwe, In: Hansen KT, Vaa M (eds): Reconsidering Informality. Perspectives from Urban Africa. Uppsala, pp 120–138

Khan MMH, Krämer A, Grübner O (2009) Comparison of Health-Related Outcomes between Urban Slums, Urban Affluent and Rural Areas in and around Dhaka Megacity, Bangladesh. In: Die Erde vol. 140 (1), pp 69–87

Kilian P, Beißwenger S, Xue D (2010) Floating or settling down? Migrant workers and mega-urban development in the Pearl River Delta, China. In: Geographische Rundschau, International Edition, 6(2), pp 50–56

Kraas F (2007) Megacities and Global Change in East, Southeast and South Asia. In: Asien vol. 103 (2), pp 9–22

Kulke E (2008) Wirtschaftsgeographie, 3rd edn. Paderborn

Kulke E, Staffeld R (2009) Informal Production Systems - The Role of the Informal Economy in the Plastic Recycling and Processing Industry in Dhaka. In: Die Erde vol. 140 (1), pp 25–43

Marshall A (1927) Industry and trade. A study of industrial technique and business organization and their influences on the conditions of various classes and nations. London

Misztal BA (2000) Social theory and contemporary practice. London New York

PCI (Pacific Consultants International) (2005) The Study on the Solid Waste Management in Dhaka City vol. 1, Dhaka

Revilla-Diez J, Schiller D, Meyer S, Liefner I, Brömer C (2008) Agile Firms and their Spatial Organisation of Business Activities in the Greater Pearl River Delta. In: Die Erde vol. 139 (3), pp 251–269

Rouse JR (2006) Seeking common ground for people: Livelihoods, governance and waste. In: Habitat International vol. 30, pp 741–753

Roy A, AlSayyad N (2004) Urban Informality: Crossing Borders. In: Roy A, AlSayyad N (eds) Urban Informality - Transnational Perspectives from the Middle East, Latin America and South Asia. Lanham/ Maryland, pp 1–6

Sinha M, Amin N (1995) Dhaka's Waste Recycling Economy: Focus on informal sector labour groups and industrial districts. In: Regional Development Dialogue vol. 16 (2), pp 173–195

Staffeld R (2007) Exportförderzonen als Entwicklungsmotor? Erfahrungen aus der Bekleidungsindustrie in Choloma, Honduras. In: Vernro: 2015 auf dem Campus vol. 2, Bonn

WCC (Waste Concern Consultants) (2006) Final Report on Composition of Plastic Waste and Market Assessment of the Plastic Recycling Sector in Dhaka City. Dhaka

Wilson DC, Velis C, Cheeseman C (2006) Role of informal sector recycling in waste management in developing countries. In: Habitat International vol. 30, pp 797–808

Chapter 14
Mega-Urbanization in Guangzhou: Effects on Water Quality and Risks to Human Health

Ramona Strohschön, Rafig Azzam, and Klaus Baier

14.1 Introduction: Megacities and their Effects on Water Resources

Due to China's economic liberalization at the end of the 1970s and the institutionalization of numerous special economic areas, Chinese agglomeration areas such as the Pearl River Delta (PRD) in southern China have recorded great economic growth in a relatively short period of time and – caused primarily by national migration – an exorbitant increase in population. Urban areas like Guangzhou, Shenzhen or Dongguan grew from small cities into giant megacities within a short time. Over the course of this development, the PRD has become one of the most dynamic and densely populated regions in China and, moreover, is among the regions in the world with the fastest rate of urbanization (Baier and Strohschön 2007). These dynamic development processes not only led to transformations of the population structure, civic economy and urban morphology, but also to considerable ecological problems and thus to changes in quality of life. In terms of the environment, the reciprocal impact of urban development and ground and surface water represents one of the most important aspects of growing cities. This is especially relevant for cities that are built atop uncovered aquifers close to the surface and/or for cities being located in a river system. To be clear, the interaction between urban development and ground and surface water is greatly influenced by the respective city's land use structure in regards to water quantity and quality. This means that the different forms of land use such as landfills, urban agriculture, industry and trade as well as diverse residential types with their corresponding wastewater systems influence the emission of pollutants in surface and groundwater, including groundwater recharge (see Fig. 14.1).

In addition to formal types of residential areas, also informal types of residential areas and housing developments assume a key role in the development of mega-urban

R. Strohschön (✉) • R. Azzam • K. Baier
Department of Engineering Geology and Hydrogeology, RWTH Aachen University, Aachen, Germany
e-mail: strohschoen@lih.rwth-aachen.de

A. Krämer et al. (eds.), *Health in Megacities and Urban Areas*, Contributions to Statistics, DOI 10.1007/978-3-7908-2733-0_14,

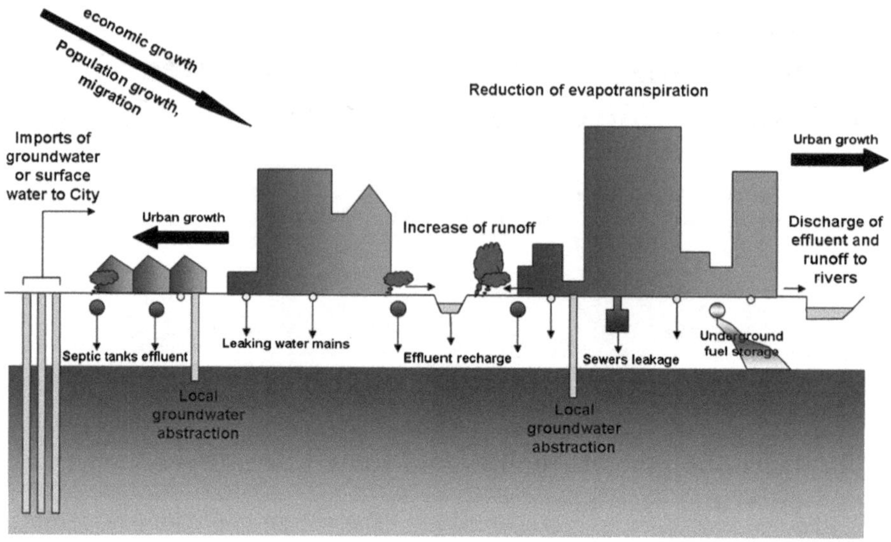

Fig. 14.1 Impacts of urbanization on the hydrology (Putra and Baier 2009)

areas and sustainable water resources management in many urban agglomerations. Thus, migrants particular often live in areas with deficient infrastructures, such as a lack of connection to the public water supply or an inadequate wastewater system.

The progressing urbanization process in China creates a huge demand for water. It is estimated that by the year 2015, there will be 109 cities in China with more than one million inhabitants. The water supply, however, is already a grave problem for many cities: between 400 and 600 cities possess only an insufficient water supply and 100 cities are already suffering from extreme water scarcity (China Daily 2003; e-fundresearch 2008). The main problems are qualitative, caused by enormously increasing consumption, an increasing amount of wastewater along with lagging capacities in the treatment of wastewater. Thus the existing wastewater systems in many cities cannot cope with the amount of wastewater resulting from economic growth and increasing populations.

Small-scale land use analyses, which will be explored in greater detail below using the example of Guangzhou, can be utilized in the initial approach to improve planning for (Chinese) megacities for the protection of water resources.

14.2 Methods: Urban Units and Water Quality Analysis

In order to be able to analyze land use types in Guangzhou, a megacity with more than 14 million inhabitants (Huang and Keyton 2010), the city was subdivided into small spatial units or *urban units* (compare Fig. 14.2). These are areas within the

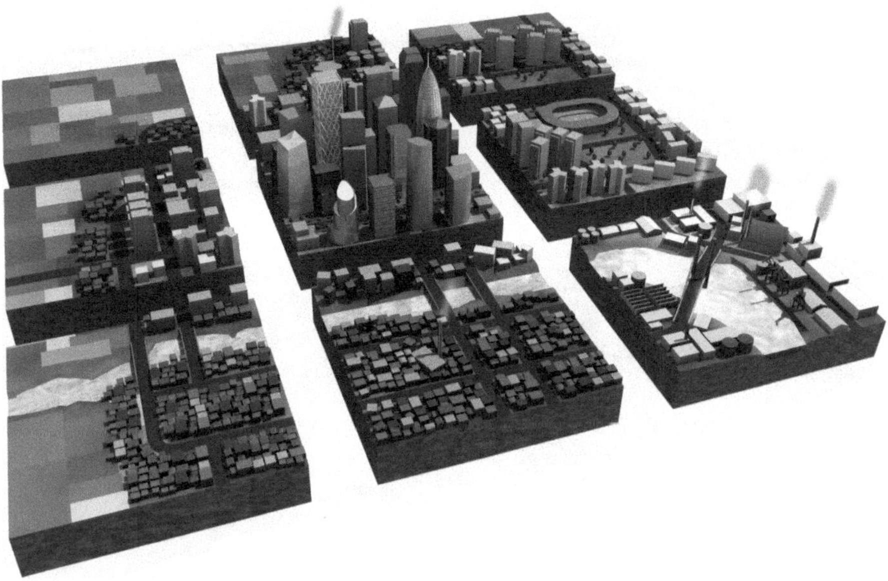

Fig. 14.2 Conceptual approach. Schematic typology and demarcation of urban pattern

cityscape, which are more or less designed morphologically homogonously within building as well as in the open space structure and thus can be clearly demarcated outward. As Guangzhou shows a wealth of various complex urban pattern, the units have to be differentiated according to a number of structural characteristics. In particular, different forms of land use in urban and peri-urban areas, such as agriculture, small business, simple village structures and highly-compact residential developments were inspected in regards to water supply and wastewater disposal. Depending on the type of land use, potential sources and types of hazardous substances were surveyed. In addition to the mentioned selection criteria, access to the investigation area as well as the opportunity to sample ground- and surface water was important for selecting the urban units.

The research shed light on general water quality as well as the possible sources of hazardous substances, such as excrement. Furthermore, evidence on the effects of urban land use on water resources in Chinese cities must be established. For this purpose, additionally residents were surveyed on noticeable changes to the optical, flavor or odor characteristics of the water quality as well as possible polluters or peculiar incidents.

Two areas under examination, the urban villages Xincun and Datang, both with a meager to middle-class standard of living, are in the urban Haizhu District. At 10,043 registered residents per square kilometer (Guangzhou city council 2007), it counts as one of the three most densely populated districts in the city, next to Yuexiu and Liwan. The third area of examination is Shibi, a still traditional village with a meager to middle-class living standard. It is located in the south, in the

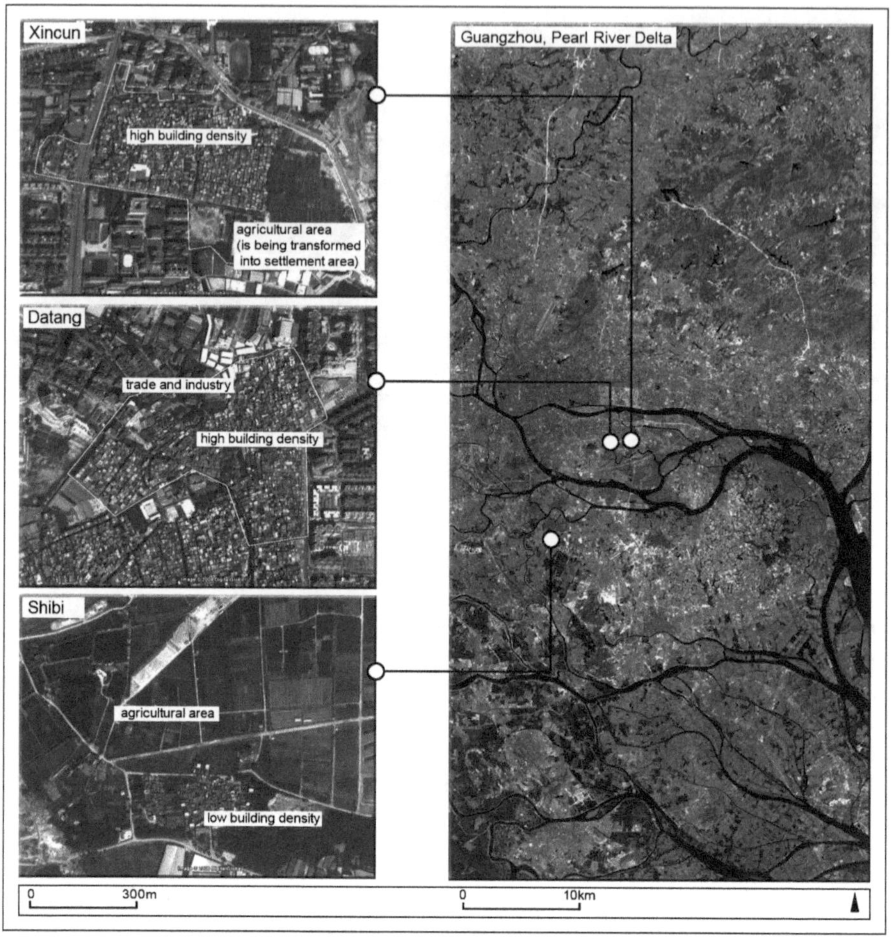

Fig. 14.3 Location of urban units in Guangzhou (based on Google Earth 2009 (left) and Landsat ETM 2000 (right))

peri-urban district of Pan Yu and has a population density of only 1,240 registered residents per square kilometer (Guangzhou city council 2007) (see Fig. 14.3).

14.2.1 Examined Parameters

In order to examine water quality in the first phase of examination in the three areas mentioned above, a total of 27 samples of tap-, ground- and surface water were taken and checked up for concentrations of total coliform bacteria, ammonium and nitrate. Additionally, temperature, redox potential, pH value and electrical conductivity were measured.

In the analysis of the data, depending on availability, the standard and limit values were taken from the Chinese environmental quality standards (SEPA 2002), the World Health Organization's Guidelines for Drinking Water Quality (WHO 2006) or the German Drinking Water Ordinance (DVGW 2001) as the basis of evaluation.

14.2.2 Results

There are many examples in Guangzhou that have demonstrated the city's vulnerability in regards to water resources. The analyses prove that the water supply and disposal infrastructure in many parts of the city is still often overwhelmed. While *access to water* seems to be standard in the urban portions of the city, there are still households in the peri-urban areas not connected to the public distribution network. It was found, moreover, that none of the tap water in the examined areas is consumed without first boiling it. Increasing contamination and an unacceptable taste were named as reasons. Water samples confirm the statements: In Datang for example, tap water is contaminated with coliform bacteria in amounts up to $7.9*10^2$ MPN/100 ml – the internationally accepted limit for drinking water set by the WHO is 0 MPN/100 ml (WHO 2006). The problem of unpurified drinking water is momentarily amplified in that humans living in peri-urban areas are using groundwater from private and public wells as part of their everyday sustenance (Wehrhahn et al. 2008). As a result, the groundwater is drank without previously being cooked because it allegedly tastes better and seems to be of better quality than the tap water. However, the sampled groundwater in Xincun as well as in Datang and Shibi was contaminated with coliform bacteria at levels of $3.3*10$–$3.3*10^5$ MPN/100 ml. The measured values of ammonium and nitrate in all units were low. But, as a result of acknowledging the problem of insufficient water quality, publicly accessible water vending machines are on the rise: conventional tap water is purified using reverse osmosis and supplied to the public for a small fee. It was noticeable, however, that these opportunities for public access are not available in all parts of the city, nor did they seem to be utilized by many residents.

In addition to water supply, *wastewater disposal* is a huge problem in Guangzhou. According to information provided by the Guangzhou Municipal Statistic Bureau (2007) 96.01% of industrial wastewater meets legal standards. Experts assume, however, that 40–60% of China's industrial wastewater is not measured (bfai et al. 2006);[1] therefore, the information regarding Guangzhou's wastewater is to be scrutinized. It became clear that the primary reason for water pollution in Guangzhou is the leading-in of untreated household wastewater into watercourses; this was the case in the inspected areas as well as in additional areas surveyed. According to other studies, only 10–25% of domestic wastewater is treated (Zhu et al. 2002; He 2005). Housing developments of meager or middle-class

[1]Bfai changed its name into Germany Trade and Invest (GTAI).

living standards are especially lacking modern wastewater disposal. A combination of drainage systems and open wastewater ditches exists in the examination areas. Human health risks are all-too obvious when children are playing around the ditches (see Fig. 14.4).

Chemical analyses of the sampled surface water (streams, feeders, fish ponds) within the units reveal measurements of $1.7*10^4$–$4.6*10^7$ MPN/100 ml for total coliform bacteria and up to 55 mg/l for ammonium and thus a heavy organic water

Fig. 14.4 Child playing at an open wastewater gutter

Table 14.1 Maximum pollutant concentrations within the investigation areas

	Total coli (MPN/100 ml)	Ammonium (mg/l)	Nitrate (mg/l)
Xincun			
Groundwater	$3.3*10^4$	8.34	3.89
Surface water	$1.1*10^7$	55	1.62
Tap water	<2	0.209	1.21
Datang			
Groundwater	$3.3*10^5$	8.06	2.9
Surface water	$4.6*10^7$	47.7	2.78
Tap water	$7.9*10^2$	0.225	0.996
Shibi			
Groundwater	$4.9*10^4$	<0.05	2.3
Surface water	$1.3*10^5$	0.884	1.86
Tap water	$4.6*10^2$	<0.05	1.66

pollution. The maximum nitrate values of 1.86 mg/l are low especially for Shibi despite intensive agricultural use (compare Table 14.1).

14.3 Discussion

The analysis of urban units like Xincun, Datang and Shibi shows that the areas do not operate as closed systems unto themselves; rather they are in direct and indirect interaction with the surrounding areas. Wastewater ends up in another area via open pipes and drainage systems, flows through, and is passed off into the neighboring areas. In terms of the dynamics of urban structures, direct dependences and interactions between the high rates of growth and change in land use and the resulting reactive informal strategies and mechanisms need to be determined. Measures such as the retroactive mounting of public water lines in many parts of the city and the individually improvised covering of wastewater ditches in Shibi or Datang for example, are proof of the residents' desire for improved living standards and environmental conditions. It is assumable that some of these strategies lead to an increase in water consumption, since access to the public distribution network is much more comfortable than a supply from a private or public well – and wastewater increases at the same time.

There is a heightened risk to the environment in particular because of Guangzhou's position in the PRD – on an unprotected and only partially protected aquifer that is close to the surface respectively – where seeping harmful substances end up in the groundwater relatively quickly due to the thin surface layers. It is not uncommon for an area's hydrological and hydrogeological basis to be negatively impacted and then subsequently destroyed as a result of the grave influence of human activities, especially concerning controlled and currently uncontrollable urbanization processes.

Even if the *coliform concept* also implicates coliform bacteria that are not of excrement origin, fecal pollution and therefore an undesirable strain on the (drinking) water supply is suspected (Tobin and Dunst 2009). If the limit value is clearly exceeded, coliform bacteria can become pathogens for health risks such as infant diarrhea and urinary tract infections. Diseases transmitted through drinking water, such as dysentery or typhus, cannot be excluded. Escherichia coli especially is an indicator organism for other pathogens in drinking water (Umweltbundesamt 2008). *Ammonium* occurs in minuscule amounts in water as part of the natural nitrogen cycle. The concentration increases greatly if the watercourse is contaminated by wastewater containing excrement and other microorganisms or fertilizer. Increased values thus suggest pollution of the water with excrement and other microorganisms. Ammonium is oxidized in the process of oxygen consumption from bacteria into nitrite and then nitrate (Wasser Wissen, no date). An increased nitrite strain is most dangerous for infants — it can trigger methemoglobinaemia due to the increased concentration of methemoglobin. Furthermore, nitrite can form nitrosamine with secondary amines in the stomach, which has been shown to be carcinogenic in animal test studies (LGL Bayern 2007).

Human health hazards are amplified by the fact that the groundwater from Guangdong Province will be increasingly used as a source for drinking water for the cities in the delta.

14.4 Conclusion

The vulnerability of megacity Guangzhou's water resources are noticeable everywhere, especially in terms of water quality. The results of the investigation substantiate the strain on tap water as well as ground- and surface water due to coliform bacteria heavily present in urban and peri-urban areas of the city. Increased ammonium values were also determined. It can therefore be assumed that coliform bacteria and ammonium are primarily of excrement origin, since large amounts of household wastewater are piped into the watercourses untreated. Drinking water vending machines reflect the fact that the problem of inadequate water quality is well known, including in regards to the water supply. Countless other examples like open wastewater ditches or eutrophicated feeders show, however, the water supply's extreme vulnerability. Activities like improvised covering for wastewater ditches express the population's need for better living and environmental standards.

A comprehensive analysis of megacity Guangzhou's water quality, the use of new technical and decentralized solutions, consequential enforcement of existing laws, sustainable ways of life and production and improved as well as forward-looking urban planning measures are all of the utmost importance for city planning and sustainable management of water resources.

Acknowledgements This research was supported by the funds of the German Research Foundation's Priority Program 1233 'Megacities – Megachallenge. Informal Dynamics of Global Change'. We also highly appreciate the editors who provided valuable comments that helped to improve the manuscript. Moreover, we would like to thank Mr Lu Lin for his invaluable support in China during the research stays and Ms Katharina Wiethoff for layout assistance of Figure 14.3.

References

Baier K, Strohschön R (2007) Informelle Dynamiken in Megastädten – Eine Strukturanalyse vor dem Hintergrund der Wasserversorgung. In: VAG Infoblatt 2: 3–4

bfai, Bundesagentur für Außenwirtschaft, Servicestelle des Bundesministeriums für Wirtschaft und Technologie (2006) Wassertechnik und Wassermanagement in der VR China. Branchenstudie, Köln

China Daily (2003) Thirsty Cities Pollute Waterways. Article published on September 17, 2003 (Accessed May 24, 2011, at http://www.china.org.cn/english/environment/75297.htm)

DVGW, Deutsche Vereinigung des Gas- und Wasserfaches e.V. (2001) Verordnung zur Novellierung der Trinkwasserverordnung vom 21. Mai 2001. Bonn

e-fundresearch.com (2008) Unabhängiger Informationsanbieter zu Investmentfonds. Chinas Wachstum droht auszutrocknen. (Accessed May 25, 2011, at http://ch.e-fundresearch.com/newscenter.php?aID=10605.)

Guangzhou city council (2007) Administrative Regions and Population. (Accessed July, 29, 2008, at http://www.gz.gov.cn/vfs/subsite/JGIN7QPB-AZE4-2MTO-EA6G-R281E8V2SFJH/category/category07.jsp?catId=5713&PageNo=5.)

Guangzhou Municipal Statistics Bureau. Guangzhou Statistical Yearbook (2007) Peking: China Statistics Press

He M (2005) The Water Household as an Example of Ecology in the Pearl River Delta. In: Ipsen D., ed. The genesis of Urban Landscape: The Pearl River Delta in South China. Kassel: Faculty of Architecture, Urban Planning, Landscape Planning. Work Report No.161; 77–84

Huang D, Keyton D (2010). Guangzhou Population Closes to 15 million. Article published on June 18, 2010. (Accessed March 11, 2011, at http://www.lifeofguangzhou.com/node_10/node_37/node_85/2010/06/18/127683154477641.shtml.)

LGL Bayern, Bayerisches Landesamt für Gesundheit und Lebensmittelsicherheit (2007) Nitrat im Trinkwasser. Erlangen (Accessed July 13, 2009, at http://www.lgl.bayern.de/lebensmittel/rueckstaende/trinkwasser_nitrat.htm.)

Putra DPE, Baier K (2009) Der Einfluss ungesteuerter Urbanisierung auf die Grundwasserressourcen am Beispiel der indonesischen Millionenstadt Yogyakarta. Cybergeo: European Journal of Geography, Environnement, Nature, Paysage, article 469. September 29, 2009. (Accessed May 25, 2011, at http://cybergeo.revues.org/22573)

SEPA, China State Environmental Protection Administration (2002) Environmental Quality Standards for Surface Water: GB 3838–2002 (Accessed May 25, 2011, at http://www.sepa.gov.cn/tech/hjbz/bzwb/shjbh/shjzlbz/200206/W020061027509896672057.pdf.)

Tobin G, Dunst M (2009). Trinkwasserhygiene auf Wasserfahrzeugen. Infoblatt, Landeshauptstadt Kiel, Amt für Gesundheit. (Accessed May 25, 2011, at http://mws.kiel.de/leben/gesundheit/hafenaerztlicher_dienst/_dokumente/Infoblatt_Trinkwasserhygiene_auf_Wasserfahrzeugen.pdf)

Guangzhou Municipal Statistics Bureau (2007) Guangzhou Statistical Yearbook 2007. Peking: China Statistics Press

Umweltbundesamt (2008) Wasser, Trinkwasser und Gewässerschutz. (Accessed May 25, 2011 at http://www.umweltbundesamt.de/wasser/themen/trinkwasser/mikrobiologie.htm.)

Wasser Wissen (2008) Das Internetportal für Wasser und Abwasser. Ammonium. (Accessed July 29, 2008, at http://www.wasser-wissen.de/abwasserlexikon/a/ammonium.htm.)

Wehrhahn R, Bercht AL, Krause, CL et al (2008) Urban restructuring and social and water-related vulnerability in mega-cities – the example of the urban village of Xincún, Guangzhou (China). Die Erde. Zeitschrift der Gesellschaft für Erdkunde zu Berlin;139(3):227–249

WHO, World Health Organization (2006) Guidelines for Drinking-water Quality. First Addendum to third edition. Volume 1 Recommendations. Genf: WHO Press

Zhu Z, Deng Q, Zhou H et al (2002) Water Pollution and Degradation in Pearl River Delta, South China. Ambio. A Journal of the human environment; Vol.31(3):226–230

Part V
Spatial Dimensions and Health

Chapter 15
A New Approach to Link Satellite Observations with Human Health by Aircraft Measurements

Britta Mey, Manfred Wendisch, and Heiko J. Jahn

15.1 Introduction

Anthropogenic ambient aerosol pollution is a worldwide obvious phenomenon and causes adverse health effects. Particularly large cities are affected. According to the World Health Organization, about 0.8 million deaths can be attributed to urban particulate air pollution globally (WHO 2002). Mostly resulting from combustion processes, aerosol particles (particulate matter, PM) can cause or exacerbate respiratory, cardiovascular diseases as well as lung cancer (Pope et al. 2002; Pope and Dockery 2006; Schulz et al. 2005). The health effects of aerosol particles are, among others, determined by particle size, their chemical composition, duration and degree of exposure and number concentration.

The terms *aerosol* and *aerosol particles* are not identical. An aerosol is a mixture of a carrier gas in which aerosol particles (liquid, solid, or liquid/solid) are suspended (not solved!), e.g. the smoke of a fire place (soot particles suspended in air). Thus the term aerosol is different to aerosol particles. The sizes of aerosol particles cover a wide diameter range of about 0.003–100 μm.

The entire size range of aerosol particles is separated into different typical modes according to their diameter (d). Firstly the particles can be classified into a fine (particle diameter $d < 1$ μm) and a coarse mode ($d > 1$ μm). A more detailed distinction separates the particles into nano particles ($d < 0.05$ μm), ultrafine particles ($d < 0.1$ μm, UFP), PM_1 ($d < 1$ μm), $PM_{2.5}$ ($d < 2.5$ μm), and PM_{10} ($d < 10$ μm) (Seinfeld and Pandis 1998). No distinctions are made between different chemical compositions. Possible sources for particles with a diameter larger than 2.5 μm but smaller than 10 μm are dust from industry, roads, or uncovered soil. Examples for

B. Mey (✉) • M. Wendisch
Leipzig Institute for Meteorology (LIM), University of Leipzig, Stephanstr. 3, D-04103 Leipzig, Germany

H.J. Jahn
Department of Public Health Medicine, School of Public Health, Bielefeld University, Bielefeld, Germany

A. Krämer et al. (eds.), *Health in Megacities and Urban Areas*,
Contributions to Statistics, DOI 10.1007/978-3-7908-2733-0_15,

sources of particles with diameters of 2.5 μm or smaller are soot particles emitted by forest fires, or particles from emissions of vehicle exhaust, power plants or industry.

The particle size is an essential parameter concerning certain human health effects since particles with smaller sizes will penetrate deeper into the respiratory tract. Aerosol particles with a diameter less than 10 μm can penetrate into the lower respiratory tract whereas $PM_{2.5}$ are able to reach the deep respiratory tract including the gas-exchange system (Brunekreef and Holgate 2002). UFP can even translocate from pulmonary alveoli into the bloodstream affecting the cardiovascular system (Duggen 2004). Therefore health effects of aerosol particles are size-dependent.

In Fig. 15.1 the deposition probability of aerosol particles in the human respiratory tract is depicted. On the right part of the figure the human respiratory tract is sketched, with different parts highlighted in different colors. On the left part of the image the deposition probability for the different parts of the respiratory tract (marked in the same colors) is shown. The total deposition probability is highest for "small" and "big" particles, where the probability is different for diverse parts of the respiratory tract. The smaller the aerosol particles the further they reach the bronchial parts.

The particle sizes of the aerosol particles suspended in an air volume can be displayed in a number size distribution (Fig. 15.2). Two exemplary distributions are presented, a distribution for clean rural air and a distribution for average urban air. Both exhibit two modes, a fine mode and a coarse mode. The difference in the coarse mode fraction is relatively small in comparison with the big

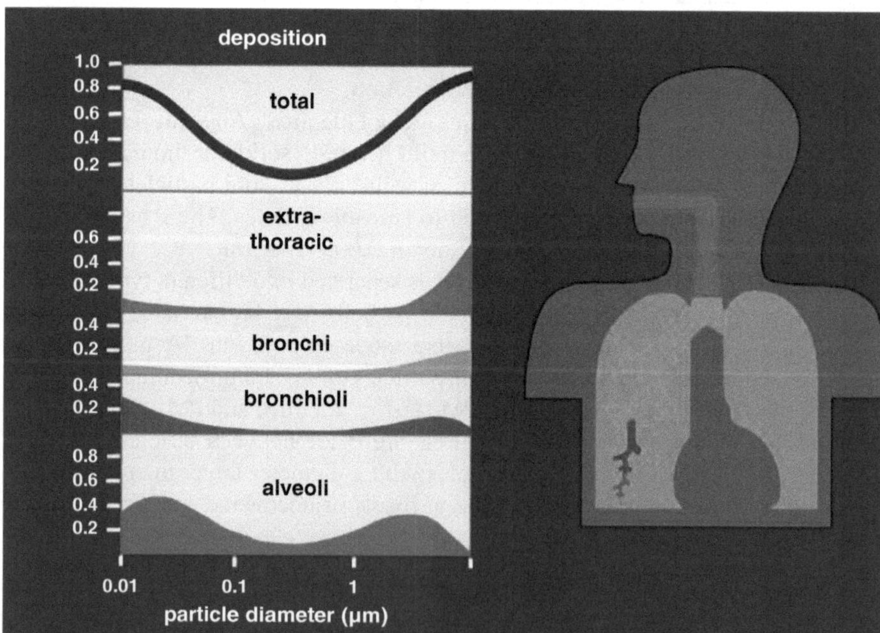

Fig. 15.1 Deposition probability of particles of different size in the human respiratory tract (Kreyling et al. 2006)

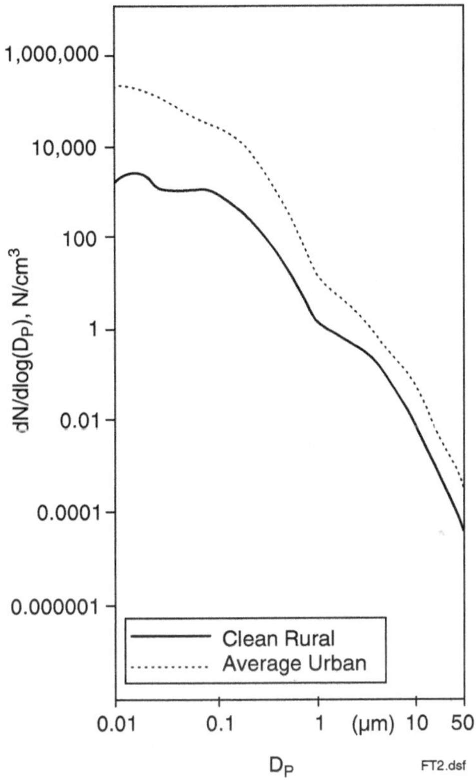

Fig. 15.2 Aerosol particle number size distributions (Beall et al. 2001)

difference in the small mode. Much smaller aerosol particles are present in urban air, mainly due to vehicle exhaust and industrial emissions in the vicinity of cities. The knowledge of the particle size distribution is not sufficient for a comprehensive discussion concerning the health effects of aerosol particles. Also the particles' chemical composition plays a role, which determines whether particlcs contain toxic substances and if they are water soluble.

All this information cannot be obtained from point measurements alone. Different spatial and temporal dimensions and limitations have to be considered and will be discussed in the following sections.

15.2 Methods

Measurement techniques for characterizing the size and chemical composition of aerosol particles are manifold and dependent on the specific scientific objectives pursued. In this section some aerosol measurement techniques are briefly described exemplarily. The instruments described below do not cover the full spectrum of aerosol measurement instruments, but present some examples of the basic techniques.

15.2.1 Ground Based and Airborne Aerosol Sampling

Ground based measurements can be more sophisticated than respective airborne techniques. This is caused by several limitations related to airborne operations. Instruments mounted on scientific aircraft have to follow specific restrictions due to limited space and electric power on an aircraft. Therefore they should be light-weight and only in the need of low electric power. On the other hand those airborne instruments also have to be robust, because of the harsh environmental conditions, as for example strong temperature and pressure changes encountered during vertical profile flights. The design of inlets for the air sampling is more complex than for ground based instruments due to the disturbance of the ambient air caused by the moving aircraft and the inlets themselves.

Ground based data, on the other hand, deliver measurements of a well defined location (point measurement), but it is not possible to get a good spatial coverage in this way.

15.2.2 Examples for In-Situ Techniques: Filters and Impactors (Ground Based)

Filters and impactors are both sampling techniques, mainly used for ground based measurements. The first technique described in this section is the aerosol particle sampling with filters. Air is sucked through a filter system where aerosol particles are deposited (Fig. 15.3).

Different kinds of filters are available for this technique. The choice of the suitable filter depends on the desired information and applied analysis method. Subsequent analyses of the particles collected on the filters include microscopic studies or chemical analyses of the filter sample to examine their shape and chemical composition.

Cascade impactors use the inertia of particles in an air flow as basic principle. Small particles (little mass, low inertia) follow the air flow, but larger, more heavy particles show a bigger inertia, collide with the impaction plate and are finally impacted (Fig. 15.4).

In this way a cascade impactor provides the possibility to separate and classify different size (or mass) fractions of aerosol particles. In each stage of the cascade impactor the air flow velocity is increased by reducing the nozzle diameter so that only smaller and smaller particles reach the next impactor stage. Impactors are used for measurements of PM_{10}, $PM_{2.5}$ and PM_1 fractions in $\mu g/m^3$ (aerosol mass concentrations). Special impactors, like the Berner impactor, are able to separate a $PM_{0.1}$ fraction from the total flow, but these are usually only used in specific field experiments. The smallest stage of the impactors used for regular monitoring stations is typically PM_1.

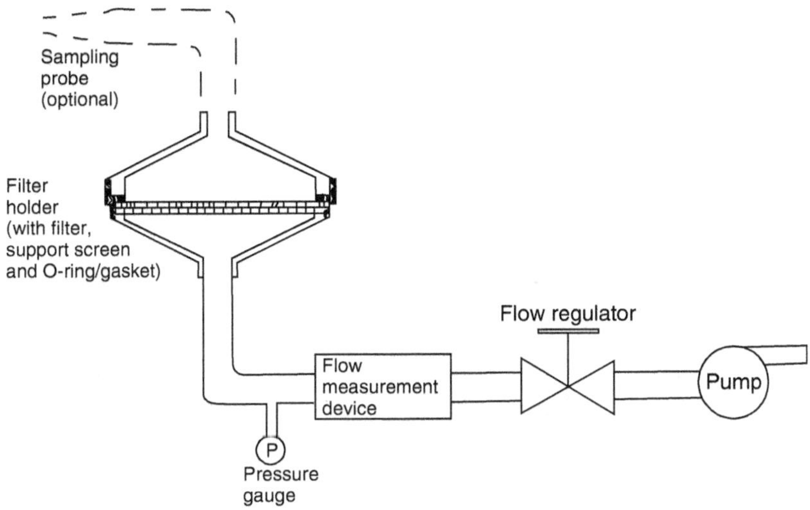

Fig. 15.3 Filter sampling. principal (Baron and Willeke 2001)

15.2.3 Remote Sensing Techniques

Remote sensing techniques use the knowledge of physical principals to gather aerosol properties (like shape and aggregate state) from the measurement of scattered and reflected radiation. Passive methods measuring the signals of scattered solar radiation and active methods with their own light source can be distinguished.

15.2.3.1 Active Remote Sensing

One example for active remote sensing is the so-called LIDAR (LIght Detection And Ranging) technique. The measurement principle uses the physical property of emitted radiation which is scattered and reflected backward by molecules and aerosol particles to gain information about the vertical distribution of microphysical properties of aerosol particles in the atmosphere. A laser pulse is emitted into the atmosphere and the backscattered signal is detected. The position of the scattering particles is calculated from the time shift between emitted and received signal (Collis 1965). LIDAR systems provide information about the optical properties of the aerosol particles (extinction, backscatter coefficient, depolarization ratio) which lead for example to the information about particle shape (spherical or non-spherical particles) (Althausen et al. 1999). Continuously measuring LIDAR systems provide the temporal development of the vertical distribution of backscattering particles (aerosol particles, cloud droplets) in the atmosphere above the instrument.

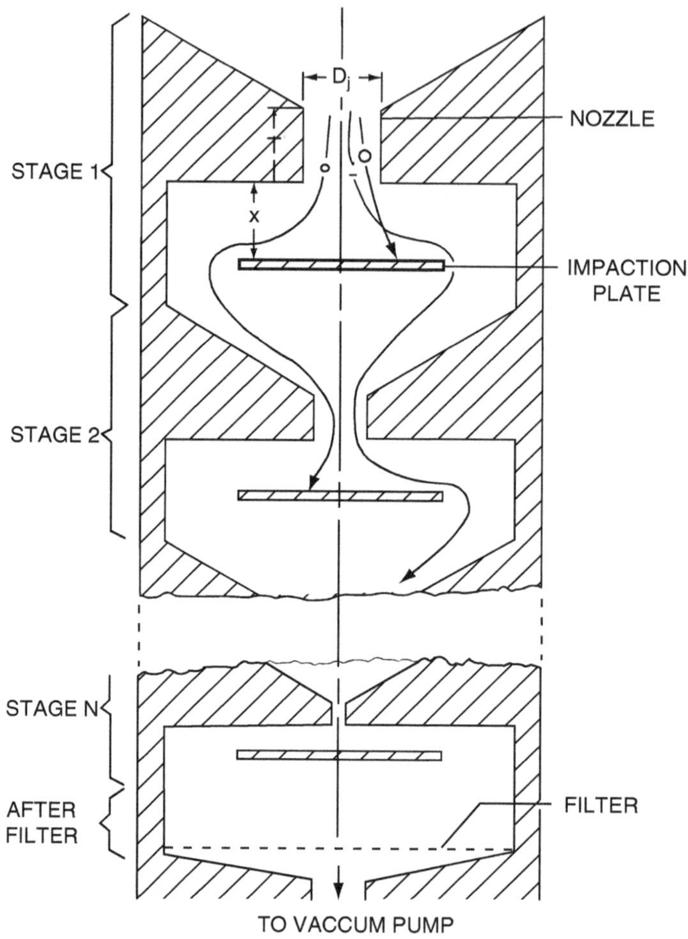

Fig. 15.4 Sketch of the measurement principal of a cascade impactor (Hinds 1999)

15.2.3.2 Passive Remote Sensing

The Spectral Modular Airborne Radiation measurement sysTem Albedometer (SMART Albedometer, previously Albedometer) (Wendisch et al. 2001) is one example for passive remote sensing instruments which enables to determine reflectivity as well as albedo of aerosol layers by measuring the incident and reflected solar electromagnetic radiation. In combination with a radiative transfer model it is possible to retrieve estimates of aerosol microphysical properties, as for example particle shape. Passive remote sensing instruments are also mounted on various satellites for monitoring different atmospheric properties (e.g. aerosol optical depth) on the global scale. The Aerosol Optical Depth (AOD) quantifies the degree to which aerosol particles absorb or scatter (solar) radiation. One example for satellite instruments which can retrieve aerosol properties is the MODerate resolution Imaging Spectroradiometer (MODIS) (Goddard Space Flight Center 1995; Kaufman et al. 1997)

15.3 Linking Satellite Aerosol Observations with Human Health Data: A New Strategy

In the combination of public health research and atmospheric science methodologies it is important to know which kind of data is available and how to overcome differences in spatial and temporal resolution on the one hand, and long time data availability on the other hand.

In public health research commonly particle mass concentrations of PM_{10} and $PM_{2.5}$ and the number concentrations of UFP are measured in order to study the resulting burden of aerosol-related diseases in a population. PM_{10} and $PM_{2.5}$ are often measured by means of in situ air quality monitoring stations to assess the daily burden of particulate matter. Sometimes also the chemical composition of the aerosol particles is analyzed at these stations.

Since it is difficult to establish everywhere an area-covering air quality monitoring network, and since a small number of stations in a city is not sufficient to provide a comprehensive picture of aerosol pollution, these point measurements frequently lack a spatial coverage.

Satellite remote sensing, however, provides spatial coverage, but is not able to measure PM_{10}, $PM_{2.5}$ or the size distribution directly. The correlation between ground based $PM_{2.5}$ measurements and respective satellite data is complicated (Liu et al. 2007). This leads to the discussion in which way satellite data might be helpful for public health purposes. Satellite data is used for global, regular observations of atmospheric conditions, where the AOD is one of the monitored parameters. In this way point sources of pollutants can be identified and monitored and long-term developments can be observed.

Satellites receive their signal partly from the ground and from the atmosphere between the Earth surface and the top of the atmosphere. Without any additional information it is difficult to derive near-surface aerosol data from satellite measurements. It is not possible to measure surface reflectivity directly with satellite borne instruments, because the incoming and reflected solar radiation is influenced (absorbed and scattered) by air molecules, aerosol particles, and cloud droplets suspended in the atmosphere. The typical approach to derive the surface reflectivity from satellite measurements is the assumption that the solar radiation at 2.1 μm is not influenced by aerosol particles and provides the surface reflectivity at this wavelength. Furthermore the surface reflectivity for 0.65 μm and 0.47 μm is calculated using tables for surface type, and experimentally derived relations between the reflectivity of these wavelengths. The surface reflectivity at 2.1, 6.5, and 4.7 μm provide the lower boundary condition for e.g. the MODIS aerosol retrieval (Levi et al. 2007). Differences between satellite retrieved AOD, and AOD obtained from sun photometer measurements at the ground (e.g. Gröbner and Meleti 2004) are discussed in scientific publications (Tripathi et al. 2005). An improvement of the surface reflectivity data quality could decrease the difference between the satellite derived AOD and the ground based derived AOD.

A new approach for linking satellite aerosol observations and human health and to improve the aerosol satellite retrieval results is a combination of airborne and ground based measurements in a specific way as described below.

Measuring albedo or reflectivity (Lubin and Masom 2006) with airborne instruments provides the albedo or reflectivity at flight level. The albedo or reflectivity at flight level can be extrapolated to surface level with the knowledge of the vertical distribution of aerosol particles and the vertical atmospheric profile (temperature, pressure, humidity) using a radiative transfer model (Wendisch et al. 2004). The vertical aerosol distribution and atmospheric profile are measured with ground based LIDAR, and radiosonde measurements respectively. The retrieved surface albedo or reflectivity is used as new boundary condition in the satellite aerosol retrieval. The usage of an imaging camera additionally to the SMART albedometer furthermore provides a spatial coverage of the reflective surface properties, whereby the inhomogeneity of urban surfaces is taken into account. It is expected that using surface reflectivity data retrieved from airborne measurements will decrease the discrepancy between MODIS AOD and sun photometer AOD, and therefore provide an improved data set for the use in public health research.

15.4 Discussion

Different techniques for the measurement of aerosol quantities, as for example PM_{10} and $PM_{2.5}$, aerosol optical depth, or size distributions, show different advantages and disadvantages with regard to public health research. Examples for such different measurement techniques are ground based filter techniques or remote sensing by satellite instruments. Ground based techniques provide data sets close to the target group for public health studies, but they show a lack of spatial coverage. The advantage of local ground based measurements is fixed only to one position, as the data values of one specific location do not count for other locations in a city.

Satellite remote sensing techniques provide the spatial coverage, but the far distance between sensor and object (atmosphere or earth surface) leads rather to a coarse spatial resolution, and offers possibilities for retrieval uncertainties, as necessary input parameters (e.g. surface reflectivity) cannot be measured directly.

Therefore using only one method with one or few instruments seems not to be suitable for describing the spatial aerosol burden. The combination of ground based measurements with computer models, calculating the propagation of the aerosol particles from their source to more distant locations, seems to be a suitable approach to overcome this difficulty.

On the other hand satellite data can be used for global long term monitoring and provides a data source for countries without dense aerosol monitoring network. The aerosol retrieval algorithms are steadily improved and compared to ground based measurements, like the sun photometers of the AErosol RObotic NETwork (AERONET) (Holben et al. 1998). Furthermore airborne measurements, as described previously, can help to improve satellite products like AOD.

The improvement of aerosol measurements, like minimizing uncertainties and errors of the instruments, refining computer models and retrieval algorithms, is part of present atmospheric research.

Acknowledgements We thank the German Research Foundation (DFG) for the funding within the priority program SPP 1233 *Megacities – Megachallenge: Informal Dynamics of Global Change*. Furthermore, we are grateful to the Institute of Remote Sensing Applications (IRSA), Chinese Academy of Sciences, in particular to GU Xingfa and YU Tao, for the fruitful cooperation.

References

Althausen D, Müller D, Ansmann A et al. Scanning 6-Wavelength 11-Channel Aerosol Lidar. JAtmOceanTechn. 1999;17:1469-1482

Baron PA, Willeke K. Aerosol Measurement: Principles, Techniques, and Applications. New York, United States, Wiley & Sons, 2001

Beall K, Grosshandler W, Luck H. AUBE '01 – 12th International Conference on Automatic Fire Detection - Proceedings, 2001

Brunekreef B, Holgate ST. Air pollution and health. Lancet 2002;360:1233-42

Collis RTH. Lidar: a new atmospheric probe. Q J R Meteorol Soc. 1965 (published online 2006);92:392:220-230

Duggen. Eine Frage der Größe. Ultrafeine Teilchen schädigen Herz und Gefäße. mensch+umwelt 2004;1:3-4

Goddard Space Flight Center. MODIS: moderate resolution imaging spectroradiometer. Greenbelt, Maryland, United States. The Administration, Goddard Space Flight Center. 1995

Gröbner J, and Meleti C. Aerosol optical depth in the UVB and visible wavelength range from Brewer spectrophotometer direct irradiance measurements: 1991–2002. J. Geophys. Res. 2004;109, D09202, doi:10.1029/2003JD004409

Hinds W. Aerosol Technology: Properties, behavior, and measurement of airborne particles 2nd ed. New York, NY: Wiley-Interscience, (1999)

Holben BN, Eck TF, Slutsker I, et al. AERONET - A federated instrument network and data archive for aerosol characterization. Rem Sens Env 1998;66, 1

Kaufman YJ, Tanré D, Remer LA, Vermote EF, Chu A, and Holben BN. Operational remote sensing of tropospheric aerosol over land from EOS moderate resolution imaging spectroradiometer. JGR 1997;102

Kreyling WG, Semmler-Behnke M, Möller W. Health implications of nanoparticles. J Nanoparticle Res. 2006;8:534–562

Levi RC, Remer LA, Mattoo S, Vermote EF, Kaufman YJ. Second-generation operational algorithm: Retrieval of aerosol properties over land from inversion of Moderate Resolution Imaging Spectroradiometer spectral reflectance, J Geophys Res 2007;112, D13211, doi:10.1029/2006JD007811

Liu Y, Franklin M, Kahn R, Koutrakis P. Using aerosol optical thickness to predict ground-level $PM_{2.5}$ concentrations in the St. Louis area: A comparison between MISR and MODIS. Rem Sens Env 2007;

Lubin D, Masom R. Polar Remote Sensing Ice sheets. Springer-Verlag Berlin, Heidelberg, New York 2006

Pope CA, 3 rd, Burnett RT, Thun MJ, et al. Lung cancer, cardiopulmonary mortality, and long-term exposure to fine particulate air pollution. Jama 2002;287:1132-41

Pope CA, 3 rd, Dockery DW. Health effects of fine particulate air pollution: lines that connect. J Air Waste Manag Assoc 2006;56:709-42

Schulz H, Harder V, Ibald-Mulli A, et al. Cardiovascular effects of fine and ultrafine particles. J Aerosol Med 2005;18:1-22

Seinfeld, JH, Pandis SN. Atmospheric chemistry and physics: From air pollution to climate change. Wiley, New York 1998

Tripathi SN, Dey S, Chandel A, Srivastra S, Singh RP, Holben BN. Comparison of MODIS and AERONET derived aerosol optical depth over the Ganga Basin, India Annal. Geophys. 2005;23:1093-1101

Wendisch M, Müller D, Schell D, Heintzenberg J. An airborne spectral albedometer with active horizontal stabilization. J Atmos Oceanic Technol. 2001;18:1856–1866

Wendisch M, Pilewskie P, Jäkel E, et al. Airborne measurements of areal spectral surface albedo over different sea and land surfaces, J Geophys Res. 2004;109:D08203, doi:10.1029/2003JD004392

WHO. World Health Report. Reducing risks, promoting healthy life. Geneva: World Health Organisation; 2002

Chapter 16
Spatial Epidemiological Applications in Public Health Research: Examples from the Megacity of Dhaka

Oliver Gruebner, Md. Mobarak Hossain Khan, and Patrick Hostert

16.1 Introduction

Public health researchers are increasingly shifting their focus from models of disease epidemiology that focus exclusively on individual risk factors to models that also consider the complex and important effects of the socio-physical environment (Geanuracos et al. 2007). The application of spatial analysis in the context of epidemiological surveillance and research has increased exponentially (Pfeiffer et al. 2009). Geographic information systems (GIS), global positioning systems (GPS) and remote sensing (RS) have been increasingly used in public health research since the 1990s (Kaiser et al. 2003). At the same time, geographers have started to extend their collaborations with public health researchers leading to the still young discipline of health geography[1] that uses geographical concepts and techniques to investigate health-related topics (Meade and Earickson 2005; Gatrell and Elliott 2009).

Space, place and location have been extensively discussed within the context of health geography (Meade and Earickson 2005) and play a fundamental role in spatial epidemiological applications. For example, information about the positional location of a household or place is essential for spatial analysis, as is the extent of a region or space and socio-economic status, ill-health, and personal perception, which can be attributed to these locations. These attributes might then vary over

[1]We refer to health geography although there is a scientific debate on the naming of this discipline. Please confer e.g., Kearns (1993), Mayer and Meade (1994), and Kearns and Moon (2002) for arguments whether to name it medical geography or the geographies of health.

O. Gruebner (✉) • P. Hostert
Geomatics Lab, Department of Geography, Humboldt-Universität zu Berlin, Berlin, Germany
e-mail: oliver.gruebner@geo.hu-berlin.de

M.M.H. Khan
Department of Public Health Medicine, School of Public Health, Bielefeld University, Bielefeld, Germany

A. Krämer et al. (eds.), *Health in Megacities and Urban Areas*,
Contributions to Statistics, DOI 10.1007/978-3-7908-2733-0_16,

space and time and environmental exposure and associated health outcomes will in turn differ with spatial and temporal scale (Kawachi and Berkman 2003; Galea et al. 2005; Robertson et al. 2010).

Research on the relationship between health and the environment requires multiple source data which need to be integrated for analysis. GIS have been widely used by public health researchers because they provide a way to link individual health data to the physical space and social community within which a person lives (Edelmann 2007). With GIS, it is possible to describe the sources and geographical distributions of disease agents, to identify regions in time and space where people may be exposed to environmental and biological agents, and to map and analyse spatial and temporal patterns in health outcomes (Cromley 2003). Understanding the spatial patterns of infectious diseases can provide insight as to their causes and controls (Ruankaew 2005). GIS supports deeper insights into how humans interact with their environment to promote better health (Ricketts 2003).

In this chapter we provide an overview of spatial applications in public health research. Focussing on both epidemiological and geographic research questions, we discuss methods and techniques that are suitable for assessing, managing, analysing and visualising spatially referenced public health data. We underline our discussion with examples from an ongoing research project focussing on health and the environment in the Megacity Dhaka, Bangladesh.

16.2 Spatial Analysis Using Geoinformation Technology

As one of the most important applications in spatial analysis, GIS cannot be regarded as one single tool or software. A GIS for public health related research for example, comprises a collection of compatible hardware, software (or algorithms) and methods for analysing spatial patterns of ill-health and their mechanisms (risk factors, pathogens, resilience), as well as for producing maps and reports of spatialised public health information (Hostert and Gruebner 2010). Besides proprietary software like ESRI's ArcGIS™, GoogleEarth™, or GoogleMaps™, there is a wealth of open source applications for different aspects of GIS, e.g. uDig, Quantum GIS, SAGA GIS, GeoDa, Open StreetMap, or spatial packages for the statistical environment R. A comprehensive overview of up-to-date open source GIS software, data, documents and projects is given by FOSSGIS (2010). The science related to implementing and applying spatial methods with GIS is referred to as "GI Science" (Longley et al. 2007). Spatial epidemiological applications within GI Science and epidemiology can be grouped in three types of study, namely disease mapping, exposure mapping, and spatial-epidemiological modelling with the aim of describing spatial patterns, identifying disease clusters, and explaining or predicting public health risk (Waller and Gotway 2004; Pfeiffer et al. 2009).

16.3 Data for Spatial Epidemiological Analysis

Manifold data sources would be appropriate for analysis in a spatial context. However, geo-referenced data, e.g. statistics from official sources, is not always available, especially when the focus is on developing countries. However, there are ways to either collect one's own data or to link different kinds of data in order to perform spatial analysis. In either case, applications and analysis methods are highly dependent on the type, scale, and quality of the data at hand. Choosing a suitable method for analysing the data is often far from trivial and common guidelines rarely exist. We here provide an overview of the most relevant data sets and related methods of analysis and discuss how this data can be integrated within GIS.

16.4 Survey Data

Often the type of data at hand determines the method to be used. For example, when public health related data is available as point data representing all disease cases in a given study area and time period, the data may be analysed in a case-control setting. This involves the spatial variation of the cases being compared to the spatial variation of a background population (control group) with the null hypothesis assuming that the risk towards ill-health is the same in all areas (constant risk hypothesis). The controls might either be assessed by conducting a census in the same area and at the same time or by considering other disease cases that do not have anything in common with the disease under investigation. For example, cases of a certain skin disease might sufficiently represent the background population when the spatial distribution of respiratory disease cases is under investigation (Waller and Gotway 2004).

A GPS-based health survey conducted in e.g. a cross sectional fashion, delivers health information of a "slice" of the investigated population assuming them (the sampled subpopulation) to be representative for the whole group. The resulting data set is thus a sampled point dataset with information on health status, individual factors and sometimes also on other physical and social environmental variables which are thought to affect the respondents' health. However, this kind of data remains rare as for privacy reasons data sets are often aggregated and health information is only available for e.g. enumeration areas, zip-codes, or administrative units. Statistical methods are therefore often chosen to suit either case-control point data representing (all) disease cases in a study area, or aggregated data, representing disease counts i.e. the number of disease cases, rates i.e. cases per person at risk, or ratios i.e. the number of cases compared to the number of controls for arbitrarily chosen area units.

16.5 Remote Sensing Based Data

Remote sensing derived information is becoming increasingly available world-wide and at multiple scales. Today, GI Science strives to better integrate remote sensing derived information in its various analysis processes. Moreover, methods from GIS are increasingly integrated with digital image processing methods for analysing remote sensing data. Lillesand et al. (2003) provide a well written introduction to remote sensing and image interpretation, also for non-remote sensing scientists. For details on digital image processing in remote sensing Richards and Jia (2005) is suggested as introductory reading.

Remote sensing from airborne and satellite platforms provides spatially continuous data at different scales. While the mere image data can serve as a cartographic base map or image backdrop for visualisation, we are usually interested in problem-specific information derived from remote sensing imagery. Raw image data is turned into information layers via image interpretation or digital image processing. This can in turn be integrated in complex analysis of health and the environment.

Typical remote sensing derived information for public health research comprises land use and land cover (see example 1), vegetation type or habitat maps, water maps or diverse structural surface properties, e.g. information on infrastructure or corridors (vegetation, airflow, etc.) to be used in exposure mapping. Also, more sophisticated products may be derived, such as – for example – indicators on housing structures in slum settlements or indirect estimates of population density, in cases where adequate statistical data is missing or erroneous. In other words: remote sensing can provide information urgently needed on the explanatory side of the equation linking health and the environment.

16.6 Data Integration for Public Health Research

In GIS, data integration is straightforward owing to the combined use of spatial databases, geo-visualisation and geoprocessing tools (Hostert and Gruebner 2010). Within spatial databases, for example, the data can be structured, documented, and also linked to data available from other sources.

Research on health and the environment in e.g. urban areas requires multi-source data that need to be combined for analysis. A spatial database can be developed for solving urban public health questions, structured to incorporate data on the social and the physical environment as well as data on individual factors determining public health. The most important benefits of a database are that it can guarantee data integrity and data redundancy, and prevent data inconsistency. Moreover, it can prevent problems with multi-user applications and data loss, and ensure data security. This can be made possible through the use of a data model[2] with

[2] A data model is an abstract model describing how data is structured. Data models are used to integrate different kinds of information, putting them into a thematic, semantic or – in the case of spatial data – in a geometric-topological structure.

database relations in the normal form.[3] While data is stored in numerous tables with atomised attributes in the GIS database, GIS-based analyses are performed via more convenient 'table views'. In such table views, attributes, e.g. health outcomes, traffic emissions or birth rates, can be rigorously combined with GIS features (i.e. a point, line or polygon) representing the location of e.g. households, streets or administrative boundaries for spatial epidemiological analysis. Furthermore, with web feature and web map services it is possible to combine the local dataset with data available from other sources via the internet. For a complete discussion on data integration within GI Science refer to Longley et al. (2007).

16.7 A Framework for Spatial Epidemiological Analysis

There is a wealth of publications that considers spatial analysis and public health research, focussing on either statistical models or applications within specific scientific fields. We draw on the work of Elliott et al. (2006), Gatrell and Elliott (2009), Pfeiffer et al. (2009), and Waller and Gotway (2004) and provide a comprehensive framework in order to structure available methods and guide researchers in finding a suitable approach for analysing their data.

Our framework includes three key pillars:

Disease mapping
Exposure mapping
Spatial epidemiological modelling

For the spatial and temporal analysis of health outcomes one can draw on the rich framework of spatial statistics which has been developed in the last 50 years (Fortin and Dale 2006, p. 25). Spatial autocorrelation analysis is one method used to discover the extent to which given observations can be regarded as spatially independent or clustered (Tobler 1970). Test statistics are used to detect the patterns of e.g. ill-health in space and time which can provide insight into causes and controls (Ruankaew 2005; Hostert and Gruebner 2010).

Spatial autocorrelation analysis is based on adjacency or distance measures and therefore depends on neighbourhood definitions. Neighbourhood definitions specify which sample points are considered to be neighbours i.e. based on adjacency or a fixed distance band, while spatial weights specify whether all neighbours should be treated the same way or whether some of them have a higher importance than

[3]A relation R is in the first normal form (1NF) if all underlying domains contain atomic values only. A relation R is in the second normal form (2NF) if it is in 1NF and every non-key attribute is fully dependent on the primary key. A relation R is in the third normal form (3NF) if it is in 2NF and every non-key attribute is non-transitively dependent on the primary key. However, two more NF exist but are rarely implemented as the data structure then often ends up in overly flat tables.

others due to shorter distances (Waller and Gotway 2004). A number of different approaches exist to specify neighbourhood relationships as well as spatial weights. Since model outcomes are sensitive to the approaches used and no theoretical guidance exists as to which approaches should be selected, we suggest following the guidance of Bivand et al. (2008), i.e. to use and compare a set of approaches including *k*-nearest neighbours and distance based approaches. Finally, a closer look at the research question at hand can help to define the appropriate method for assessing neighbourhood relations. For example, if our research targeted an infectious disease comparing the spatial variation of the disease between a city area and a sub-urban area, a spatial weights neighbourhood (i.e. distance based) relation should be chosen with a greater importance being given to those neighbours that are closer in space than others. This allows the best representation of the spatial variation of the background population at risk of contracting a certain disease, as the population is heterogeneously distributed with higher concentrations in the city than in the surrounding areas.

Fundamental to the analysis of spatial patterns is consideration of the spatial processes creating the pattern. Related concepts are stationarity, isotropy, first order (trend), and second-order (local) spatial effects. In brief, a spatial process is termed (non-)stationary if the dependence between measurements of the same variable across space is the same (different) for all locations in an area. The distance dependency of the variance of the variable under study might vary with direction – in this case the process is called anisotropic otherwise isotropic. First-order effects describe large-scale variations in the mean of the outcome of interest due to location or other explanatory variables, while second-order effects describe small-scale variation due to interactions between neighbours (Pfeiffer et al. 2009).

16.7.1 Statistical Significance Tests

Statistical significance of most tests for spatial autocorrelation can be assessed by a randomisation procedure (Monte Carlo test). If specific assumptions are met, a normal approximation distribution test could be used as an alternative. During the randomisation procedure, data values are reassigned among N locations, providing a randomisation distribution against which one can judge the observed value. If the observed value of the test statistic lies in the tails of this distribution, it could be stated that there is significant spatial autocorrelation in the data to reject the assumption of independence among the observations. Another option is to compare the Z-score (standard deviation) to a standard normal distribution since the Z-score can be assumed to have an approximately normal distribution (Cliff and Ord 1973, 1981). In general however, the randomisation test is the preferred procedure.

16.8 Disease Mapping

Disease mapping may be defined as the spatial and temporal estimation and presentation of health outcomes with the aim of e.g. cluster detection, assessing health inequalities, generating hypotheses, and estimating spatial variability in underlying risk towards ill-health (Elliott et al. 2006).

The risk towards ill-health represents, for example, the probability of a person contracting a disease within a specified time period. Risk is an attribute of a person and is determined by endogenous characteristics such as age, gender, and education, and by exogenous socio-physical environmental factors such as occupation, living conditions, and social network, amongst others. Risk is an unobserved and dynamic quantity to be estimated.

In the following we focus on global measures that enable tests for spatial autocorrelation over the whole study area, and on local indicators of spatial associations (LISA), which provide information about the type of clustering and the locations of clusters (Anselin 1995). While we can only present an overview of existing methods here, the interested reader is referred to Bivand et al. (2008), Pfeiffer et al. (2009) and Waller and Gotway (2004).

16.8.1 Global Estimates of Spatial Autocorrelation

Global autocorrelation methods are used to assess whether significant spatial patterns are apparent throughout the study area, however, these do not help to identify the location of spatial patterns. The null-hypothesis is that no spatial pattern exists (Pfeiffer et al. 2009).

Having the population at risk and corresponding neighbourhood relations in mind, several methods of spatial autocorrelation analysis are available. Depending on the data at hand, the k-nearest neighbour test (Cuzick and Edwards 1990), the K-function (Ripley 1977), and the Cumulative Sum statistic (Rogerson 1997) are qualified for case-control point data. For aggregated data and also for sampled point data, the Moran's I (Moran 1948; 1950), Geary's c (Geary 1954), Tango's Index (Tango 1995), and Whittemore's method (Whittemore et al. 1987) are suitable. When trying to determine whether a disease is infectious, considering space-time clustering detection methods is important. It is crucial to gain knowledge about whether disease cases that are close in space are also close in time and vice versa. Global space-time clustering detection tests include the space time k-function (Diggle et al. 1995), Ederer-Myers-Mantel test (Ederer et al. 1964), Mantel's test (Mantel 1967), Barton's test (Barton et al. 1965) and Jacquez's k-nearest neighbour test (Jacquez 1996)

16.8.2 Local Estimates of Spatial Association

While global measures enable testing for spatial patterns over the whole study area (first order effects), local indicators of spatial associations (LISA) test for statistically significant local spatial patterns (Anselin 1995). Hence, local methods of spatial association define the type, location and extent of spatial patterns, e.g. clusters, and describe second-order effects in the data. The primary goal of such methods is to determine where the observed value, rate, or ratio differs significantly from the value, rate, or ratio observed over the rest of the study area (Waller and Gotway 2004). Some authors further divide local methods into focused and non-focused tests for detecting clusters. None-focused tests identify the location of all likely clusters in a study area, while focused tests investigate whether there is an increased risk around a pre-determined point, such as a source of air pollution (Pfeiffer et al. 2009). Local estimates of spatial associations can be further divided into distance and adjacency based approaches like the global measures, and into moving window based approaches, which are typically local measures for case control data (Elliott et al. 2006). With moving window based approaches, usually circular windows of varying radii are applied over the study area in order to compare the observed number of disease cases with the expected number of cases, assuming that the process under investigation follows a Poisson process (Pfeiffer et al. 2009).

For case-control point data, moving window based approaches like the Geographical Analysis Machine (Openshaw et al. 1987), the Cluster Evaluation Permutation Procedure (Turnbull et al. 1990), the Spatial Scan Statistic (Kulldorff 1997), Besag and Newell's method (Besag and Newell 1991) and Rushton and Lolonis Disease Mapping and Analysis Program (Rushton and Lolonis 1996) are suitable for detecting clusters. Local measures for point data also involve focused tests for including explanatory variables in a model to explain health status of the population by distance from a putative health threat. Health data (the disease) and exposure data (a certain point in space contributing to the disease) is modelled in order to investigate the association of the point source, distance and the corresponding disease. Focused tests for detecting local clusters include Stone's test (Stone 1988), Lawson and Waller's Score test (Waller et al. 1992; Lawson 1993), Bithell's Linear risk score test (Bithel et al. 1994; Bithel 1995), and Diggle's test (Diggle 1990). For the detection of local clusters in space and time Kulldorf (2001) proposed the Space-Time Scan Statistic.

For aggregated data and also for sampled point data, the moving window based methods above will also be suitable. Additionally, distance and adjacency based approaches like the Anselin Local Moran's I (Anselin 1995), and the Getis Ord Gi* (Getis and Ord 1992) are found to be qualified for spatial pattern detection (Waller and Gotway 2004; Elliott et al. 2006; Pfeiffer et al. 2009).

With local spatial autocorrelation tests second-order effects of the spatial process can be described, such as small-scale variation due to interactions between neighbours. However, the moving window based local spatial pattern detection methods work with circles and thus assume that disease clusters are circular. Moreover, another limitation of window based spatial autocorrelation tests is the *a priori* choice of cluster size, as testing for a variety of cluster sizes results in problems of multiple inferences, although this can be accounted for in a Bonferoni correction (Pfeiffer et al. 2009).

16.8.3 Spatial Variation of Risk Towards Ill-Health

The final goal of disease mapping is often to provide maps showing the spatial variation of the population at risk towards ill-health. These maps provide important evidence for disease causes and controls and thus can ideally inform policy in a synoptic way. The information presented may be e.g., the density of disease cases or standard mortality/morbidity ratios (SMR). The objective is to show the important spatial effects present in the data. The resulting smoothed map should have increased information content without introducing significant bias. Again, the data at hand determines the methods used for producing the maps. For point data e.g., containing locations of disease cases, kernel smoothing methods would be best in order to facilitate visual assessment of the pattern. Bayesian methods are best qualified for aggregated data e.g., SMR in order to account for the uncertainty of local measurement and spatial dependence between neighbouring measurements (Pfeiffer et al. 2009). For a more in-depth discussion on how these methods work, the reader is referred to Elliott et al. (2006), Lawson et al. (2003), Lawson (2009), Pfeiffer et al. (2009), and Waller and Gotway (2004).

16.9 Exposure Mapping

Exposure modelling and mapping can be defined as the spatial and spatio-temporal estimation and presentation of factors from the social or physical environment which are (supposed to be) associated with health outcomes (Elliott et al. 2006).

Various methods exist with which to analyse the manifold factors contributing to ill-health. While spatial autocorrelation analysis may also be applied to exposure data, we here focus on methods for the geoprocessing of exposure data and present the most typical methods available in GIS.

16.9.1 Topological Analysis

Topological analysis refers to the fact that each feature (point, line or polygon) "knows" its geographic coordinate and its neighbouring features (adjacency). This is achieved by incorporating topological information into the spatial data model.

It is then easily and quickly possible to track all connected features, e.g. shared borders between adjacent areas. For example, when a point is connected with a line and this line is connected with another point, topology infers that both points are also connected (connectivity) and analysis procedures can use this information in an automated way. Topology can also be used to detect features lying in a polygon (containment). Additionally, attribute data can be used in a spatial database query, like: "Which housing is adjacent to an industrial area with recorded blood cancer or respiratory disease cases?"

16.9.2 Overlay Analysis

Spatial relationships between different data sets can be discovered and new layers with a higher level of information content created from their joint analysis. Vector (topological) overlay functions are, for example, "intersect", "union", or "clip". "Intersect" is used to combine two data sets and to preserve those features and attributes falling in the spatial extent of both layers, while "union" generally keeps all features of both data sets. With "clip", one portion of a layer is cut by using another layer as a kind of "cookie-cutter". For detailed information on how these functions work and how they are used refer to Longley et al. (2007). Raster overlay deals with cell values from different raster grid scenes that can be combined via mathematical operations (map algebra) to generate new values of cells at corresponding positions in a new grid layer (Boulos et al. 2001; Boulos 2004).

16.9.3 Interpolation

Interpolation methods deriving spatially continuous information from spatially discontinuous point data are needed to relate point sampling data on diseases with explanatory information sampled at different locations. In order to achieve a spatially continuous map surface from a point data set with information on phenomena that are also spatially continuous, i.e. precipitation measures or ozone values, interpolation can help to estimate the values for in between locations at which no measurements are available. "Kriging", for example, is an established method for producing such continuous map surfaces from point data sets (Cromley and McLafferty 2002). Kriging uses the existing underlying spatial structure of the sample data (distances among samples or observations) to estimate parameters to describe the spatial structure of the data. This distance based functional relationship is then used in a weighted moving average approach to predict values and standard errors for no-sample locations.

16.9.4 Proximity Analysis

Proximity analysis works with so-called buffers that are drawn around a point, line or polygon to measure e.g. distances to known pollution sources or to quantify the population at risk. By including thematic information, it can be used to stratify data. The population at risk may be divided into age groups or quartiles of age distributions may be used to derive buffer zones with equal proportions of certain age groups.

16.10 Spatial Epidemiological Modelling

We define spatial epidemiological modelling as quantifying and predicting the spatial distribution of a particular health outcome by a set of explanatory exogenous variables from the socio-physical environment and endogenous individual level factors.

Analytical models are a means with which to quantify the effects of the explanatory variables on health outcomes while simulation models are used for predicting health outcomes. In this section we review solely analytical models while the interested reader is referred to Maguire et al. (2005) for a profound discussion of simulation models.

A wide variety of analytical model approaches exists. Linear and generalised linear models are the most widespread type of models used to describe empirical relationships between health outcome and explanatory variables. Depending on the properties of the data, Gaussian, Poisson, negative Binomial, Binomial and other kind of distribution families can be used to properly fit the model to the data (Waller and Gotway 2004).

Multivariate models are applied to provide both a means of quantifying first-order effects and, when first order effects have been considered, for second-order effects. If only one explanatory variable is to be used, the bivariate Moran's I statistic can be considered. The statistic is good at gaining information on the extent to which values for the outcome variable observed at a given location show a systematic (more than likely under spatial randomness) association with another variable observed at the "neighbouring" locations. This bivariate spatial correlation can be considered in addition to or instead of the usual (non-spatial) correlation between two variables at the same location (Anselin et al. 2002).

Residuals from multivariate regression models can be examined for evidence of spatial autocorrelation to identify the presence of second-order effects. If there is no evidence of autocorrelation in the residuals, the data is most likely not spatially structured and a non-spatial model should provide a satisfactory description of the data. However, there may also be some non spatial effects. Regression approaches for spatially independent data include linear regression models or additive models (GAM) (Waller and Gotway 2004; Pfeiffer et al. 2009).

If a second order spatial pattern is evident in the residuals, the model has to be extended to account for the spatial dependency in the data. Mixed models for example, provide a means to account for dependencies as they consist of a fixed part and a random part. The fixed part describes the response variable as a function of the explanatory variables. The random part contains components that allow e.g., for heterogeneity,[4] nested data (random effects), spatial or temporal correlation, and a real random term (Zuur et al. 2009). Thus the random part allows for a correlation of the response variable. Mixed modelling approaches take into account the clustered structure of the data, assuming e.g. that individuals are nested within households, households are clustered within neighbourhoods, neighbourhoods within settlements and so on. Subjects from within the same cluster will be more similar than subjects from different clusters due to their shared environment. In this way, mixed effects models regard spatial proximity as a form of multilevel clustering (Pfeiffer et al. 2009).

For a profound discussion on how these models work, the interested reader is referred to Dormann et al. (2007), Waller and Gotway (2004), and Zuur et al. (2009).

16.11 Examples

The following examples from the Megacity Dhaka, Bangladesh are based on the distinguished framework for urban health developed by Galea et al. (2005). They assume that the social and physical environments that define the urban context are shaped across scales, from global to municipal level. Local factors are accordingly embedded in this multi-level framework.

16.11.1 Example 1: Remote Sensing Based Meta Indicators

Griffiths et al. (2010) present an approach to map urban land-use change from multi-sensoral data, exemplified for the Dhaka megacity region in Bangladesh between 1990 and 2006.

Imagery from the Landsat series of satellites is a great asset for such an analysis due to its synoptic coverage of large urban areas as well as its unique historical archives. In their approach, they solve problems of spectral ambiguities and seasonal phenological dynamics through incorporating multi-temporal imagery for each monitoring year (1990, 2000, and 2006) and by extending spectral information from Landsat with synthetic aperture radar (SAR) data. They use a support vector machine (SVM) classifier and post classification comparison of

[4]Heterogeneity, the violation of homogeneity, happens if the spread of the data is not the same at each X value, and this can be checked by comparing the spread of the residuals for the different X values (Zuur 2009).

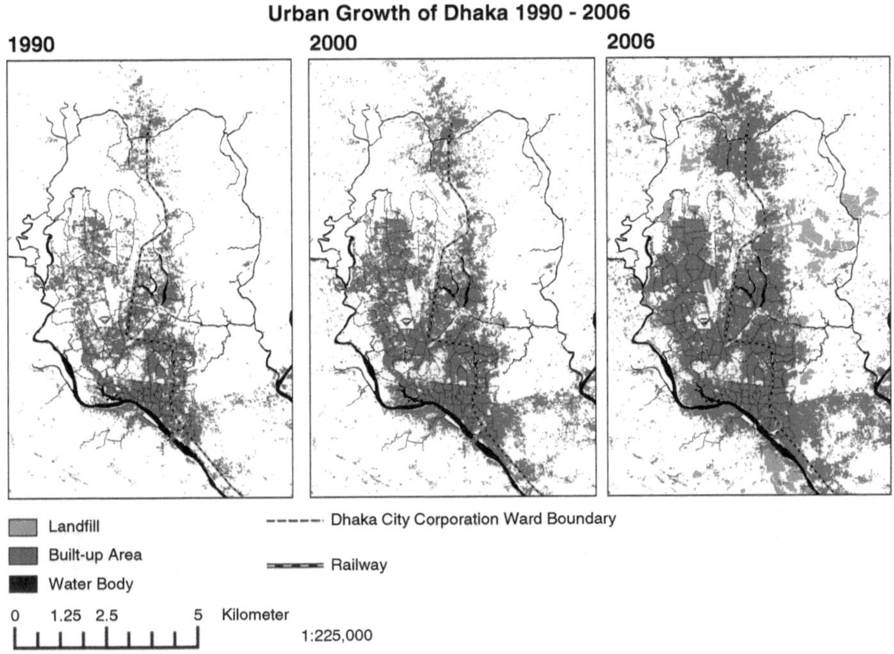

Fig. 16.1 Land use and land cover change between 1990 and 2006 in the Megacity Dhaka after Griffiths et al. (2010)

the three maps to reveal spatio-temporal patterns. The study unveils a profound expansion of urban areas at the expense of rural ecosystems, i.e. prime agricultural areas and wetlands (Fig. 16.1). Deprived ecosystems and ecosystem functions induced by urbanisation and associated impervious surfaces lead to e.g., increased surface runoff after heavy rain or an extended urban heat island effect (Alberti 2009), thus affecting urban health. Moreover, the study provides relevant land use information for subsequent spatial analysis of geo-referenced public health data. Maps of urban fabric, landfill, urban green and park areas, as well as water bodies provide spatially explicit information, which can be used for exposure mapping (c.f. example 2) as well as for spatial statistical models that explain the relationship of health and the environment (c.f. example 3).

16.11.2 Example 2: Geoprocessing for Exposure Mapping

In this example we show how to generate valuable additional information with geoprocessing methods from various geo-referenced data.

From March to April 2009 we conducted a health survey in several slums of Dhaka. In total, 1932 slum household members were interviewed and the residences were geo-referenced via GPS. In the following, we concentrate solely on one slum.

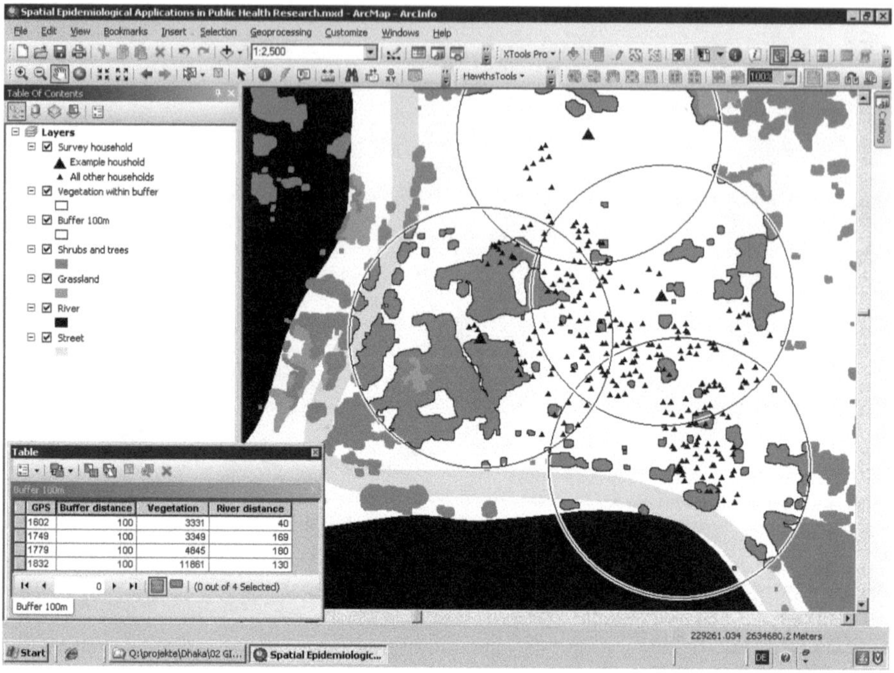

Fig. 16.2 Here, geoprocessing in ArcGIS is visualised. Variables representing the physical environment are geoprocessed by proximity measures and overlay analysis. The triangles on the map represent interviewed households in the settlement of Bishil/Sarag. A buffer of 100 m around each household was created exemplified by four circles in order to calculate the share of vegetation (grey patches). Furthermore, the distances from each household to the nearest river (in black) was measured. In the table the share of vegetation in m^2 and also river distances in meters are provided exemplarily for four surveyed households

In Fig. 16.2 the slum of Bishil/Sarag, the interviewed households (black triangles), and location based environmental information (water bodies and urban green) from example 1 are visualised. Within GIS, the calculation of physical environmental variables such as share of vegetation in 100 m around each sampled household and distances to the next river is now possible. In ArcGIS™ (ESRI 2011), we calculated buffer of 100 m around each household (exemplified by four circles). Within the extent of each buffer, we calculate the area of vegetation extracted from satellite data (cf. example 1) using an overlay approach. In this way all polygons representing vegetation are merged with the buffer layer (i.e. through intersect). For every buffer, area calculations are then performed to gain the number of square meters of vegetative area in each buffer. Furthermore, for each household the distance to the nearest river is calculated in a proximity approach by calculating the Euclidean distance in meters. Having assembled this valuable information through geoprocessing, it can now be used to calculate household risk indexes in order to map household exposure (not shown). However, we use the gained information as input for further analysis in a spatial statistical model (c.f. example 3).

16.11.3 Example 3: Spatial (Auto-)Correlation Analysis

With this example, taken form Gruebner, Khan et al. (2011), we explain how geo-referenced data can be analysed in simple disease mapping (mental well-being), exposure mapping (housing quality), and epidemiological modelling approaches (mental well-being related to housing quality).

The explanatory covariates for this example were taken from the above mentioned survey and from geoprocessing of satellite data (cf. example 2). We subsequently extracted 14 principal components from these covariates (not shown), representing the socio-physical urban environment (amongst others housing quality or population density) along with individual attributes like age, gender, education, marital status, and migration background. The WHO-5 Well-Being Index was used as a measure for self-rated mental well-being. The brief screening instrument assessed the indicators of depression by five questions rated on a 6-point Likert scale (Likert 1932), from 0 to 5. The rates were summed to a range from 0 to 25. Within that range, a raw score of <13 suggested poor well-being (WHO 2010).

Spatial autocorrelation analysis is applied to summarise the degree to which persons with similar health status (WHO-5 scores) or households with similar socio-physical environmental factors (housing quality) tend to occur next to each other, i.e., form spatial clusters. We detect spatial clusters of poor well-being, and

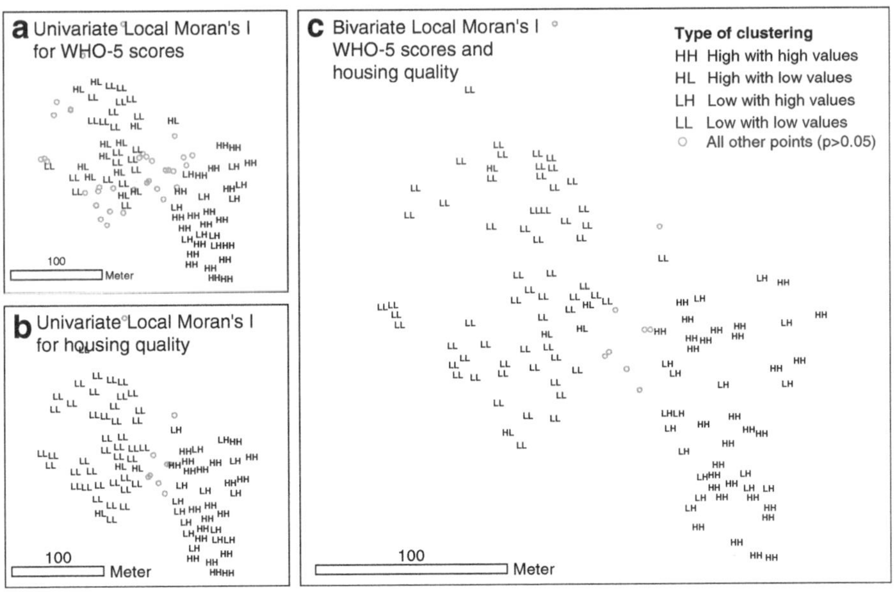

Fig. 16.3 (**a**) Local spatial auto-correlation analysis of the response variable "mental well-being" (measured in WHO-5 scores). (**b**) Local spatial auto-correlation analysis of the explanatory variable "housing type". (**c**) Bivariate local spatial correlation analysis between "mental well-being" and "housing type". The neighbourhood definition for all three analyses was a 60 m fixed distance band (cf. Gruebner et al. 2011)

also spatial clusters of good well-being most significantly among males in Bishil/ Sarag (Fig. 16.3a). We also find spatial clustering of good and poor housing quality within this settlement (Fig. 16.3b).

We further apply the bivariate Moran's *I* statistic to gain information on the extent to which values for the well-being of one person (WHO-5 scores) observed at a given location show a correlation with another variable observed at the "neighbouring" locations. Thereby, well-being is found to be most strongly positively spatially correlated with housing quality (cf. Fig. 16.3c).

16.12 Summary

Spatial epidemiology within GI Science is an emerging field and can be based on established concepts and methods from the area of public health, epidemiology, spatial statistics and geography. We provided examples which gave evidence on how to apply remote sensing and to enrich survey based geo-referenced data in GIS. We further showed how geoprocessed data can be analysed in a simple disease and exposure mapping, as well as in a simple spatial epidemiological approach. We thereby showed that mental well-being at one location is spatially dependent on the mental well-being and other socio-physical environmental factors (e.g. housing quality) prevalent at neighbouring locations. As we assume similar spatial structures to be found in other studies focussing on health and the environment, we would like to emphasise that a spatial epidemiological approach helps to avoid violating the assumption of data independence (e.g. through spatial autocorrelation analysis). We hope that we could give some new ideas to colleagues from related research fields for analysing their data. Collaborative efforts between epidemiologists, biostatisticians, environmental scientists, GI Science experts, and health geographers are needed to realise the full potential of spatial epidemiology in environmental health research. This may then lead to innovative solutions to complex questions.

Acknowledgements We would like to thank the German Research Foundation (DFG) for funding the project Dhaka INNOVATE under the priority programme 1233 "Megacities-Megachallenges". We further thank Tobia Lakes, Sven Lautenbach and Daniel Müller for thoughtful comments on the manuscript.

References

Alberti, M. (2009). Advances in Urban Ecology: Integrating Humans and Ecological Processes in Urban Ecosystems. New York

Anselin, L. (1995). "Local Indicators of Spatial Association – LISA." Geographical Analysis **27** (2): 93–115

Anselin, L., I. Syabri, et al. (2002). Visualizing multivariate spatial correlation with dynamically linked windows. New Tools for Spatial Data Analysis: Proceedings of the Specialist Meeting, Santa Barbara, Center for Spatially Integrated Social Science (CSISS), University of California

Barton, D. E., F. N. David, et al. (1965). "A criterion for testing contagion in time and space." Annals of Human Genetics **29**: 97–103

Besag, J. and J. Newell (1991). "The Detection of Clusters in Rare Diseases." Journal of the Royal Statistical Society. Series A (Statistics in Society) **154**(1): 143–155

Bithel, J. F. (1995). "The choice of test for detecting a raised disease risk near a point source." Statistics in Medicine **14**: 2309–2322

Bithel, J. F., S. J. Dutton, et al. (1994). "Distribution of childhood Leukemias and non-Hodgkin's lymphomas near nuclear installations in England and Wales." British Medical Journal **309**: 501–505

Bivand, R. S., E. J. Pebesma, et al. (2008). Applied Spatial Data Analysis with R. Berlin, New York, Springer

Boulos, M. N. K. (2004). "Towards evidence-based, GIS-driven national spatial health information infrastructure and surveillance services in the United Kingdom." International Journal of Health Geographics **3**(1): 1

Boulos, M. N. K., A. V. Roudsari, et al. (2001). "Health Geomatics: An Enabling Suite of Technologies in Health and Healthcare." Journal of Biomedical Informatics **34**: 195–219

Cliff, A. D. and J. K. Ord (1973). Spatial Autocorrelation. London

Cliff, A. D. and J. K. Ord (1981). Spatial Processes: Models and Applications. London

Cromley, E. K. (2003). "GIS and Disease." Annual Review of Public Health **24**(1): 7–24

Cromley, E. K. and S. L. McLafferty (2002). GIS and Public Health. New York

Cuzick, J. and R. Edwards (1990). "Spatial Clustering for Inhomogeneous Populations." Journal of the Royal Statistical Society. Series B (Methodological) **52**(1): 73–104

Diggle, P. (1990). "A point process modelling approach to raised incidence of a rare phenomenon in the vicinity of a prespecified point." Journal of the Royal Statistical Society Series A **153**: 349–362

Diggle, P., A. G. Chetwynd, et al. (1995). "Second-order analysis of space-time clustering." Statistical Methods in Medical Research **4**: 124–136

Dormann, C., F., J. McPherson, M., et al. (2007). "Methods to account for spatial autocorrelation in the analysis of species distributional data: a review." Ecography **30**(5): 609–628

Edelmann, L. S. (2007). "Using geographic information systems in injury research." Journal of Nursing Scholarship **39**(4): 306–311

Ederer, F., M. H. Myers, et al. (1964). "A statistical problem in space and time: Do Leukemia cases come in clusters?" Biometrics **20**: 626–638

Elliott, P., J. C. Wakefield, et al., Eds. (2006). Spatial epidemiology: methods and applications. Oxford, Oxford University Press

ESRI (Environmental Systems Research Institute) (2009). ArcGIS 9.3.1. Redlands, USA, www.esri.com

ESRI (Environmental Systems Research Institute) (2011). ArcGIS 10. In. Redlands, USA: www.esri.com

Fortin, M.-J. and M. Dale (2006). Spatial Analysis. A Guide for Ecologists. Cambridge

FOSSGIS e.V. (2010). "FreeGIS.org." Retrieved April, 8., 2010, from http://www.freegis.org/

Galea, S., N. Freudenberg, et al. (2005). "Cities and population health." Social Science & Medicine **60**(5): 1017–1033

Galea, S., Freudenberg, N., & Vlahov, D. (2005). Cities and population health. Social Science & Medicine, 60, 1017–1033

Gatrell, A. C. and S. J. Elliott (2009). Geographies of Health: An Introduction. Oxford

Geanuracos, C. G., D. D. Cunningham, et al. (2007). "Use of Geographic Information Systems for Planning HIV Prevention Interventions for High-Risk Youths." American Journal of Public Health **97**(11): 1974–1981

Geary, R. C. (1954). "The contiguity ratio and statistical mapping." The Incorporated Statistician **5**: 115–145

Getis, A. and J. K. Ord (1992). "The Analysis of Spatial Association by Use of Distance Statistics." Geographical Analysis **24**: 189–206

Griffiths, P., P. Hostert, et al. (2010). "Mapping megacity growth with multi-sensor data." Remote Sensing of Environment **114**(2): 426–439

Gruebner, O., Khan, M.M., Lautenbach, S., Muller, D., Kramer, A., Lakes, T., & Hostert, P. (2011). A spatial epidemiological analysis of self-rated mental health in the slums of Dhaka. International Journal of Health Geographics, 10, 36

Hostert, P. and O. Gruebner (2010). Geographic Information Systems. Modern infectious disease epidemiology - Concepts, methods, mathematical models, public health. A. Krämer, M. Kretzschmar and K. Krickeberg. Berlin, Springer: 177–192

Jacquez, G. (1996). "A k-nearest neighbour test for space-time interaction." Statistics in Medicine **15**: 1935–1949

Kaiser, R., P. B. Spiegel, et al. (2003). "The Application of Geographic Information Systems and Global Positioning Systems in Humanitarian Emergencies: Lessons Learned, Programme Implications and Future Research." Disasters **27**(2): 127–140

Kawachi, I., & Berkman, L.F. (Eds.) (2003). Neighborhoods and health. Oxford, New York: Oxford University Press

Kearns, R.A. (1993). Place and Health: Towards a Reformed Medical Geography. The Professional Geographer, 45, 139–147

Kearns, R., & Moon, G. (2002). From medical to health geography: novelty, place and theory after a decade of change. Progress in Human Geography, 26, 605–625

Kulldorff, M. (1997). "A spatial scan statistic." Communications in Statistics - Theory and Methods **26**(6): 1481–1496

Kulldorff, M. (2001). "Prospective time periodic geographical disease surveillance using a scan statistic." Journal of the Royal Statistical Society Series A **164**: 61–72

Lawson, A. (1993). "On the analysis of mortality events associated with a prespecified fixed point." Journal of the Royal Statistical Society Series A **156**: 363–377

Lawson, A. (2009). Bayesian disease mapping: hierarchical modeling in spatial epidemiology, CRC Press

Lawson, A., W. Browne, et al. (2003). Disease mapping with WinBUGS and MLwiN, J. Wiley

Likert, R. (1932). "A Technique for the Measurement of Attitudes." Archives of Psychology **140**

Lillesand, T. M., R. W. Kiefer, et al. (2003). Remote Sensing and Image Interpretation. New York, Wiley

Longley, P. A., M. F. Goodchild, et al., Eds. (2007). Geographic Information Systems and Science. New York, Wiley & Sons

Maguire, D., M. Batty, et al., Eds. (2005). GIS, Spatial Analysis, and Modeling. Redlands, California, ESRI Press

Mantel, N. (1967). "The detection of disease clustering and a generalized regression approach." Cancer Research **27**: 209–220

Mayer, J.D., & Meade, M.S. (1994). A Reformed Medical Geography Reconsidered. The Professional Geographer, 46, 103–106

Meade, M. S. and R. J. Earickson (2005). Medical Geography. New York

Moran, P. A. P. (1948). "The interpretation of statistical maps." Journal of the Royal Statistical Society Series B **10**: 243–251

Moran, P. A. P. (1950). "Notes on Continuous Stochastic Phenomena." Biometrika **37**(1–2): 17–23

Openshaw, S., M. Charlton, et al. (1987). "A Mark 1 Geographical Analysis Machine for the automated analysis of point data sets." International journal of geographical information systems **1**(4): 335–358

Pfeiffer, D. U., T. P. Robinson, et al. (2009). Spatial Analysis in Epidemiology. New York, Oxford University Press

Richards, J. A. and X. Jia (2005). Remote Sensing Digital Image Analysis. Berlin, Germany, Springer

Ricketts, T. C. (2003). "Geographic information systems and Public health." Annual Review of Public Health **24**: 1–6

Ripley, B. D. (1977). "Modelling spatial patterns (with discussion)." Journal of the Royal Statistical Society Series B **39**: 172–212

Robertson, C., Nelson, T.A., MacNab, Y.C., & Lawson, A.B. (2010). Review of methods for space-time disease surveillance. Spatial and Spatio-temporal Epidemiology, 1, 105–116

Rogerson, P. A. (1997). "Surveillance systems for monitoring the development of spatial patterns." Statistics in Medicine **16**: 2081–2093

Ruankaew, N. (2005). "GIS and epidemiology." Journal of the Medical Association of Thailand **88**(11): 1735–1738

Rushton, G. and P. Lolonis (1996). "Exploratory spatial analysis of birth defect rates in an urban population." Statistics in Medicine **15**: 717–726

Stone, R. A. (1988). "Investigations of excess environmental risks around putative sources: statistical problems and a proposed test." Statistics in Medicine **7**: 649–660

Tango, T. (1995). "A class of tests for detecting 'general' and 'focused' clustering of rare diseases." Statistics in Medicine **14**: 2323–2334

Tobler, W. R. (1970). "A computer model simulation of urban growth in the Detroit region." Economic Geography **46**(2): 234–240

Turnbull, B. W., E. Iwano, et al. (1990). "Monitoring for clusters of disease: applications for leukemia incidence in upstate New York." American Journal of Epidemiology **132**: 136–143

Waller, L. and C. Gotway (2004). Applied Spatial Statistics for Public Health Data. Hoboken, New Jersey

Waller, L., B. W. Turnbull, et al. (1992). "Chronic disease surveillance and testing of clustering of disease and exposure: Application to Leukemia incidence and TCE-contaminated dumpsites in upstate New York." Environmetrics **3**: 281–300

Whittemore, A. S., N. Friend, et al. (1987). "A test to detect clusters of disease." Biometrika **74**(3): 631–635

WHO (World Health Organization) Collaborating Centre for Mental Health Frederiksborg General Hospital. (2010). "WHO-Five Well-being Index (WHO-5)." Retrieved 27.04.2010, from http://www.who-5.org/

Zuur, A. F., E. N. Ieno, et al. (2009). Mixed Effects Models and Extensions in Ecology with R. New York, Springer

Chapter 17
Health Inequities in the City of Pune, India

Mareike Kroll, Carsten Butsch, and Frauke Kraas

17.1 Urban Fragmentation and Health Inequities

Since the so-called urban turn in the year 2008, more than half of the population worldwide is living in cities. This is leading to a growing number of people whose health is being influenced by urban living conditions. Whereas a third of the urban population worldwide is currently living in small urban centres with a population of below 100,000, 16% lives in so-called emerging megacities (five to ten million inhabitants) and megacities (above ten million inhabitants); two thirds of these megacities are located in poor or newly industrializing countries (UN 2010). Megacities – especially in Asia and Africa – have grown at a very high pace over the last decades with many implications for the urban population.

Rapid changes in the physical environment (e.g. infrastructure development, land use changes, environmental degradation) and the social environment (e.g. social pluralization, lifestyle changes, dietary habits, socio-cultural and political conflicts) that go hand in hand are posing new challenges for human health (Bork et al. 2009). Concerning the social implications of urbanization, the increasing social fragmentation of urban populations (Coy and Kraas 2003) is jeopardizing social justice within societies. While economic growth – closely linked to the new global patterns of investment as well as to the new division of labour – is creating a new urban middle class in many cities with new consumption patterns and lifestyles, urban poverty remains a problem which is coming to the fore in form of mushrooming slum settlements. Steep socio-economic gradients – often on a small spatial scale – have far-reaching consequences for the differentiation of the urban health situation (Shukla 2007).

Health as an important prerequisite for human development is very closely connected to social justice, yet the terms "health inequity", "health inequality" and "health disparity" are not consistently defined in the literature (Bravemen 2006). The term "health inequality", often used synonymously with "health

M. Kroll (✉) • C. Butsch • F. Kraas
Institute of Geography, Cologne University, Cologne, Germany

A. Krämer et al. (eds.), *Health in Megacities and Urban Areas*,
Contributions to Statistics, DOI 10.1007/978-3-7908-2733-0_17,

disparity", mainly refers to the description of health differences between population groups or individuals at different socio-economic levels. Socio-economic status is usually calculated by using variables measuring educational status, income and occupation (Bravemen 2006). There are only few studies on health inequalities or disparities which focus on gender, ethnicity or geographical location (Carter-Porkas and Baquet 2002) as influencing factors of human health.

The term "health inequity" goes beyond the sheer description by adding a normative dimension. Health inequity can be defined as "avoidable disparities in health and its determinants – including but not limited to health care – between groups of people who have different underlying social advantage or privilege, i.e. different levels of power, wealth or prestige due to their positions in society relative to other groups" (Bravemen 1998: 10). Other authors also stress that health inequities are – in comparison to health inequalities – not only considered as unnecessary and avoidable but also as unfair and unjust (e.g. Whitehead 1991). Further distinctions can be made between definitions which unilaterally address differences in health status or focus on access to health care. Many definitions also integrate both perspectives. Since they are logically linked, they will both be addressed in this paper.

The correlation between socio-economic status and health has been intensively studied in Western societies (for Germany e.g.: Richter and Hurrelmann 2006; Mielck 2005; Bauer et al. 2008). However, the findings of these studies can hardly be transferred to the fast growing agglomerations of newly industrializing countries such as in India as they differ clearly regarding their social, economic, ecological and political conditions. Furthermore, differences in health system design, distinct epidemiological patterns and cultural influences on health behaviour prohibit an easy transfer of concepts and analytical frameworks. Against this background, this paper aims to address inequities in health status and access to health care services in the emerging megacity Pune in India.

The findings are based on two interlinked PhD projects of the University of Cologne which are conducted in close collaboration with the Bharati Vidyapeeth University's Institute for Environmental Research and Education. In the first section the research area and the methodology will be described. Then, findings on inequities in health status and in access to health care in Pune will be presented. Based on these results, some possible implications that health inequities have for health governance will be discussed in the conclusion.

17.2 Setting and Research Area

The Indian city of Pune as an emerging megacity is currently undergoing a rapid expansion and structural change which is laying the ground for fundamental transformation processes. In the last two decades it has grown substantially in terms of population, mainly due to high economic growth, resulting in massive urban sprawl. Today, Pune is India's seventh largest and one of its fastest growing cities.

It comprises an urban agglomeration of roughly five million inhabitants with its adjacent twin city Pimpri-Chinchwad (UN 2010). It is located approximately 150 km east of the Indian economic centre Mumbai. Pune's proximity to Mumbai as well as its long-standing reputation as an intellectual centre ("Oxford of the East") hosting several universities and research centres, have been favourable to its development as a centre for the automotive industry since the 1950s. The liberalization of the Indian economy since the 1990s has put forth Pune as an emerging information technology centre and has induced an economic diversification (Bapat 2009).

The rapid, predominantly unregulated expansion and structural change of this urban agglomeration, which has gone hand in hand with a doubling of the population in the last two decades (UN 2010), has far reaching consequences for the physical and social environment of the city. The process of mega-urbanization is accompanied by a growing differentiation and fragmentation of urban society which is reflected in the increasing number of inhabitants living in formal and informal slum areas in close proximity of the newly developing mostly semi-gated "housing societies" of the upper middle class. According to the Municipal Corporation, currently more than 40% of the total population of Pune is living in slum areas (Pune Municipal Corporation PMC 2008). It can be assumed that the widening gap between rich and poor is also inducing a rise in health inequities.

Due to the complexity of the urban environment, individual health-determining parameters are affected by manifold interwoven influencing factors. Air quality, for example, depends on a city's size and growth patterns, the steering capacity of the municipality, geographical location, the climate and many other factors. In Pune which was renowned as a hill station with a pleasant climate and surroundings three decades ago, the environment has become increasingly degraded in the course of the city's expansion and due to the population's growing demand for resources (Pune Municipal Corporation PMC 2008).

New qualities and quantities of household and industrial waste, untreated sewage (33% of the total wastewater discharged), increasing traffic (the number of vehicles has doubled within a decade) (Pune Municipal Corporation PMC 2008) and sealing of open spaces have - among other factors – contributed to the environmental degradation of the once green and healthy city. Also, the local climate has changed over the last decades, with changing monsoon patterns and rising temperatures, especially in the summer months. Moreover, poverty levels in urban Maharashtra have increased since 2000 despite some economic growth (Mishra et al. 2008: 14).

Changing health determinants in Pune's physical environment (e.g. contaminated air, water and soils, traffic congestion and noise pollution) as well as in its social environment (e.g. changing social networks and diet patterns, rising income disparities) cannot be unilaterally linked to the prevalence of single diseases due to the fact that the impact of influencing parameters can hardly be quantified exactly.

Thus, the increasing prevalence of certain diseases rather has to be explained by a "web of disease" (Jenkins 2003: 19). Nevertheless, some diseases are closely associated with certain developments: The increasing number of cases of chronic obstructive pulmonary disease (COPD), for example, can be associated with rising levels of air pollution, in the case of Pune especially indoor air pollution; the increasing prevalence of high blood pressure and diabetes can be associated with changing dietary patterns, rising stress levels, a lack of physical exercise, but also demographic changes, mainly an increasing life expectancy. The latter leads to an increasing prevalence of chronic diseases in general. Diseases of the circulatory system were the major cause of death in Pune in 2004, accounting for 26.4% of all registered deaths (State Bureau of Health Intelligence and Statistics and Vital Statistics (SBHI and VS), not dated). However, the rising burden of chronic diseases in Pune does not mean that infectious diseases have been successfully controlled.

Pulmonary tuberculosis was the third highest single cause of death in 2004 in Pune, accounting for 7.6% of all deaths (SBHI and VS, not dated). Even if some fatal diseases such as small-pox or poliomyelitis have been eradicated in India (Gupte et al. 2001), the re-emergence of "old" infectious diseases such as tuberculosis and the emergence of "new" diseases such as HIV/AIDS or H1N1-influenza ("swine flu") have shown the vulnerability of the urban population towards communicable diseases (for TB, HIV see Gupte et al. 2001, for H1N1 see The Hindu 2009). The aim of the first perspective, which focuses on health status, is to get a better understanding of the epidemiological patterns of different socio-economic groups and the factors which influence these patterns.

The comparatively low advances in the overall health status due to the still high burden of communicable diseases and the growing burden of chronic diseases is also influenced by the access to health care services. While some aspects of urban life, such as life styles, dietary patterns and exposition to pollutants are mostly viewed as negative influences on health status, the opportunities offered by preventive, curative or rehabilitative health care services usually are addressed as positive influences (Galea and Vlahov 2005). Penchansky and Thomas (1981) defined access to health care as the degree of fit between the demand side and the health care system, which can be measured in five dimensions: "availability" (volume and type of services), "accessibility" (location of supply in relation to the location of clients), "accommodation" (organisation of services, e.g. opening hours), "affordability" (relation of prices of services to the ability to pay) and "acceptability" (influence of attributes as religious group, gender, type of facility, etc.). Access to healthcare not only depends on the population's needs (health status and disease patterns), but also on the framework of the (local) health care system.

The Indian health care system as a whole consists of an overburdened public and an unregulated private health care sector (Butsch 2008). In 1947, guidelines for the development of a comprehensive public health care system were established by a committee chaired by Sir Joseph Bhore (Gangolli et al. 2005). With these guidelines which focused on the provision of basic preventive and curative services, India became the forerunner of the primary healthcare approach. Unfortunately, this visionary system never became fully functional as financial resources for the public health care sector remained scarce. Although the tertiary treatment institutions are mostly located in cities, traditionally investment in rural primary health care services has been given more importance. In 2005, with the launch of the National Rural Health Mission, the Indian government again devoted a substantial part of the public health budget to rural areas.

Due to the lack of public services, an unregulated private health care sector emerged in urban India, which offers top class services at the upper end of the scale and services offered by self trained, i.e. untrained and uncertified, medical personnel at the lower end. The national health care system's framework lays the basis for the diverse patchwork of healthcare providers in Pune. The second approach which focuses on access to health care services has the aim of understanding barriers and facilitators which have an impact on the decisions different population groups make on which type of health care provider to patronize. Another aim of this approach is to explore whether they receive adequate health care.

17.3 Study Sites and Survey Design

Several field work campaigns were conducted in order to analyze health inequities from both of the above mentioned perspectives. One of the campaigns was a joint initiative which encompassed aspects from both approaches, using some common data collection tools. In both projects a set of qualitative and quantitative methods was applied.

Firstly in a jointly conducted household survey respondents from 450 households living in six different research sites were interviewed. 75 households were chosen in each research area using random walk sampling (Diekmann 2003): each field assistant followed a default route through the research site selecting every 7th, 10th or 12th household depending on the total number of households in each research site. Three slum areas and three middle class areas were initially selected according to the building structures and assumed socio-economic status, which later was verified in a pre-test. Further, the research areas are situated in three different parts of the city: one slum and one middle class area are situated in the old city centre (Somwar Peth), one upper middle class area and one slum area, consisting of three neighbouring construction worker camps, are located in the former British cantonment (Koregaon Park), three neighbouring slum pockets (counted as one slum) surround an upper middle class area at the edge of the urbanized area (Kondhwa) which has developed over the last two decades. Although each research site shows some heterogeneity in situ, the differences between the six areas are much more relevant. In each area not only migration status and socio-cultural background, but also households' economic status vary significantly, as indicated by the wealth index (Fig. 17.1). Interestingly, the wealth index not only shows a different distribution of quintiles between middle and lower class areas, but also between the three slum areas.

Secondly, qualitative methods were applied including expert and in-depth-lay interviews. Expert interviews (82 interviews) with medical practitioners, academics, experts from non-governmental organizations which deal with health related issues, especially concerning the urban poor, and staff from the Pune Municipal

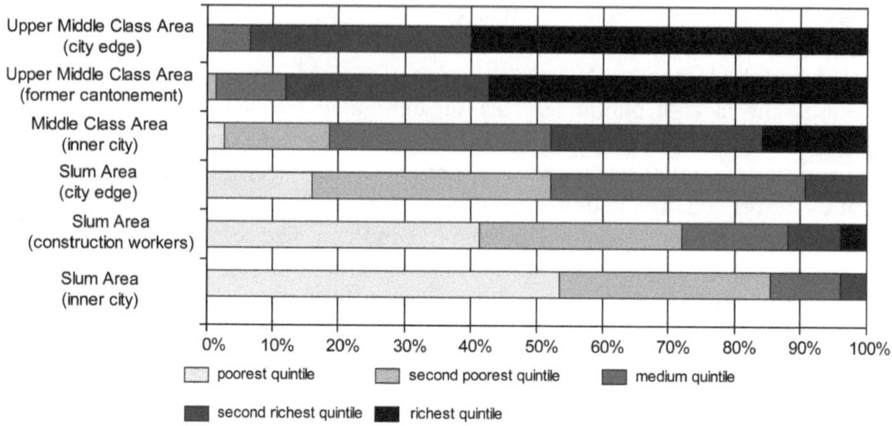

Fig. 17.1 Distribution of wealth index in the six research areas
(*Source*: own survey by C. Butsch and M. Kroll, a total of $6 \times 75 = 450$ households were interviewed)

Corporation were performed. In-depth lay interviews (72 interviews) were conducted in the six research areas with citizens who had already participated in the household survey.

Thirdly in and around the six research sites the health infrastructure and health-affecting factors (e.g. sources of pollution) were mapped. In the following section the results from the household survey and from in-depth interviews with experts and lay people will be discussed.

17.4 Inequities in Health Status

Due to the absence of a comprehensive health monitoring system in Pune, reliable prevalence rates of chronic and infectious diseases only exist for a few diseases such as tuberculosis. These data only exist since they are collected through national surveillance programs (for India: see Misra et al. 2008). Further, it is difficult to link existing morbidity and mortality data to socio-economic indicators such as income or profession, which would allow for conclusions on health inequities in Pune. Regular surveys such as the governmental National Family Health Survey, which also collects data on income and education, compare urban and rural settings but do not make any spatial or socio-economic differentiation within cities. The general focus on the urban-rural divide in India as well as the lack of appropriate data impedes conclusions on health inequities in Pune.

Preliminary results from the household survey and expert interviews show that especially in the poorer sections of slum areas the prevalence of diseases such as gastroenteritis, which are associated with the poor quality of drinking water and poor hygienic conditions, are still common. There are also huge differences in terms of health-determining factors such as hygiene, housing quality, quality of water supply and water availability, the availability of food as well as social networks, income and education within the settlements officially declared as slums.

The prevalence of other infectious diseases such as tuberculosis is difficult to assess in a survey because of their association with social taboos. Tuberculosis is still prevalent in slum areas but can also occur in higher income classes, especially if the immune system of the population is weakened through extremely high stress levels or nutritional deficiencies. A medical specialist used the following example during an expert interview to illustrate this point: Managers with high stress levels are often immunocompromized and thus more vulnerable to infectious diseases such as tuberculosis. For example, they could contract tuberculosis from their driver who probably lives in a slum area and might be carrying the bacillus. The prevalence of tuberculosis in urban Maharashtra was 3.7 per 1,000 people in 2006 (International Institute for Population Sciences IIPS and Macro International 2008). Further, multi-drug resistant tuberculosis is on the rise due to improper or interrupted treatment which is often associated with a lack of knowledge about health and other factors such as alcohol abuse. Although there has been a decrease in cases of tuberculosis in the last years in Pune, medical specialists are expecting a rise in connection with the spread of HIV/AIDS.

However, people from the lower socio-economic strata also increasingly suffer from chronic and degenerative diseases. Diseases such as high blood pressure and diabetes, which often co-occur and are considered to be lifestyle diseases, do not only appear in the middle and upper middle classes, as shown in Fig. 17.2. Whereas high prevalence of diabetes and high blood pressure in the upper middle class is usually connected to lifestyle changes, increasing incidence levels, especially in the poorer segment of the population, can be accounted for by physical predispositions such as abdominal obesity or low birth weight (Barker hypothesis) (Siegel et al. 2008). Though, lifestyle factors such as poor eating habits and lack of physical

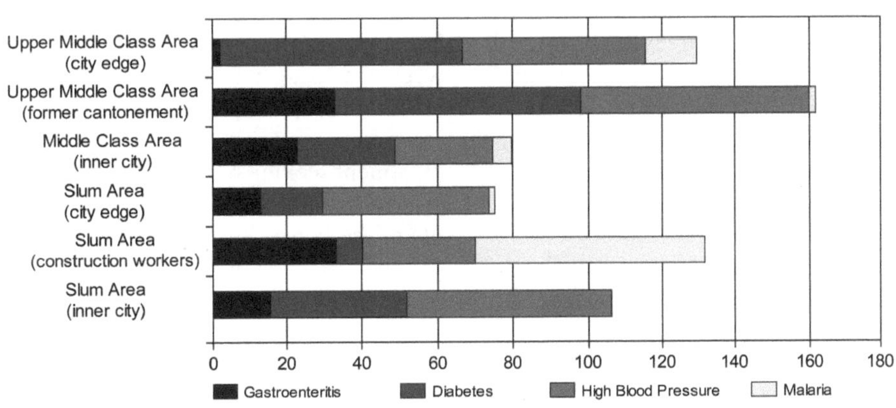

Fig. 17.2 Age-standardized prevalence rate of gastroenteritis, diabetes, high blood pressure and malaria in the six research areas
(*Source*: own survey by M. Kroll, a total of 3875 individuals were interviewed from 6×150 = 900 households)

exercise also obtain relevance in slum areas in households which are above the poverty level.

Vector borne diseases such as malaria, dengue fever and chikungunya, which have been the targets of several long term extinction campaigns, are partially on the rise again. The incidence of malaria in the construction worker camps is striking in Fig. 17.2. This can partly be explained by a higher number of puddles of water in this area which are usually not sealed and which may serve as breeding grounds for mosquitoes. They are also located near a river which falls dry in the dry season, leaving stagnant pools of water in the river bed. Besides, the municipality does not spray against mosquitoes in this area because this is only done in "housing societies" and formal slum areas. These uncontrolled hot spots can contribute to a re-expansion of vector-borne diseases in Pune which is affecting all strata of society.

Health-determining factors such as the quality of housing and awareness about nutrition and health differ in many respects in the six different research areas. However, especially the case of diabetes shows that differences in health status do not necessarily become apparent in the type of diseases which are prevalent in the local population. It is rather the severity of a disease which often makes differences in health status unjust: diabetes and high blood pressure are silent diseases which can lead to a lot of complications if they remain undetected over a period of time or if they are not properly treated. For example, a woman in a slum area said during an in-depth interview that she was suffering from diabetes but got cured by a healer. Therefore there is no need for any further treatment from her point of view. The lack of education, knowledge about health and financial assets make the lower socio-economic strata much more vulnerable towards health risks (Sakdapolrak 2007).

17.5 Inequities in Access to Health Care

Both the qualitative and the quantitative methods revealed the existence of barriers and facilitators in all of the aforementioned five dimensions of access to health care. These are a product of the interplay between the local health care system (i.e. supply side) and population characteristics (epidemiological profile, ability to pay, etc., i.e. demand side). The differences in access to health care in Pune are illustrated by the clear discrepancies in the treatment seeking behaviour of different population groups as shown in Fig. 17.3.

Even a very superficial look at the results of the household survey reveals that the public sector does not play a major role in the provision of health care in Pune. When asked about their usual health care providers, the overwhelming majority of respondents stated facilities from the private health care sector as their primary health care providers. In total, 83% of the respondents named a private facility as their household's first choice. A closer look at Fig. 17.3 shows that preferences differ between the six research areas. The highest proportion of households using governmental services can be found in the inner city slum area, the lowest in the middle class area in the former cantonment. In the latter a comparatively high share

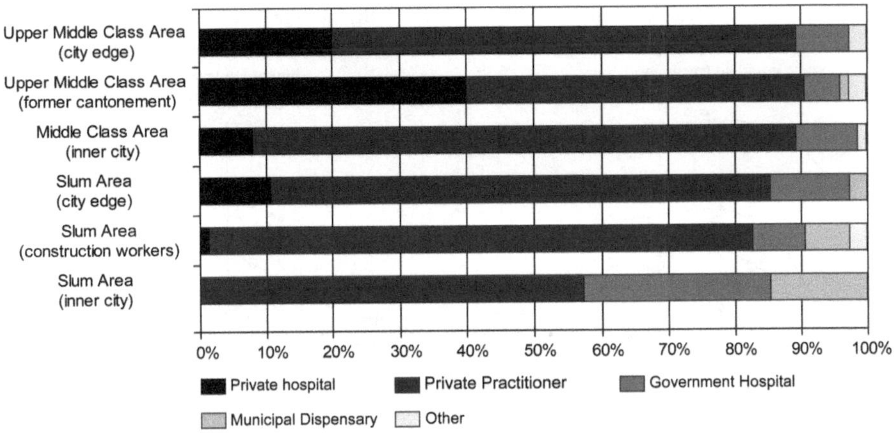

Fig. 17.3 Treatment seeking behaviour in different areas of Pune: To which treatment facility does your household usually go to?
(*Source*: own survey by C. Butsch, a total of $6 \times 75 = 450$ households were interviewed)

of households goes to private hospitals, because these hospitals also offer the services of out-patient departments.

The other four research areas are situated between these two extremes. Fig. 17.3 shows that in areas with a larger proportion of households with higher socio-economic status a) the percentage of people who rely on public health care is lower and b) the number of private hospitals is higher. An obvious explanation for this connection is the difference in the prices of health care provision. Services offered by public health care providers are subsidized and service charges are much lower than those of private health care providers. People who live below the poverty line even have access to public health care services free of charge.

Private practitioners on the other hand offer slightly more expensive services – with varying treatment costs depending on the qualification of the health care provider – and the private hospitals' outpatient departments offer the most expensive services. Therefore the utilisation rates of public services correlate with the share of poor households.

However, as Fig. 17.4 shows, the differences in the treatment costs per outpatient consultation, which were computed using the results of the household survey, are only minor in the case of acute diseases. Contrary to that, the differences in the costs for inpatient treatment are much higher. This can be linked to the fact that the total costs for the treatment of acute diseases which do not require inpatient treatment include a minor share of treatment costs and a major share of costs of consumables (e.g. medicine, dressing material etc.).

Due to the shortfalls in the public health care sector, these consumables can hardly be provided at primary or secondary health centres. Therefore, the financial gain of consulting a public facility is negligible. This also explains the relatively low share of respondents who named public primary health care centres as their usual source of treatment. Because of opportunity costs (waiting times, travel

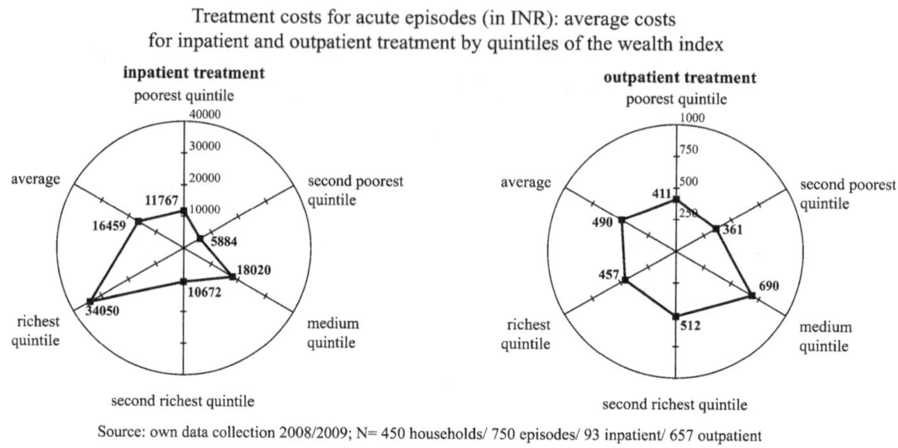

Fig. 17.4 Mean costs for inpatient and outpatient care in case of an acute disease (*Source*: own survey by C. Butsch, a total of 6 × 75 = 450 households were interviewed)

costs), public primary health care can in fact even be more expensive than private health care, particularly if additional medicine has to be bought on the market. Nevertheless, exclusively focusing on the "affordability" would be too easy since barriers and facilitators in the other dimensions also have an influence on the population's access to health care.

"Accessibility", for example, is another important dimension which should be considered when attempting to explain the differences in the utilization of different treatment facilities. While private practitioners are spread all over the city, there is only a limited number of public health care facilities, which are concentrated in the older parts of the city. Especially in newly developed areas there is a lack of public health care facilities, as mapping and expert interviews which were conducted in the course of our research revealed. This also explains the higher utilization rates of public services in the inner city slum as compared to the two other slum areas. This does not mean that public health care facilities are inaccessible, especially since there is a public transportation system in the city. However, the accessibility of private providers is much higher, which clearly is a facilitating factor.

In terms of "availability" hardly any limitations exist in Pune, as the following quotation from an expert interview shows:

Interviewer: "Would you say the coverage of health care has changed?"

Expert: "Yes, coverage has changed. Many people have increased the machineries and newer instruments. Whatever is launched in say US or UK, next day or in that day itself it is coming to these hospitals now. So patients they don't have to go outside to treat themselves".

Most of the interviewed experts expressed similar views on the availability of health care. Especially because of the increase in solvent patients in the middle class since the economic reforms, health care providers have invested in advanced investigation and treatment facilities. However, due to financial barriers these facilities are not evenly accessible, especially since the public health care sector cannot keep pace with the city's growth.

In the dimension "accommodation" again barriers and facilitators are very unevenly distributed between the private and the public sector. The following quotation from an interview with a lay person from a middle class area shows that severe barriers exist in the public sector:

Respondent: "I will go to these guys in the private hospital but I will not go to Sassoon [the public tertiary hospital]. If the same doctors at the Sassoon had their own practicing clinics I would go to them there. You know, it's a question of time lack... the waiting period... the crowds... the clipping of quality treatment ...that would be reason which keep me away from there."

As mentioned implicitly in this quotation, nearly all experts and lay people agreed that the private sector is much more accommodating than the public sectors as the opening hours are much longer, waiting times are shorter and service in general is more comprehensive.

The fifth dimension introduced by Penchansky and Thomas (1981), "acceptance", addresses mainly socio-cultural and behavioural factors which influence the relationship between practitioner and patient. Indian society is highly complex, and factors related to this dimension seem to reinforce people's decision which health care provider to go to rather than functioning as an access-barrier:

Respondent: "...if most people of your social and cultural background go to this kind of certain hospital maybe one might find it embarrassing to go to a government hospital or a cheap facility. So even if you can't afford it you might feel the need to go to ... pretty expensive things..."

As this quotation shows, the influence of social networks on the decision which health care provider to consult seems to be important. However, all respondents agreed that the discrimination of certain segments of the population do not represent major problems regarding access to health care.

17.6 Conclusion and Recommendations

The above-mentioned findings which incorporate both perspectives, i.e. health status and access to health care, show that numerous factors contribute to health inequities in Pune, which in turn are either linked to contextual or compositional factors. While the context is determined by environmental factors such as the local health care system and the general mega-urban setting, compositional factors include individual parameters such as income, educational status, occupation, etc. (Smyth 2008). As the WHO commission on social determinants of health underlined: "...a toxic combination of poor social policies, unfair economics, and bad politics is responsible for much of health inequity. [...] [This] toxic combination is also responsible for the social gradient in health in those who are above the level of material deprivation but still lack the other goods and services that are necessary for a flourishing life" (WHO 2008: 35).

Based on the presented findings it can be concluded that people from the lower socio-economic strata, who mostly live in slum areas, face access problems basically in terms of quality of health care services and the financial consequences of treatment. In terms of health status, infectious diseases still pose an unsolved problem, especially in unregistered slums which even lack basic amenities such as sanitation and safe drinking water. However, the infrastructure in authorized slums has been improved over the last decade (Bapat 2009), resulting in an increase of the availability of basic amenities and an improvement of the general hygienic situation.

Chronic diseases pose a "new" challenge, especially for the urban poor, as they do not have access to adequate treatment and lack the knowledge to assess their own health status. The lower economic strata are therefore more vulnerable to a deterioration of their health status due to a higher exposure to health risks and less coping capacities. In the middle class, most hygienic problems such as waste disposal or water supply have been over the last two decades. Besides environmental health risks, the exposure to social health risks is comparatively lower, too, as traditional norms and values are still deeply rooted in this segment of society. People from the middle class tend to go to smaller private health care facilities. Social networks have a strong influence on the choice of the treatment facility. However, if inpatient treatment is required (including expensive surgery) the costs of treatment can create serious difficulties for a household. The ability to handle day to day health problems creates an illusive safety, which masks the danger of a crisis triggered by severe medical problems (World Bank 2002).

For the upper middle class in Pune, major problems linked to access – also to high quality health services – do not include issues related to availability; access in this context is rather determined by soft factors such as social networks. Further, this socio-economic group is prone to irrationalities at another level, leading to an overconsumption and overprescription of medicines and treatments. A positive development which can be observed in this group is a growing utilization rate of preventive health measures including e.g. regular health checkups, which are increasingly offered by private hospitals in Pune. Regular medical consultation increases the chance of an early detection and diagnosis of chronic diseases, such as diabetes or high blood pressure, and therefore improves therapeutic success.

However, this group is not safe from infectious diseases due to the complex pathways of disease transmissions in a highly fragmented society. Though, in contrast to people of a lower socio-economic status, they can afford preventive measures such as filtering systems for safe drinking water supply, mosquito nets or vaccinations. Further, access to adequate treatment averts or mitigates long-term consequences for their health. In contrast to members of the middle class, they are not forced to take a high financial risk. This shows that even the higher socio-economic strata are exposed to various health threats, although their ability to cope is much higher, which results in an improved health status. However, even this group is affected by the still unsolved problem of high prevalence rates of infectious diseases in Pune. A comparison of the six different research areas illustrates that Pune's social fragmentation is reflected in rising health inequities, which indirectly affect the whole urban population.

Several studies have revealed that in less egalitarian societies the lower socio-economic strata suffer from direct negative impacts on their health status. However, even higher income groups face worse outcomes in terms of health as compared to more equitable societies which have e.g. a welfare system to balance disparities (see e.g. Butsch and Sakdapolrak 2010). Megacities and emerging megacities are extreme social-ecological systems in which complex interdependencies and the simultaneousness of processes cause the emergence of various new (health) problems. Health inequities in such complex systems do not only influence the overall health outcome negatively. It is also very likely that new health problems emerge from these conditions as examples in the recent past have shown. The plague outbreak in Surat in 1994, for example, hit a city that had been devastated by social turmoils and a flood wave earlier and that was facing severe governance shortfalls (Sinha 2000). The SARS epidemic in 2003 became a global threat after having evolved as an urban disease in Hong Kong from where it entered the global network of metropolises (Keil and Ali 2007).

Reducing health inequities is therefore not only a philanthropic cause but also an essential matter of public health. In order to reduce health inequities in Pune, primarily infrastructure deficits related to sanitation and overall hygienic and environmental conditions in slum areas have to be improved. Access to health care services has to be increased by reducing the financial barriers, either through facilitating pooling mechanisms or by increasing the availability of public health care facilities. Further, there is a need for regulation in the health care system, such as the approbation of medical practitioners and compliance with treatment standards in order to ensure minimum treatment standards.

Awareness building is essential for health protection (e.g. on nutrition, alcohol abuse, etc.) and to enable patients to responsibly utilize medical services. A comprehensive health monitoring system has to be established to supervise epidemiological trends. This is crucial in order to adjust the health system to the population's needs. Until present, these requirements remain unmet due to a lack of steering capacity of the municipality, which lacks financial and regulative power. Non-governmental organizations have started to act as mediators between marginalized population groups and the public health care system; they also provide health care services and conduct health awareness campaigns. However, this civil society engagement cannot constantly bridge the gap in health governance.

In most low and middle income countries and in development cooperation, improving rural health is still the focus of political agendas and action plans. With the ever increasing number of urban and mega-urban dwellers a paradigm shift is urgently needed to prevent South Asia's cities from future health crises – not only the ones which are reported in the global media but also the silent day-to-day crises.

Acknowledgements This research project is carried out in close cooperation with the Institute of Environment and Education of Bharati Vidyapeeth University in Pune/India. We would especially like to warmly thank Prof. Dr. Erach Bharucha and his research team for their very valuable support and advice.

References

Bapat M (2009) Poverty lines and lives of the poor. Underestimation of urban poverty – the case of India. London

Bauer U, Bittlingmayer U, Richter M (2008) Health Inequalities. Determinanten und Mechanismen gesundheitlicher Ungleichheit. Wiesbaden

Bork T, Butsch C, Kraas F, Kroll M (2009) Megastädte: Neue Risiken für die Gesundheit. Deutsches Ärzteblatt; 106: 1877–1881

Bravemen P (1998) Monitoring Equity in Health: A policy oriented approach in low and middle income countries. Equity Initiat. Pap. No. 3. World Health Organization, Geneva

Bravemen P (2006) Health disparities and health equity: concepts and measurement. Annu. Rev. Public Health; 27:167–94

Butsch C (2008) Access to healthcare in the fragmented setting of India's fast growing agglomerations – a case study of Pune. Source; 10: 62–72

Butsch C, Sakdapolrak P (2010) Geographien von Gesundheit in Entwicklungsländern. Geographische Rundschau; 6–7: 12–17

Carter-Porkas O, Baquet C (2002) What is a "health disparity"? Public Health Reports; 117: 426–434

Coy M, Kraas F (2003) Probleme der Urbanisierung in den Entwicklungsländern. Petermanns Geographische Mitteilungen; 147: 32–41

Diekmann A (2003) Empirische Sozialforschung. Grundlagen, Methoden, Anwendungen. Hamburg

Galea S, Vlahov D (2005) Urban Health. Populations, Methods, and Practice. In: Galea S, Vlahov D, eds. Handbook of Urban Health. New York, Springer Press: 1–15

Gangolli L, Duggal R, Shukla A (2005) Review of health care in India. Mumbai

Gupte M D, Ramachandran V, Mutatkar R K (2001) Epidemiological profile of India: historical and contemporary perspectives. J. Biosci; 26: 437–64

International Institute for Population Sciences (IIPS) and Macro International (2008) National family Health Survey (NFHS-3), India, 2005–06: Maharashtra. Mumbai

Jenkins C (2003) Building better health. A handbook of behavioral change. Washington D.C.

Keil R, Ali H (2007) Governing the Sick City: Urban Governance in the Age of Emerging Infectious Disease. Antipode; 39: 846–873

Mielck A (2005) Soziale Ungleichheit und Gerechtigkeit. Einführung in die aktuelle Diskussion. Bern

Misra R, Chatterjee R, Rao S (2008) India Health Report. Oxford University Press

Mishra S, Duggal R, Lingam L, Pitre A (2008) A report on health inequities in Maharashtra. Pune

Penchansky R, Thomas JW (1981) The concept of access. Medical Care XIX (2): 127–140

Pune Municipal Corporation (PMC) (2008) Environmental status report. Pune

Richter M, Hurrelmann K (2006) Gesundheitliche Ungleichheit. Grundlagen, Probleme, Perspektiven. Wiesbaden

Sakdapolrak P (2007) Water related health risks, social vulnerability and Pierre Bourdieu. Source; 6: 50–95

Shukla A (2007) Key Public Health Challenges in India: A Social Medicine Perspective. Social Medicine; 1: 1–7

Siegel K, Narayan KM, Kinra S (2008) Finding a policy solution to India's diabetes epidemic. Health Affairs; 27: 1077–90

Sinha H (2008) Plague: A Challenge for Urban Crisis Management. Journal of Contingencies and Crisis Management; 8: 42–54

Smyth F (2008) Medical geography: understanding health inequities. Progr Hum Geogr; 32: 119–172

State Bureau of Health Intelligence and Statistics and Vital Statistics, SBHI and VS. (not dated) Annual Report 2003 and 2004. Pune

The Hindu (2009) Swine flu claims its 91st victim in Pune. http://www.thehindu.com/news/national/article38581.ece. 26.10.2009

UN (2010) World Urbanizations Prospects: The 2009 Revision, New York (Accesses June 15, 2010 at: http://esa.un.org/unpd/wup/Documents/WUP2009_Highlights_Final.pdf)

Whitehead M (1991) The concepts and principles of equity and health. Health Promotion International; 6: 217–28

WHO CSDH (2008) Closing the gap in a generation. Geneva

World Bank (2002) Better Health Systems for India's Poor. Washington

Part VI
Urban Livelihoods, Urban Food and Health

Chapter 18
Urban Development and Public Health in Dhaka, Bangladesh

Sabine Baumgart, Kirsten Hackenbroch, Shahadat Hossain, and Volker Kreibich

18.1 Public Health as an Objective of Urban Planning

The history of urban development and the advance of planning as a tool for guiding urban growth have always been closely linked with public health issues. Recognising the links between a city's urban layout and its public health status has in the past often been the beginning of the establishment of planning principles, whether in Mesopotamia to improve air circulation, in medieval times to protect cities from fire hazards, or in the industrial age to fight epidemics such as cholera by adhering to sanitary guidelines. The interdependencies between socio-spatial development and the public health status of urban settlements are especially important nowadays in the fast growing cities of developing and transforming regions, where city authorities often have problems accommodating rapidly growing populations and an expanding economy in such a way that public health risks are minimised. In light of the severity of public health risks, especially in growing cities, the WHO (http://www.who.int) in 2010 launched the campaign 'Urban planning as a critical link to building a healthy 21st century'. The campaign emphasises promotion of urban planning, improvement of urban living conditions, and participatory governance so as to enhance the resilience of cities to disaster and provide a better urban living environment (ibid.). The megacity Dhaka, the subject of our research, is a striking case showing the deterioration of the urban environment under uncontrolled and rapid urbanisation with detrimental effects on public health.

This paper starts by taking a historical perspective, explaining how public health issues have influenced urban development and provided one of the starting points for the regulatory urban planning framework of today. Subsequently, the paper aims

S. Baumgart (✉) • K. Hackenbroch • S. Hossain
Department of Urban and Regional Planning, Faculty of Spatial Planning, TU Dortmund University, Dortmund, Germany
e-mail: sabine.baumgart@udo.edu

V. Kreibich
Faculty of Spatial Planning, TU Dortmund University, Dortmund, Germany

A. Krämer et al. (eds.), *Health in Megacities and Urban Areas*,
Contributions to Statistics, DOI 10.1007/978-3-7908-2733-0_18,

to present the challenges for planning and public health that are found in contemporary megacities growing under conditions of poverty and with failed urban planning institutions. Based on empirical findings from four years of research in Dhaka, the paper seeks to explain the importance of the spatial set-up of large urban agglomerations for public health. The linkages between spatial development and public health will be analysed on two different levels: the planning regulations for Dhaka City and their implications for public health, and local experiences with public health deficits that apparently result from strategic planning and management shortcomings in two urban low-income settlements.

18.2 The Issue of Public Health in Urban Planning: a Brief Historical Reading

Public health issues are not a new phenomenon in urban planning. They were important goals for urban planning and design even in ancient and medieval towns. In Mesopotamia's (3000 BC) capital Ur the Sumerians cooled and ventilated the city by aligning its main axis according to the prevailing wind direction. In the first century BC Vitruvius, in his first of ten books on architecture, laid down climate oriented rules intended to provide healthy settlements. Some decades later the medical scientist Hippocrates (450–350 BC) emphasised public health goals by demanding restrictions on urban growth in his 'Essay on air, water and locality'. The lay-out of ancient towns in Greece was based on a holistic planning concept known today as classic urban design (Mumford 1980). In European medieval towns (in about the twelfth/thirteenth century), where strong walls acted as defined boundaries against dangers in the open landscape, the first statutory building codes to protect against fire and regulate tannery sites were established.

The issue of public health and the wellbeing of city dwellers were the most important considerations of the *Vastu Shastra* principle, a traditional Hindu system of design based on directional alignment, which was applied when planning Jaipur city in India in 1727. The planning idea was to take advantage of a balanced relationship of the five elements of nature (land, water, air, fire and sun) and to create a conjugal living and working environment to ensure wellbeing and enhance health, wealth, prosperity and happiness (Shastri 1996; Puri 1997).

In Europe, industrial development from the early nineteenth century onwards saw increasing building and population densities in the inner cities and increasing threats to the public health of the inhabitants. The first Public Health Act of 1848 led to spatial planning regulations aimed at the renewal of slums in order to improve the environmental conditions of settlements in the European period of industrialisation. These objectives were considered by a newly established local board of health and thus institutionally rooted. It was Patrick Geddes, a Scottish biologist (1854–1932), who became an important catalyst for the development of an integrated perspective on health and spatial aspects not only in Europe but also in Asia, including Dhaka. He criticised the disregard of architecture and housing for basic hygienic needs in the nineteenth century, when rents and revenues were the only criteria for urban design (Mumford 1980).The renewal of substandard housing to form healthy neighbourhoods was introduced as a principle objective of planning during these decades (Platt 2007).

The policy, strongly supported by English reformers, led to improvements in hygiene through trunk infrastructure – gas, drainage, water – and regulations for the use of public land as well as the establishment of social infrastructure such as hospitals, boarding houses and public green spaces. Civil engineers like George Eugene Haussmann in Paris (1853/70), Ildefonso Cerda in Barcelona (1859) or James Hobrecht in Berlin (1862) played leading roles in introducing modern land use planning with public hygiene based on trunk infrastructure, mobility based on railways, and fire protection through building regulations. They set restrictions for building heights and standards for high quality construction to counter land speculation. The location and design of public green spaces became important to compensate for the loss of private gardens (Grassnick and Hofrichter 1982).

In 1898 Ebenezer Howard published his diagram of a garden city as a model linking the benefits of the urban with the rural landscape and guiding growth and the location of public infrastructure. The rather paternalistic concept focussed on the welfare of industrial workers. Maximum population size of a city was based on health considerations and restricted to 30,000 inhabitants (Hotzan 1994).

Two decades later the local planning concepts were extended across municipal boundaries. Core goals of regional planning focussed on a wide range of problems such as increasing transport and traffic, housing for low income families, and the safeguarding of open space and its accessibility. The 1933 Congrès Internationaux d'Architecture Moderne (CIAM) in Athens, Greece, created the idea of a 'functional city' to address overcrowding and unhygienic conditions in industrialised cities. Proposals for improvement were based on strict zoning for different functions such as housing, working, leisure, and transport, separated by green belts. This contributed to the establishment of a normative base for planning regulations in the north-western hemisphere with (i) building codes for security such as fire protection, spaces between buildings and public roads for prevention of hazards; (ii) planning regulations for land use and land tenure to safeguard the climate, restrict emissions from obvious pollutants and facilitate re-use of existing buildings to restrict the growth of cities (Mumford 1980; Grassnick and Hofrichter 1982; Benevolo 1984).

In 1948 Le Corbusier applied the modern European city planning principle of CIAM to the foundation of Chandigarh in India. The new city of Chandigarh was to be located between two rivers to cope with the summer heat, and the planning of the city was guided by the need for an adequate supply of water, easy drainage and a suitable climate. To ensure these urban necessities and thus public health, Le Corbusier introduced a division of urban functions (zoning), an anthropocentric layout and a hierarchy of road and pedestrian networks. Through basing his ideas on the CIAM planning principle, he wanted to make the city different from existing Indian towns that suffered from overcrowding and unhygienic conditions (Prakash 2002; Von Moos 2010).

Public health was also the most important issue that took Delhi through three distinct phases of city planning (Priya 1993:824). In 1912, a team of British town planners introduced the garden city concept to symbolise British imperial power in colonised India. New Delhi was planned with wide roads, green spaces and houses for government officials, and was separated from the old city by vast stretches of land. Lack of attention to the problems of the old city resulted in its gradual transformation into a large slum area through deterioration and dilapidation with huge threats to public health (ibid.). In 1937 the Delhi Improvement Trust was therefore founded to address the issue of public health with a focus on the renovation of old Delhi. Following the partition of India and in response to the dramatic increase of population, the Delhi Development Authority was formed in 1957 to prepare a Master Plan, ensure guided development and thus improve public health and functionality of the city.

Similarly, Patrick Geddes, a friend of the Bengali poet and Nobel Prize winner Rabindranath Tagore, developed the first Town Plan for Dhaka in 1917. It was based on the garden city concept and covered important urban planning issues like the geographical and social context, a survey of the city, provision for housing, open space, industrial development, the introduction of canals as local transportation routes and the issue of public health (Geddes 1990). The municipality, however, paid no attention to this important document, thus paving the way for slum-like living conditions with overcrowding, congestion and severe pollution. The Architects and Town Planning Consultants of London that were commissioned with the preparation of a Master Plan for Dhaka in 1960 described the appalling environmental and hygienic conditions:

"Shops, commercial premises, warehouses, workshops and small factories are often intermixed with houses; narrow and tortuous streets, never designed for motor cars and buses, are so overcrowded that traffic is seriously delayed, and to be a pedestrian is a precarious experience" (DIT 1959:2).

Initiatives to improve public health in Asia, particularly in the Indian subcontinent, have informed urban planning - the influence of the British planning system is still evident today – but they could not eradicate public health risks. In Asian cities, the plans, prepared mostly by foreign consultants, were often difficult to implement in the local context. In explaining the 'ill-planned urbanization', Fendall (1963:569) indicated the problem precisely by stating that urban planning practice in the megacities is largely based on "preconceived ideas or the transposition of planning methods" and that it fails to consider "the political, economic, demographic, and cultural factors of the region in which urbanization is occurring". Such planning practice rarely meets the needs of the urban majority. Rather it disrupts the physical state of the urban environment, crudely disturbing the living and working conditions of the people, disorganising their access to facilities and thus threatening their health, well-being and quality of life.

18.3 Public Health as a Challenge to Urban Growth Under Conditions of Poverty

Especially in Africa and Asia urban growth rates continue to be high: all new megacities projected to emerge by 2025 will be located either in Africa or in Asia (UN-Habitat 2009:6). Large cities in the developed countries experienced a steady urbanisation process until reconstruction and development after World War II. However, it was only after the Second World War that all large urban centres of the global South first embarked on their urbanisation process that then proceeded very rapidly within a few decades. In poor countries, this process can be characterised as urbanisation under conditions of poverty. For example, the mega-city of Dhaka experienced growth rates of up to 10% during the 1960s and 1970s, until in the early 2000s the growth rate slowed down to 3% (Islam 2005:12–14).

Urbanisation under conditions of poverty stems to a large extent from rural–urban migrants coming to the cities in search of income opportunities and social services, including health care facilities. City governments are seldom prepared to cater for high levels of immigration, with the consequences of exorbitant residential densities, haphazard settlement layouts and the occupation of hazardous sites for residential purposes.

While urban population growth and the underlying consequences invariably affect public health outcomes there are two sides of the coin. On the one hand, inhabitants are actively involved in creating public health risks by increasing densities in existing settlements. For example, the average residential density in the urban core of Shanghai is 26,000 persons per km^2, compared to the London urban core with 4,800 persons per km^2 (Burdett and Sudjic 2007:198–199). The average residential floor area per capita can also serve as an indicator for healthy living conditions: about 45 m^2 per person in Germany compared to 16 m^2 per person in Shanghai (ibid.); in Dhaka it amounts to only 2.3 m^2 per person (Nazem 2007).

On the other hand, these public health risks are willingly accepted by urban dwellers, because the option of not migrating to the urban centres is no alternative in terms of livelihood outcomes. Thus in order to achieve a desirable livelihood outcome, public health is sacrificed through accepting sub-standard and unhealthy living and working conditions in an urban agglomeration. A case in point is Keko Mwanga – a saturated informal settlement in Dar es Salaam, Tanzania which in 2002 had 17,000 inhabitants on an area of 34 ha (Sheuya 2004:69). Kombe and Kreibich (2000:90) assessed the settlement and concluded that it provides the poor with walking access to the city centre in exchange for crowded housing, highly deficient services, environmental deficits and public health hazards. A survey in two urban villages in the Chinese city of Ningbo found that due to high densities – one of the urban villages has an estimated density of over 400,000 inhabitants per km^2 – dwellings became multifunctional spaces for living, sleeping, storing and cooking, and 55% of inhabitants, mostly the women, even had to take a bath in their bedroom using a water container (Changqing et al. 2007:37). Furthermore, mental stress caused by the threat of eviction is often accepted in order to live in proximity to jobs and services and thus to reduce opportunity costs (Payne 2002; Hackenbroch et al. 2008).

Increasing disparities and fragmentation in megacities pose considerable public health risks to those who cannot afford to move to 'low-risk areas'. The huge gap between rich and poor population groups leads to socio-spatial segregation expressed in land use, zoning and urban design practices and indicated by density, available open space, social and technical infrastructure supply, and – last but not least – by land prices (Burdett and Sudjic 2007:190). The middle and high income groups tend to leave behind the health-threatening city, moving to the suburbs where the real estate market happily provides them with condominiums and gated communities. Here public health threats are low, even though these developments often do not correspond with sustainable planning strategies and thus lead to an accumulation of public health risks in places exposed to environmental risks.

The morphology of informal settlements of low income households has developed without consideration of security regulations, prevention of accidents, exposure to pollution and land tenure. These deficits are aggravated by the fact that neither inner city gentrification processes nor urban sprawl are under the control of a functioning institutional authority at the local and regional level. Rather, large parts of the cities are facing a process of institutional 'unmapping' (Roy 2003) and are no longer considered as part of planning processes, except in cases of demolition. For example in Mumbai "many middle-class neighbourhood organisations [...] interpret the sanitising of urban space through a logic of demolition rather than one of improvement of informal settlements" (McFarlane 2008:92).

This raises questions about the notions of citizenship that are employed, indicating that the urban poor's 'right to the city' is grossly disregarded. Slum and squatter citizens constitute a considerable amount of the population of megacities but adequate housing and utility provision that would considerably minimise public health risks for these citizens is often not high on the agenda of city governments.

18.4 Dhaka: Spatial Organisation, Urban Livelihoods and Their Impact on Public Health

With today about 12 million inhabitants and 21 million inhabitants projected for 2020, Dhaka Metropolitan Area is one of the world's largest cities. Almost 40% of the population (3.4 million) of the smaller administrative unit of Dhaka City Cooperation with 9 million inhabitants live in 5,000 slum and squatter settlements/clusters (CUS 2006). With an estimated current annual growth rate of 3% (Islam 2005:14), providing shelter and adequate urban utilities for poor rural–urban migrants has been and still is an enormous challenge in Dhaka. Additionally, many settlements of the middle and high income groups are undergoing rapid transformation and densification processes or are being newly developed in flood retention areas; both processes put the functionality of the city and its neighbourhoods at risk. The resulting deficient spatial organisation, combined with the need for urban livelihoods and often dangerously mixed land uses, impacts greatly on public health. Increasing densities in the inner city areas and uncontrolled growth in the periphery result in multifarious problems in the city like poor drainage and sanitation, regular flooding after rainfall, high levels of water and air pollution, the development of heat islands, inadequate urban utilities, and the deterioration of law and order, adding significant threats to public health in the city.

18.4.1 Planning Control in Dhaka City and its Implications for Public Health

In Dhaka city planning exercises are frequently administered in violation of the legal planning documents. Plans prepared to consider important public health issues often bring little improvement in the urban environment because of their violation in practice. The Dhaka Improvement Trust (DIT) was founded in 1956 under the provision of the 'Town Improvement Act 1953' and with the authority to improve the physical and urban conditions of Dhaka. It prepared the Dhaka Master Plan in 1959 as a guiding document for development control and land management in the city, intended to accommodate an increasing population and maintain or improve living conditions. The plan specified areas for different uses (zones), proposed the extension of the road network and assigned areas for housing, open spaces, urban facilities and industrial development. Land was to be made available mainly through the clearance of slum settlements in Old Dhaka, the filling of inner city low-level land and canals, and the extension of the city primarily towards the north (DIT 1959).

However, a major part of the plan was not implemented due to a number of difficulties. Most importantly, the specific needs of the urban poor were not reflected in the planning documents, especially in Old Dhaka where mixed use inside the buildings and open spaces has a historical relevance and tradition. It therefore became difficult to relocate the inhabitants of Old Dhaka as proposed in the plan and to generate the space required for planned facilities. Moreover, due to the absence of any in-situ development proposal, there was no improvement to the slum-like living conditions in the area.

A major focus of the plan was on improving the transportation system through upgrading only the surface road network. The statement that "there is little evidence that this waterway is now of any appreciable commercial value to abutting premises, except for the transport of timber" (DIT 1959:7) demonstrates the plan's negligence of the potentials of waterways in Dhaka and its interest in filling existing canals to create new building land. Furthermore, the River Buriganga was described only in relation to its function for the surrounding industries and no importance was attributed to water related pollution control and air ventilation in the city.

The plan suggested accommodation of the increasing population of the city on the proposed land developed through land filling in the low contour areas. DIT was given the overall authority for housing and especially land management including development control, while the implementation responsibilities were kept open for organisations with qualified staff. The plan thus provided scope for private sector involvement in housing for the city. However, by the end of the 1980s living conditions in the city had deteriorated considerably due to problems with traffic, drainage and water supply. At that time *Rajdhani Unnayan Kartipakkha* (RAJUK, Capital Development Authority) emerged as a new governmental institution with a view to improving the urban environment through urban planning, land development and building control in the city and its periphery (www.rajukdhaka.gov.bd).

In 1995, RAJUK prepared the three-tier Dhaka Metropolitan Development Plan (DMDP) to provide for guided urban development for the period up to 2015 with a projected population of 16 million (RAJUK 1997). While the Structure Plan and the Urban Area Plan of the DMDP became legal documents on 3rd August 1997, the preparation of the Detailed Area Plans (DAP) was only completed through governmental gazette notification on 23rd June 2010 following huge pressure from groups of professionals and environmentalists (The Daily Star 24.06.2010). DMDP re-categorised land into different uses. This involved, for instance, the preservation of surrounding wetlands and low-lying land for flood retention and agricultural uses, areas where development is prohibited in order to allow urban facilities like rainwater drainage to function properly (RAJUK 1997). Difficulties, especially in controlling unplanned development in prohibited areas (flood flow zones and demarcated agricultural areas) and urban growth in violation of DMDP guidelines, may have been exacerbated by the long delay in plan preparation. Often there were demonstrations and press-briefings by urban professional groups trying to force the government to implement the DAP immediately and thus protect the wetlands from further land development (Photo 18.1).

One of the major developments in recent years is the direct involvement of RAJUK in the implementation of housing projects in addition to development control. Like housing and land development projects run by private organisations, a number of government projects have been developed in violation of DMDP guidelines and thus contribute negatively to the city environment. Of the 17 housing and land development projects found to be located in flood flow zones, harming the urban environment and thus illegal according to the DMDP guidelines, a Review Committee formed by the Ministry of Housing and Public Works found that two projects were being implemented by the development control authority RAJUK itself (The Daily Star 24.06.2010). The committee also recommended relocation of at least five other government projects (e.g. terminals and waste dumping zones) away from DMDP specified flood flow zones (ibid.).

Although different departments, agencies and ministries at the national level are responsible for dealing with public health issues, weak interdepartmental cooperation on the local level seems to be one of the major reasons for the urban

Photo 18.1 A demonstration calling for immediate implementation of DAP to save flood flow zones and wetlands of Dhaka (*Source*: The Daily Star 17.07.2010)

problems Dhaka is currently facing. As many as 19 ministries and 40 government organisations are involved in the planning and development of Dhaka with practically no coordination between them (Islam et al. 2000; Siddiqui et al. 2000). Besides the failure to protect the wetlands, RAJUK, obviously often in violation of existing regulations, is allowing increase in floor space and non-conforming land uses in many formerly planned settlements like Dhanmondi and Gulshan. The assumption has to be that the approval process includes no adequate consideration of available facilities like urban utilities, transport and air ventilation or the issue of heat islands in the city. The semi-autonomous utility authorities (Dhaka Water Supply and Sewerage Authority, Dhaka Electricity Distribution Company and Gas Distribution Company) also extend their services to settlements developed in violation to DMDP, thus quasi-legalising them.

18.4.2 Local Experiences: Public Health and Urban Settlements

The structural properties of urban neighbourhoods and their effects on public health in the absence of planning has combined with the haphazard transformation of formerly planned areas lead to unhealthy living conditions and threats to public health in many of Dhaka's settlements. Based on research in two consolidated low-income settlements, Korail and Islambag, and on a newspaper review of relevant events in other areas of Dhaka, discussion now turns to the effects of high density and lack of accessibility, low-quality building structures, zoning and mixed land uses, and tenure insecurity on public health.

Islambag, a city ward that had about 60,000 inhabitants in 2001 (BBS 2007:123), is located next to Old Dhaka, the oldest part of the city which developed along the shores of the Buriganga River. Housing development started here in the mid 1970s; today it is a consolidated settlement characterised by high building density with many multi-storey buildings and an infrastructure that was designed for considerably lower densities. Industrial plastic recycling provides the economic base of the area. The socio-economic structure is rather heterogeneous, including low-income households who labour in the plastic recycling industry and lower middle and middle-income households who are engaged as businessmen in the same industry (Hackenbroch et al. 2009:54).

Korail is a squatter settlement which started to expand in the 1990s. It is located in proximity to high-income neighbourhoods that developed during the rapid expansion of the city from the 1960s. Korail is currently the largest slum in Dhaka with an estimated 110,000 inhabitants living mostly in one-storey tin houses in an area of 60 ha (CUS 2006): a density of 180,000 inhabitants per km^2. The majority of households belong to low-income groups and are mostly engaged in providing services to the surrounding high-income areas, while a few – often those involved in local committees – belong to the lower-middle income group (Hackenbroch et al. 2009:58; Hossain 2010:615–616).

18.4.2.1 High Density and Lack of Accessibility

According to the DAP, the aim of RAJUK is to produce living spaces with maximum densities of 62,000–87,000 inhabitants per km^2 (RAJUK 2010:89, 97; two different maximum densities are found in the same document, hence the range of maximum densities). According to the Centre for Urban Studies some of the Dhaka slum and squatter settlements reach densities of up to 500,000 inhabitants per km^2, while the average density in these areas amounts to 220,000 inhabitants per km^2 (CUS 2006:40). Islambag, on a 35 ha area, has an average density of 170,000 inhabitants per km^2 with an inadequate street and footpath network in relation to these densities. Only 10% of the study area comprises open spaces, i.e. streets, footpaths and squares (authors' fieldwork in 2007). Most of the internal roads are hardly accessible by car due to their limited width, thus prohibiting access for ambulances or fire fighting vehicles in case of emergencies. Accessibility is further hampered by the extension of livelihood activities into street spaces, for example through storing goods, extending workshop premises or setting up mobile or semi-mobile stores (Hackenbroch et al. 2009:56, see Photo 18.2). While these activities reduce accessibility, they are but a necessary reaction to the dysfunctional urban structure. Due to the lack of adequate space indoors or outdoors, streets and other open spaces are blocked in the pursuance of livelihoods. If an adequate layout had been implemented and densities had been controlled, it could perhaps be possible for today's inhabitants to perform their livelihood activities without hampering the functionality of the settlement. Blockage of the street network and especially of the drainage system results especially in the monsoon period in increased water logging and increased risk of water-borne diseases.

Photo 18.2 Narrow pathways in Islambag used as storage space by the plastic recycling industry, prohibiting vehicular movement (Picture: K. Hackenbroch, July 2009)

High densities and the economic necessities of poor households force inhabitants to adapt to a new concept of privacy and publicness, especially with regard to gender issues. Traditionally it is not common for women to work in places with high public visibility, as reflected in the following statement by a woman who considers herself as living in a middle-income household in relation to the range of socio-economic profiles in Islambag:

"If the work can be done inside the house, I do it. But I do not go outside. It is a matter of shame for us. We have never worked outside. Now if we go outside, we will lose our prestige. That is why we accept hardship but still remain inside the house" (focus group discussion, 03.04.2010).

On the other hand, women in households that are less economically well-off or stable in their livelihoods increasingly use urban public space for selling goods, crushing brick stones or sorting plastic waste. One resident even commented on the picture of a woman cooking in a street in a slum area:

"She is cooking in a dirty place. After preparing *ruti* [bread] and cooking rice she keeps these along the street side. She also allows her children to eat there, in the street. [. . .] She is very poor and living in great misery" (interview with participant of solicited photography, 08.04.2008).

Between the two study settlements perceptions of publicness seemed to differ, indicated by the formality of the culturally determined dress-codes. While the internal area of Korail, the squatter settlement, seemed to be perceived as a semi-public entity, the outdoor spaces of Islambag were considered to be public from the doorstep. The differences between the settlements can be attributed to the different socio-economic background of their inhabitants as well as to the higher integration of Islambag into the city context. It is, however, very difficult to assess what performing livelihood activities in public spaces implies for women's physical and mental health.

18.4.2.2 Low-Quality Building Structures

All over Dhaka the lack of control of building codes and the highly dynamic development of informal settlements, especially slums and squatter settlements, have lead to building structures having low ventilation and insufficient daylight. The lack of both building regulations and their enforcement encourages the development of high rise buildings as the transformation process in the inner city areas from one-storey to multi-storey buildings progresses. Construction of buildings is often poor with regard to structural soundness or the use of materials that withstand flooding and earthquakes. The consequences of the lack of enforcement of building codes and regulations became obvious once again on 1st June 2010 when in the Tejgaon area a five-storey-building collapsed on nearby tin-shed housing causing the death of 20 people. An article in The Daily Star commented:

"The entire city is virtually swarming with buildings crossing the six-storey mark in even the most cramped of areas. Majority of these buildings were either constructed without any authorisation form [sic] Rajuk or are in total violation of the proper construction rules causing repeated incidents of collapses" (The Daily Star 04.06.2010).

The editor also calls into question the role of RAJUK "in ensuring that proper building rules and codes are followed by builders, contractors and owners of the buildings in the city" (The Daily Star 03.06.2010). The collapsed building had been constructed in 2001 as a three-storey, it was just one week before the collapse that the fifth storey was added. Only a few days later another six-storey building tilted dangerously to one side in the same area (Photo 18.3, The Daily Star 04.06.2010).

Photo 18.3 Mix of different building structures, Islambag (Picture: K. Hackenbroch, March 2010)

18.4.2.3 Zoning and Mixed Land Uses: Residential Quarters and Hazardous Economic Activities

Due to pressure from up-market real estate developers to develop all available land for housing and economic activities, river banks, canals and flood plains are being transformed mostly through land grabbing and filling. Consequently, the land available for housing projects of low and middle-income households has been considerably reduced. Grabbing of water lands also occurs every day on a smaller scale. All these processes go on largely unabated, despite the fact that they represent a violation of the DAP (see above). The inefficiency of land use regulations enables private middlemen to acquire and occupy reserved land through informal arrangements with residents who have no alternative if they wish to secure living space in the city. The intervention of the authorities, RAJUK and Bangladesh Inland Water Transport Authority (BIWTA), is sporadic and depends on public opinion and the power constellations of the encroachers. The 'Save rivers, save Dhaka' campaign launched by the local media in May 2009 caused BIWTA to conduct an eviction drive to free river banks from encroachment (The Daily Star 01.06.2009). Sustainable solutions to prevent land encroachment have thus not yet been found. Furthermore, it is often only the urban poor and their informal settlements that are blamed for the situation and thus that suffer most from encroachment-grounded evictions. In contrast, government authorities tend to generously overlook encroachments conducted by developers and higher income groups.

There is also a lack of approved land use regulations with regard to mixing only compatible land uses that do not lead to increased public health risks. The dangers stemming from unregulated mixed land uses can be seen all over the city. In Islambag the problem is the integration of the plastic recycling industry into a residential area (Hackenbroch et al. 2009), which causes very high air pollution and severe public health risks to the residents (Burkart and Endlicher 2009:100–102). In Hazaribag it is the leather industry that causes inhabitants to suffer from air and water pollution resulting in skin and respiratory diseases, and that also pollutes the River Buriganga for residents downstream. Although relocation of the leather industries has been under discussion for about 10 years, nothing has so far been done to disentangle this highly dangerous mix of land uses (Sharif and Mainuddin 2003:10). The hazards from unregulated mixed land uses became tragically obvious when a fire broke out in Nimtoli, Old Dhaka, and killed about 120 people on 3rd June 2010. The fire spread fast due to chemicals that were being stored in the ground floors of the otherwise residential buildings. The danger this poses to public health quickly became part of public discussion: "Thousands of Old Dhaka residents live close to grave danger as many warehouses store inflammable substances and industries use these in residential areas in violation of environmental rules" (Photo 18.4, The Daily Star 08.06.2010).

Unregulated mixed land uses do not only pose a threat to residents of lower middle-income or slum and squatter settlements, as can be seen by the effects generated by the uncontrolled concentration of educational institutions in a few areas of Dhaka, for example in the high income area of Dhanmondi. The rapid increase of educational institutions as well as medical facilities (Mahabub-Un-Nabi and Hashem 2007:36), combined with increased car ownership in the middle and high-income groups and roads designed for a primarily residential area, causes traffic jams making Dhanmondi rather inaccessible at the start and end of the school day and contributing to air pollution.

Photo 18.4 Burning of remnants from the plastic recycling industry in narrow pathways with mixed residential and industrial land uses, Islambag (Picture: K. Hackenbroch, February 2010)

18.4.2.4 Tenure Insecurity of Informal Settlements

Most of the slum and squatter settlements identified by the Centre for Urban Studies lack security of tenure. Despite various commitments by government agencies and initiatives by non-government organisations, evictions of slum dwellers continue to be frequent (Hackenbroch et al. 2008). Insecure tenure and the fear of forced eviction result in considerable mental stress for residents, in addition to their often already deprived economic and environmental situation. The following quote indicates the health risks residents have to deal with after eviction:

"We are living in a temporary house and sometimes in the rain and cold it was really hard to survive. The sanitation facilities of the area are horrible, causing health problems. All our good toilets were demolished during the eviction" (interview, June 2008, with a female resident who was evicted in January 2008 by RAJUK because of a planned infrastructure project. She resettled on the same land due to a lack of alternatives close to her work place).

While a considerable number of settlements violate the strategic planning guidelines of the Dhaka Master Plan and DAP, evictions most often threaten the livelihoods of the urban poor rather than the encroaching developers working for middle and high-income groups. Indeed, evictions are carried out for the benefit of the latter groups, whether they are undertaken as 'urban cleansing' of prime housing sites (Ghafur 2008) or with the broader aim of 'city beautification'.

18.4.2.5 Limited and Expensive Access to Utilities: the Example of Water Supply

Urban utility supply in Dhaka is limited and expensive. While there is a huge deficit of utility supply in the existing settlements, provisions are extended only selectively to the growing parts of the city. The water supply by the Dhaka Water Supply and Sewerage Authority (DWASA) is limited by its dependency on ground water extraction as the primary source (85% of the total production). The supply in the whole city is therefore often unreliable, limited and varying in quality. A serious shortage of water occurs in the hot season when the electricity supply is erratic and water fails to reach most of the houses in the city at the normal delivery time, it may even be unavailable for a whole day. Frequent replacement of household connections, illegal tapping and unofficial negotiations with DWASA staff are a few of the many strategies the inhabitants use to cope with the water shortage situation (Hossain 2011a). The consequence of such informal practices is a drastic reduction in both the quantity and quality of supply. The shortage of safe water thus poses a serious threat to public health, whether in terms of a lack of drinking water, water for cooking, or water for bodily hygiene.

DWASA does not officially supply piped water to the informal settlements of Dhaka where more than one third of the total population live (see above). Though a few non-government organisations are active in supplying water through community groups and negotiation with DWASA staff, their services are very limited and selective, helping in only a few settlements. The need for water in informal settlements is met by local water vendors under very different conditions and at a very different level of expense. Our study in Korail found that about 45 water vendors were involved in supplying water to about 7,700 households in the settlement (Hossain 2010). The water vendors maintain unofficial negotiations with DWASA field staff to enable illegal taping of water from DWASA main lines and follow a number of different strategies using various affiliations to permit the

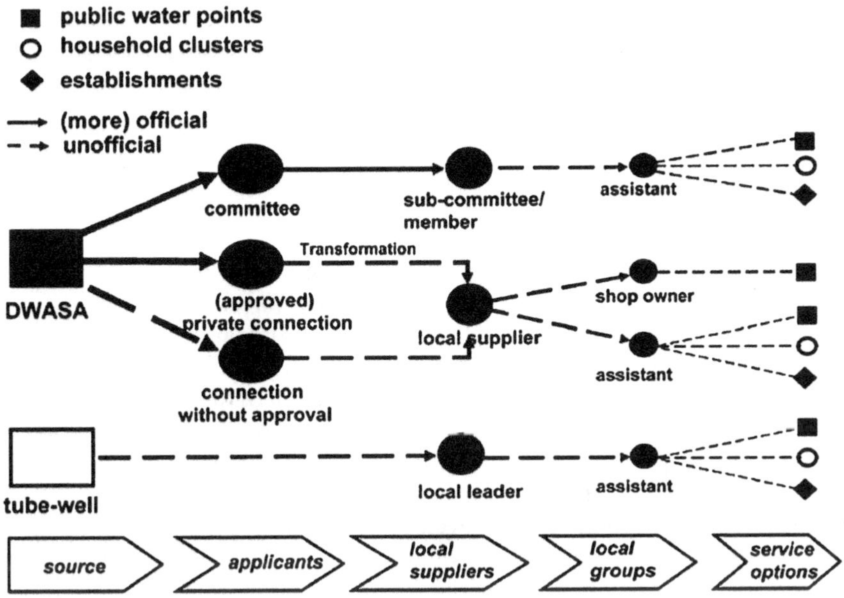

Fig. 18.1 Actors and institutions in water supply in Korail (*Source*: Hossain 2010)

uninterrupted operation of their businesses (Hossain 2010, 2011b). Figure 18.1 presents the actors and institutions involved in water supply in Korail.

Though the diagram shows multiple water access options for the inhabitants of Korail, the supply is very expensive and limited. The water vendors make water available through illegal tapping. This involves connections being frequently displaced during official monitoring and the employment of unskilled technicians and improper materials, like unsafe rubber pipes being used to collect water from the other side of a lake. The quality of water deteriorates greatly as it moves through the rubber pipes.

Though our study found that 85% of all Korail households have access to a water supply in their housing compounds, only one third of the total compounds have in-house water reservoirs (authors' field work in 2009). The remaining households either store water in pots or other available containers, or carry water from water vending points to meet their additional daily water needs. The maintenance of reservoirs is poor, mainly due to the financial crisis and negligence by house owners, thus further increasing threats to public health due to the use of contaminated water. Insufficient control and monitoring of safe drinking water is evident in urban as in rural areas in Bangladesh (Khan et al. 2007).

Due to the restricted water supply the price of water in Korail is about 10–15 times higher than the unit price of DWASA piped water (Hossain 2010). This creates a situation where inhabitants are forced to limit their access to water to an average duration of an hour per room-cluster inhabited by eight to ten households.

As they are unable to increase consumption of the expensive water, the inhabitants sacrifice their minimum water needs for domestic use and bodily hygiene.

18.5 Conclusions

Urban planning for Dhaka, whether the DIT Master Plan or DMDP, considers planning to be merely a set of technical guidelines for development control. It thus fails to integrate a public health framework into its urban planning strategy and has developed no appropriate planning guidelines with which to address increasing urban problems including public health. Transformation of the inner city as well as peripheral development therefore progresses without consideration or control of the availability and capacity of infrastructures and utilities, creating large threats to public health. Severe difficulties are faced when trying to impose planning guidelines, e.g. corruption in the development authority, political support and protection of violators, and unwillingness on the part of residents to bear the costs and difficulties inherent in adhering to planning provisions.

The above findings show the need for the implementation of appropriate spatial and environmental planning instruments to improve an urban environment that is currently characterised by high densities, a lack of land use regulations, social and spatial exclusion and a lack of adequate infrastructure. Further research is required, particularly to focus on finding a proper balance between financial expenditure aiming to develop healthy urban environments and consequences for affordability and access after implementation. Upgrading processes in these fields can lead to gentrification and segregation, widening socio-spatial disparities.

However, planning cannot be successful in achieving positive public health outcomes if it is only understood as a technical exercise of urban design – as became evident when the planning approaches formulated by the Charta of Athens and implemented in a top-down approach did not produce the desired outcomes on the ground. 'Planning the city' is not only a technocratic task, it should be informed by the needs of all citizens. Successful consideration of public health impacts, for example via environmental assessments of strategic planning documents, requires not only an appropriate set of indicators. It also demands the early participation of public and private institutions and stakeholders at every stage of the planning process, as well as continuous participation throughout the coordination and implementation phases. The consideration of the context specific power structure and the socio-political culture is of utmost importance as they condition participation and thus largely determine the success of any participatory planning process in an urban environment defined by deep difference and diversity. While cross-sectoral planning integrates public health issues, policies on public health have to equally address the socio-spatial setting, the securing of households' livelihoods and the related governance framework.

References

Bangladesh Bureau of Statistics (BBS) (2007) Population Census-2001. Community Series, Zila: Dhaka. Dhaka

Benevolo L (1984) Die Geschichte der Stadt. Campus Verlag, Frankfurt New York

Burdett R, Sudjic D (2007) The endless city. London. The urban age project by the London School of Economics and Deutsche Bank's Alfred Herrhausen Society. Phaidon, London

Burkart K, Endlicher W (2009) Assessing the Atmospheric Impact on Public Health in the Megacity of Dhaka, Bangladesh. Die Erde 140:93–109

Center for Urban Studies (CUS) (2006) MEASURE Elevation; National Institute of Population Research and Training (2006): Slums of Urban Bangladesh: Mapping and Census, 2005. Dhaka

Changqing Q, Kreibich V, Baumgart S (2007) Informal Elements in Urban Growth Regulation in China – Urban Villages in Ningbo. ASIEN 103:23–44

Dhaka Improvement Trust (DIT) (1959) Master Plan for Dacca. Dhaka

Fendall NRE (1963) Public Health and Urbanization in Africa. Public Health Reports 78 (7):569–584

Geddes P (1990) Dhaka Nogor Unnayan Parikalpana 1917 (Report on Dhaka City Planning 1917) translated by Abdul Mohaimen, Dhaka: Dhaka City Museum

Ghafur S (2008) Spectre of product fetishism. Reviewing housing development proposal for Dhaka. In: The Daily Star (16.8.2008):1–7

Grassnick M, Hofrichter H (1982) Stadtbaugeschichte von der Antike bis zur Neuzeit. Vieweg, Braunschweig Wiesbaden

Hackenbroch K, Baumgart S, Kreibich V (2009) The Spatiality of Livelihoods: Urban Public Space as an Asset for the Livelihoods of the Urban Poor in Dhaka, Bangladesh. Die Erde 140:47–68

Hackenbroch K, Hossain MS, Rahman MA (2008) Coping with Forced Evictions: Adaptation Processes of Evicted Slum Dwellers in Dhaka. TRIALOG 98 3:17–23

Hossain S (2010) Dividing the ordinary: Negotiating water supply in an informal settlement of Dhaka. In: Rawani AA, Kettani H, Ting Z (eds) Proceedings of 2010 International Conference on Humanities, Historical and Social Sciences. World Academic Press; England, pp 614–619

Hossain S (2011a) Informal dynamics of a public utility: Rationality of the scene behind a screen. Habitat International 35(2):275–285

Hossain S (2011b) The production of space in the negotiation of water and electricity supply in a Bostee of Dhaka. Habitat International 35(4):(in press)

Hotzan J (1994) dtv-Atlas zur Stadt. Von den ersten Grundzügen bis zur modernen Stadtplanung. Dtv, München

Islam N (2005) Dhaka Now. Contemporary Urban Development. Bangladesh Geographical Society, Dhaka

Islam N, Khan MM, Nazem NI, Rahman MH (2000) Reforming governance in Dhaka, Bangladesh. In: Islam N (ed) Urban Governance in Asia: Sub-regional and City Perspectives. Pathak Shamabesh, Dhaka, pp 135–162

Khan MH, Khandoker A, Kabir M, Mori M (2007) Determinants of drinking arsenic-contaminated tubewell water in Bangladesh. Health Policy and Planning 2007 22:335–343

Kombe WJ, Kreibich V (2000) Informal Land Management in Tanzania. Dortmund SPRING Research Series 29

Mahabub-Un-Nabi ASM, Hashem M (2007): Trends of Development in Dhanmondi. In: Jahan S, Maniruzzaman KM (eds) Urbanization in Bangladesh - Patterns, Issues and Approaches to Planning. Bangladesh Institut of Planners, Dkaka, pp 36–42

McFarlane C (2008) Sanitation in Mumbai's informal settlements: state, 'slum', and infrastructure. Environment and Planning A 40 (1):88–107

Mumford L (1980) Die Stadt. Geschichte und Ausblick, Band 1. dtv, München

Nazem NI (2007) Informal settlements or slums – what are we talking about. Presentation within the Cooperation Workshop in Dhaka 22.11.2007

Payne G (2002) Tenure and Shelter in Urban Livelihoods. In: Rakodi C, Lloyd-Jones T (eds) Urban Livelihoods: A People-centred Approach to Reducing Poverty. Earthscan, London Sterling, pp 151–164

Platt HL (2007) From Hygeia to the Garden City. Bodies, Houses, and the Rediscovery of the Slum in Manchester, 1875–1910. Journal of Urban History 33:756–772

Prakash V (2002) Chandigarh's Le Corbusier: The Struggle for Modernity in Postcolonial India. University of Washington Press, Seattle

Priya R (1993) Town Planning, Public Health and Urban Poor: Some Explorations from Delhi. Economic and Planning Weekly 28 (17): 824–834

Puri BB (1997) Applied vastu shastra in modern architecture. Vastu Gyan Publication, New Delhi

RAJUK (1997) Dhaka Metropolitan Development Plan (1995–2015), Vol. I & II, Dhaka

RAJUK (2010) Final Report - Preparation of Detailed Area Plan (DAP) for DMDP Area: Group-C. Dhaka

Roy A (2003) City Requiem, Calcutta - Gender and the Politics of Poverty. Globalization and Community 10. University of Minnesota Press, Minneapolis New York

Sharif MI, Mainuddin K (2003) Country Case Study on Environmental Requirements for Leather and Footwear Export from Bangladesh. Bangladesh Centre for Advanced Studies, Dhaka

Shastri T (1996) Bharatiya vastu shastra (principles of vastu shastra). The Banaras Mercantile Bholanath Pustakalaya, Calcutta

Sheuya S (2004) Housing Transformations and Urban Livelihoods in Informal Settlements. The Case of Dar es Salaam, Tanzania. Dortmund

Siddiqui K, Ahmed J, Awal A, Ahmed M (2000) Overcoming the governance crisis in Dhaka City. University Press, Dhaka

The Daily Star (01.06.2009) Death of a lifeline. www.thedailystar.net

The Daily Star (03.06.2010) Begunbari building collapse - Put the lessons to use.

The Daily Star (04.06.2010) 6-storey building tilts at Begunbari.

The Daily Star (08.06.2010) Old Dhaka at chemical risk.

The Daily Star (17.07.2010) Implement DAP, save Dhaka city.

The Daily Star (24.06.2010) DAP now official. All recommendations find place on gazette notification; fate depends on implementation.

UN-Habitat (2009) State of the World's Cities 2008/2009 - Harmonious Cities. Earthscan, London Sterling

Von Moos S (ed) (2010) Chandigarh 1956: Le Corbusier, Pierre Jeanneret, Jane B. Drew, E. Maxwell Fry. Scheidegger and Spiess, Zürich

Chapter 19
Urban Food Security and Health Status of the Poor in Dhaka, Bangladesh

Wolfgang-Peter Zingel, Markus Keck, Benjamin Etzold, and Hans-Georg Bohle

19.1 Introduction: Hunger and Vulnerability After the Urban Turn

Amartya Sen, in his seminal work on food entitlements and deprivation (1981), has effectively demonstrated that food security is first and foremost a question of access to food rather than of general availability. Furthermore, research has shown that not only the rural populations are vulnerable to food insecurity, but that it is a significant challenge to urban dwellers as well (Sen 1981: 32; Pryer and Crook 1988; Watts and Bohle 1993). This is particularly true after the so-called "urban turn" – more than half of the world's population now live in urban habitats (UN 2008). The global food price hike of 2007 and 2008 again has taught national governments and the international aid community that an undisturbed supply of and access to food are the basic prerequisites for urban food security where basically all urban populations depend on food markets to access food.

According to Michael Watts (1983) the lack of access to food can be regarded as "silent violence"; it constitutes a 'normal' crisis for the poor, despite the fact that sufficient amount of food is available in the country to avoid hunger, although on a low average level. This paradox will be examined in this chapter in the light of the concept of social vulnerability. Generally, social vulnerability refers to "exposure to contingencies and stress, and difficulty in coping with them" (Chambers 1989: 1). Vulnerability is a dynamic, multilayered and multidimensional social condition, which is structured by intersecting social, political, economic and ecological forces in specific places at specific times (cf. Watts and Bohle 1993). Social vulnerability

W.-P. Zingel (✉) • M. Keck
South Asia Institute, University of Heidelberg, Heidelberg, Germany
e-mail: h93@ix.urz.uni-heidelberg.de

B. Etzold • H.-G. Bohle
Geography Department, University of Bonn, Bonn, Germany

A. Krämer et al. (eds.), *Health in Megacities and Urban Areas*,
Contributions to Statistics, DOI 10.1007/978-3-7908-2733-0_19,

is rooted in actors' (or communities') capacities to cope with and recover from all kinds of stressors, among which are environmental hazards, global economic transformations and personal misfortunes (Bohle 2008: 39). Vulnerability to urban food insecurity has been examined in detail (cf. Ruel et al. 1998; Maxwell 1999; Bohle and Adhikari 2002; Pryer 2003; Community-Studies-Team 2007) and acknowledged in 'flagship reports' of international organisations, e.g. by FAO (FAO 2004: 18f) or UN-HABITAT (2006: 104ff). Experts agree on the fact that urban food security is first and foremost a question of income and of households' (HH) social access to food. The fact that the urban poor spend a greater portion of their income on purchasing food than the middle and upper strata makes them particularly susceptible to food price hikes; low and unstable incomes, in turn, seriously hamper the satisfaction of their nutritional and health-related needs (cf. Lam 1982: 53, Pryer and Crook 1988: 26ff, Maxwell 1999: 1945, Bohle and Adhikari 2002: 411, Pryer 2003: 141).

This chapter on urban food security and health in the megacity of Dhaka examines food security in terms of the availability, accessibility and utilization of food and its specific health outcomes. For this exercise, a food entitlement perspective is employed, and linked to the ideas of social vulnerability. The focus is on the urban poor living in selected slums of Dhaka. While most academic work deals with the consequences of food scarcity on a national level only, in this chapter we look at local impacts, and also take the considerable social stratification of the urban poor into account. It is assumed that prices of food as well as income and entitlement opportunities of the poor are the factors that count, and that locality contexts and differentiations within the poor are highly important aspects of food security and health status of the most vulnerable populations in the megacity.

With these assumptions in mind, we present the Coping Strategy Index (CSI) developed by Maxwell (1996) as a methodology to measure food security and related health risks. In the first section, the global food price hike is examined as a stressor for Bangladesh's food supply. In the next sections the vulnerability of the poor to food and health insecurity in Dhaka is analysed against the background of the Coping Strategy Index, taking a comprehensive food and health survey as a base. The respective problems of food availability, accessibility and outcomes in terms of food security and health are discussed in empirical terms. The last section, which is again based on the CSI, examines the coping strategies and the capabilities of the poor to deal with the food price hikes and food insecurities in the megacity. In conclusion, we stress that global food crises have to be investigated as local-level experience of the poor and vulnerable, particularly in the growing megacities and their increasing number of slum dwellers. We also point to the considerable social differentiation within these slum areas, with extreme vulnerabilities to food insecurity and health risks for female-headed households. We also show that the urban poor, in order to cope with the crisis, heavily rely on their informal social networks.

19.2 Methodology: The Coping Strategy Index (CSI)

In order to investigate the vulnerability to food insecurity and health risks of low-income groups in Dhaka's slums we rely on the work of Daniel Maxwell (1996), who developed the Coping Strategy Index (CSI), a tool to measure food insecurity on the household (HH) level that is frequently applied by the World Food Programme. The CSI is based on the assumption that the way how people deal with insufficient food in critical times provides a general indication for the food security status of HHs. It combines context specific coping strategies, people's perceived severity of these strategies and the frequency of their use into a composite index.[1]

The coping strategies that were used in a standardized Food Consumption Survey in nine slum areas in the megacity of Dhaka (April–June 2009; $n = 205$)[2] encompass information on food access, dietary change and rationing strategies within HHs. As access to food in urban areas largely depends on the HH's income, the assessment of coping strategies was based on the simple question "What do you do if you do not have enough money to buy sufficient food?" In order to rank the different strategies, the perceived severity and frequency of having to cope with insufficient food was assessed. It was asked "How severe are the following ways of dealing with insufficient money to buy food?"[3] and "How often did your HH had to manage in the following ways in the last week?"[4] The coping strategies mentioned by the interviewees were the following:

- "We try to work and earn more than before to make up for higher expenses."
- "We buy food in the local food stall or in the grocery shop on credit."
- "We eat less preferred but less expensive food (e.g. less meat or fish)."
- "We borrow food or money from relatives or neighbors."
- "The mother eats less in order to ensure that children have enough food."

[1]We are grateful to Patrick Sakdapolrak from the Geography Department, University of Bonn, for introducing the methodology to us. He applied the CSI-method in a vulnerability study of slum households in Chennai, India (Bohle and Sakdapolrak 2009; Sakdapolrak 2011)

[2]Nine slum settlements were selected for the survey, located in different parts of Dhaka, six within the Dhaka City Corporation (Begunbari-Tilatek, Pallabi; Bishil and Sarang Bari Bastee, Mirpur; Bhuiapara road, Khilgaon; Kunipara, Tejgaon; Adabar No-10 Bastee, Mohammadpur; Natun Jurain Bastee, Alambagh, Shyampur), and three within Dhaka Union (Kamranginchar; Abdullapur, Dakshin Khan; Harirampur, Turag). 18–31 household interviews were carried out in each of the slums. Their population ranged from 3,000 to 30,000. The households were randomly selected from a sample that was drawn at the same time at the very same study sites by the INNOVATE research consortium from the Universities of Bielefeld and Humboldt at Berlin. Their Public Health Survey was conducted under the supervision of Dr. MMH Khan and O. Grübner. We, hereby, would like to thank Dr. Khan and his colleagues for the co-operation in conducting the research and for letting us use parts of their data set

[3]Perceived severity of the respective coping strategy: 1 = 'not severe', 2 = 'little severe', 3 = 'severe', 4 = 'very severe'

[4]Frequency of applying the respective coping strategy within the last week: 0 = 'never (0 days)', 1 = 'hardly at all (1 day)', 2.5 = 'once in a while (2–3 days)', 5 = 'pretty often (4–6 days)', 7 = 'all the time (everyday)'

Table 19.1 Validity of the Coping Strategy Index (CSI) (based on a food consumption survey in nine slums in April–June 2009, $n = 205$; the higher the score, the worse is the food security situation)

"Do you sometimes go to bed hungry?"	Share of all HH (%)	CSI-Score
"Yes, a few times a week"	7	68
"Yes, but seldom"	53	62
"No, never"	40	54
Total/mean	100	59

- "We save less and send less money to our family in rural areas than before."
- "We eat less/We eat smaller portion sizes of meals."
- "We sell personal goods (e.g. jewelleries) to get enough money for food."
- "We eat prepared food from a local food stall/roadside shops instead of cooking at home."
- "We skip two meals a day."
- "We do not eat anything a whole day."
- "We send our children to eat with relatives or neighbors."
- "We buy cheaper food from BDR shops/Open Market Sales (Public Food Distribution)."

In order to assess the food (in)security status of each of the interviewed HHs, the respective severity score of each coping strategy was multiplied with the respective frequency score. If one adds up the respective weighted scores for all the 13 coping strategies, every HH gets a total coping strategy index score. In case of our survey in Dhaka, the lowest CSI-score of all 205 respondents was 17, while the highest was 132. The higher the CSI-score, the higher is the vulnerability of a household to food insecurity. A simple test of the CSI-score against a 'classical' question in food security research ("Do you sometimes go to bed hungry?") shows the validity of the CSI-method for assessing the food security status of a household (see Table 19.1).

In the following sections of this chapter, the CSI-score is used as a proxy indicator for food security.

19.3 The Global Food Price Hike as a Stressor for Bangladesh's Food Security

In most of South Asia rice is the major staple of the people, regardless of their economic or social class, caste or ethnic background. In late 2007 the price of the major rice exporters showed a pronounced increase with a peak in mid-2008; by the end of 2008 prices finally started to fall. The price of high quality Thai rice quadrupled (!) within a period of 1 year, only. In the period of August 2004 until November 2006 it had risen from 250 US$ per metric ton (MT) to 300 US$/MT, while afterwards it jumped up to more than 1,000 US$/MT in mid-2008. From

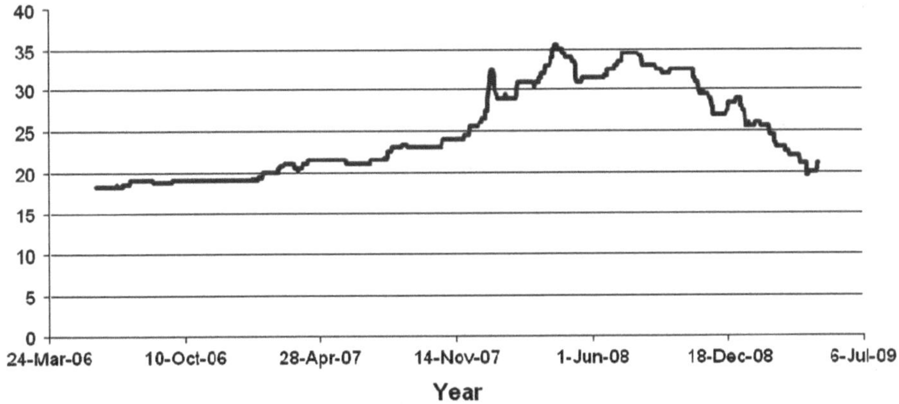

Fig. 19.1 Daily retail rice prices in Dhaka, in Bangladesh Taka (Source: Ministry of Agriculture, Government of the People's Republic of Bangladesh (2009))

the end of 2008 onwards prices fell and remained at a high level of around 600 US$/MT.[5]

Global price hikes have a damaging effect in countries that have to import at any price. If the government is not interfering in between and subsidizes food imports (directly or indirectly) those who have to rely on the market have to pay the higher prices, often impossible for the urban poor, who to a large extent have to depend on their exchange entitlements for accessing adequate amounts of food (Sen 1981). Figure 19.1 provides data on low quality (coarse) rice that is mainly consumed by low income groups in Dhaka. Here, prices more than doubled in the mentioned period, i.e. from 16.5 Bangladesh Taka (BDT) in January 2006 to 35 BDT in December 2008. Since the beginning of the year 2009 prices have come down again to a rate of 20 BDT to 25 BDT.

Experts identified several factors being responsible for rising global prices, such as the increasing world energy prices (of oil and gas) that in turn raised the prices of major agricultural inputs such as fertilizer and water (via diesel for pumps and tractors); an increasing demand for rice in developing countries with a high income elasticity like China and India; more land used to grow fuel crops; the weak US-Dollar; and massive price speculation in agricultural commodities (Cohen and Garrett 2009). But how does the particular situation look like in Bangladesh?

Bangladesh's rice granaries are located in north-western and northern regions of the country. Rice is grown on small farms of less than 2.5 acres (1 ha). Up to three harvests are possible in a year, i.e. *aman*, *boro* and *aus* (Ahmed 2001: 2; Dorosh et al. 2004: 14). In the fiscal year 2006/07 55% of the total rice production stemmed from the *boro* season, whereas *aus* lost its importance and plays only a marginal role now with hardly 6% (FPMU 2009: 1). Bangladesh has been able to more than triple rice (paddy) production since Independence in 1971. Domestic production

[5]According to www.indexmundi.com, accessed: 20.08.2009

rose from 14.9 mio MT in 1971 to 43.1 mio MT in 2007 and to a record harvest of 46.9 mio MT in 2008. This success was possible only because yields increased from 1,602 kg/ha in 1971 to 3,995 kg/ha in 2008. Since all arable land is cultivated for long, the area harvested increased only from 9.3 mio ha to 11.7 mio ha, mostly by multiple cropping and irrigation. Today, Bangladesh holds the fourth position among the world's rice producing countries (FAOSTAT 2010).

But despite all impressive production increases, Bangladesh is still not in a position to feed its population. More than 3 mio MT foodgrains had to be imported in 2007: 2.4 mio MT wheat, 0.6 mio MT (milled) rice and 0.2 mio MT maize (FAOSTAT 2010).[6] For many years around one-tenth of all food-grains consumed in Bangladesh had to be imported; thanks to marked improvements in production that ratio has fallen. Due to the poor monsoon rains of 2009 rice imports are projected at 0.6 mio MT in 2010 after 0.4 mio MT in 2009. In 2008 they were well over 1 mio MT as in the year before (FAO 2009b: 25). Food imports not always reflected food requirements. This never came out as clearly as in 1974 when emerging shortages were seen too late and donors were hesitant to help. The crisis resulted in a famine in the same year and political turmoil in the following one. Food imports initially fully depended on donors' preparedness to help and on the availability of foreign exchange for commercial imports. 1999 saw another poor harvest: a record 2.2 mio MT of milled rice had to be imported on top of 2.4 mio MT of wheat after severe floods and harvest losses (FAOSTAT 2010). In 2007 floods and the cyclone Sidr brought devastation and much of the paddy crop was destroyed. Consequently the country had to import large quantities of grain just when world market prices started to rise to unprecedented heights.

Bangladesh's market liberalization of the 1990 s coincided with India's removal of export restrictions. India's rice exports increased dramatically from 0.9 mio MT in 1994 to 4.9 mio MT in 1995. India started dominating the rice imports of her neighbour, thus replacing Thailand as the major source of rice imports to Bangladesh (Dorosh and Murshid 2004: 109). One has to bear in mind, however, that India had been exporting rice to Bangladesh also before, although illegally and unrecorded. The long and winding border was never effectively controlled and whenever prices differed in the two neighbouring countries large quantities would be smuggled across. But also after India started to liberalize its foreign trade, she continued to look at her consumers first: In times of high world market prices, India restricts her exports rather than allow her domestic prices to rise. As a consequence, India's rice exports were reduced from 5 mio MT in 2007 (FAO 2008: 32) to 3.57 mio MT in 2008 (FAO 2009a: 29) at a time when Bangladesh needed to import. One of the peculiarities of the world cereal market is that rice is traded much less than wheat or maize; less than one-tenth of the world rice production is internationally traded. Consumers also do not easily change their eating habits;

[6]According to the FAO (2008) 1.4 mio MT of rice were imported in 2007. Note: FAOSTAT differentiates between paddy (unmilled rice), (milled) rice, broken rice and other varieties. As paddy loses weight in the milling process (in the order of one third), quantities cannot be easily added up

making bread from wheat flour is also very different from cooking rice. The world market, thus, reacts sharply to changes of supply and demand of rice. Otherwise it could not be explained how upward changes of world market prices of rice could affect a country so much, that imports only a few percent of its total consumption.

19.4 Vulnerability of the Poor to Food Insecurity in Dhaka

19.4.1 Food Availability in the Megacity of Dhaka

The most food insecure areas of Bangladesh are the North-West Region (Dinajpur, Rangpur), northern *char* islands (sand bars emerging as islands within Jamuna and other river channels), the 'drought zone' in the West (western parts of Nawabjanj, Rajshahi, Noagaon), the Sylhet *hoar* basin (a wetland ecosystem in north-eastern Bangladesh), the southern coastal belt and Chittagong Hill Tracts in the South-East (GOB/WFP 2004). In these regions, subsistence production is often insufficient to feed the families of small farmers or to provide work to landless labourers throughout the year, in particular during the *monga* period (a season of poverty and hunger in some areas of Bangladesh prior to the major rice harvest); additional resources are necessary to buy food from local markets. Rural food insecurity, widespread poverty, and a general lack of employment are thus among the most important driving forces of migration to Dhaka and partly cause of its rapid population growth.

Because of the limited absorption capacity of industry and government service in the capital, this rural exodus contributes to an ever growing 'informal' economy. People from the aforementioned areas settle down in one of Dhaka's countless slums (makeshift huts rather than run-down inner city houses). They sustain their livelihoods by means of self-employment, e.g. as rickshaw pullers, street food vendors (Etzold et al. 2009; Hackenbroch et al. 2009), or by taking up jobs under often dangerous and unhygienic conditions, e.g. in construction, in the plastic recycling and processing industry (Kulke and Staffeld 2009), the garment or the emerging pharmaceutical sectors.

Less rice was traded in Dhaka's six major markets in early 2009 as compared to early 2008, e.g. 1.424 mio kg per day and 1.527 mio kg.[7] These markets can be considered to be fairly representative for Dhaka: At a rate of 0.5 kg of rice per person per day[8] they supply rice for three million people or one-third to one-half of

[7]Surveys were conducted in February and March 2008 and again in the same months in 2009 in the Mirpur 11, Malibag, Jatrabari North, Mirpur 1, Kochuket and Babubazaar/Badamtuli

[8]The net availability of rice in Bangladesh was 188.4 kg in 2004–2005; it was a little less in the year before (BBS 2008:411)

the population of the 'inner city' that comes under the administration of the Dhaka City Corporation (cf. Keck et al. 2008: 30).[9] A reduction of 7% from 2008 to 2009 meant on average 35 g or 120 Kcal[10] less per person and day – a substantial amount for people who often have a food energy intake of 2,200 Kcal per day.[11]

19.4.2 The Urban Poor and Their Access to Food

Livelihood groups in Dhaka can be distinguished by occupation (Pryer 2003). A look at the CSI-scores of the respective labour groups[12] shows that workers with a permanent job in private services or in agriculture enjoy the highest level of food security (mean CSI-score 40 and 46) (see Table 19.2). Moreover, their HH income as well as their food expenditure are far above the respective averages. On the contrary, HHs depending on incomes from working as domestic servants or security guards (household services) have the lowest income, spend the smallest amounts of money on food, are most likely undernourished and are the least food secure according to the CSI-score (mean CSI-score 63). Self-employed people in retail (e.g. street food vendors) or transport sectors (e.g. rickshaw pullers), and factory workers are situated in between the above mentioned extremes. Among all, cooks and day labourers (in the group of 'others') have the highest mean CSI-score with 74 and 66 respectively. It becomes clear from these numbers that urban vulnerability to food insecurity is largely determined by (gainful) employment.

Table 19.3 backs this statement. It shows a clear connection between the level of income and food security and health. HHs in the poorest income quintile have less than half of the average income at their disposal. Consequently, there is less money available per head to purchase the required amount of food and the HHs are less food secure (as indicated by a high CSI-score). There exist, however, marked disparities within each income group.[13] This brings additional factors into play. The fact that 28% of the HHs within the poorest quintile are female headed

---

[9]The last population census was in 2001. 5.3 mio people were counted in the area under DCC (BBS 2008: 94). At an annual growth rate of 5% their number would have increased to 8.2 million in 2010

[10]Rice in Bangladesh on average has a nutrient content of 347 Kcal per 100 g (BBS 2008: 398)

[11]In urban areas average per capita calorie intake per day was around 2,200 Kcal since the late 1980 s; it was 2,193 Kcal in 2005 (BBS 2008: 397)

[12]Labour groups have been identified by the economic sector in which the head of the household earns his/her main income

[13]This might be explained by the fact that the CSI-Method brings out the perceived sensitivity of households and their coping behaviour. If a HH with a relatively higher income has to cut down its expenditure on meat and fish in order to ensure a provision with 'good' rice, the deteriorated food situation might result in a higher CSI-score as compared to a family just surviving on rice, oil and some vegetable and without having been used to eating meat and fish of good quality

Table 19.2 Occupational groups, income, food expenditure, food security and health in Dhaka's slums (based on a food consumption survey in nine slums in April–June 2009, $n = 205$)

Economic Sector, in which main income earner of household works	Share of all HH (%)	Share of HH headed by women (%)	Share of HH in lowest income quintile (%)	HH income per person in HH, per month[a] (BDT)	HH food expenditure per person in HH per month (BDT)	CSI-Score	% of HH with underweight Body Mass Index <18.5 kg/m^2	Health status
Finance and private service industry	7	7	27	1,662	1,416	40	40	3.5
Agriculture	3	0	0	1,898	1,403	46	57	3.4
Construction	13	4	4	1,449	1,120	58	23	2.9
Manufacturing (incl. RMG sector)	8	12	24	1,799	815	60	35	2.9
Transport (incl. rickshaw puller)	25	4	16	1,538	1,551	60	41	2.9
Trade/Retail (incl. street vendors)	25	6	16	1,619	1,464	61	35	3.0
Household services (incl. maid)	5	40	50	789	876	63	50	3.1
Other	13	19	33	1,177	1,198	67	40	3.6
Total/mean	**100**	**9**	**19**	**1,515**	**1,316**	**59**	**37**	**3.1**

Body Mass Index and Health Status were calculated on the basis of data provided by the Bielefeld-Berlin INNOVATE consortium's Public Health Survey (cf. Khan et al. 2009). According to the World Health Organisation a Body Mass Index (BMI; weight/height) of below 18.5 is an indicator of underweight, a BMI of 18.5–25 indicates normal weight, a BMI of 25–30 overweight and a BMI above 30 obesity. The question regarding the health status was: "In general, how do you rate your health?": excellent = 1; good = 2; so-so = 3; fair = 4; poor = 5. The higher the mean value, the worse the average perceived/self-reported health status of that group

[a]In Bangladeshi Taka; at the time of the survey (April–June 2009) 100 BDT = 1.062 EUR

highlights the importance of social factors that also determine access to employment and income. Divorced (and to some extent also widowed) women are especially stigmatized and socially excluded, with severe consequences for all their family members' food security and health status. Social mechanisms like these need to be considered when thinking of fighting food insecurity in Bangladesh.

19.4.3 Utilization of Food and Health Outcomes

Fifty six percent, more than half, of all income is spent on food in Bangladesh; breads and cereals account for 50% of food expenditures; another 20% of food expenditure is spent on other foods, beverages and tobacco (USDA 2010). Dhaka's

Table 19.3 Relative income quintiles, food expenditure, food security and health in Dhaka's slums (based on a food consumption survey in nine slums in April–June 2009; $n = 205$)

Relative Income Group based on total HH income per month	Share of all HH (%)	Share of HH headed by women (%)	HH income per person in HH, per month (BDT)	HH food expenditure per person in HH, per month (BDT)	CSI-Score	Share of HHs with underweight BMI <18.5 kg/m^2 (%)	Health status
Poorest quintile	19	28	791	980	60	46	3.4
Poor quintile	17	3	1,038	1,168	60	38	2.7
Middle quintile	36	8	1,380	1,277	58	38	3.0
'Rich' quintile	7	0	2,109	2,057	61	13	3.3
'Richest' quintile	21	0	2,643	1,595	53	31	3.1
Total/mean	**100**	**9**	**1,515**	**1,320**	**58**	**36**	**3.1**

See FN 19: data provided by the Bielefeld-Berlin INNOVATE consortium (cf. Khan et al. 2009)

poor spend even more on food: sometimes even more than they earn. Spending 127% of household income on food alone as has been reported by respondents in the poorest quintile (see Table 19.4) is only possible if the household income is measured by earnings only and if consumption can also be funded by debt. As poor people do not have access to bank loans, family members and friends are the major source of credit. Of course, there are micro credit schemes in the country of the Peace Nobel Price laureate Mohammad Yunus. But micro credits are meant to finance investments for self-employment rather than consumption and daily survival. But even if they were available, it seems that the outcome still is the same, i.e. that poor families become more and more indebted.

The largest share of food expenditure is spent on rice (see Fig. 19.2). On average, a slum HH consumes about 12 kg of rice per week and spends about 71 BDT on every person that needs to be fed. The most vulnerable HHs, in turn, spend more than half of their food budget on rice, almost exclusively of the lower quality (coarse rice such as *guti, pari, mota*, or *lali*). In contrast, the families that are relatively better-off not only buy more rice per person, but also better rice.

Important to note is that the better-off a HH is, the more money it spends on fish. While the poorest almost exclusively buy small fishes (*choto mach*) in only small quantities, the more affluent buy bigger species, mostly carps such as *rui* or *katol*, and the rather expensive but highly valued fish *hilsha*, i.e. the 'national fish' of Bangladesh. The same applies to meat: the poorest can hardly afford it and only spend 2% of their food budget on small quantities, whereas, the more affluent families spend up to 9% on meat. For dairy products, poor slum dwellers spend just 1%, while the richest slum HHs can afford to spend 4% on milk, cheese, curd and other dairy products. Another interesting aspect is that while both income groups spend more than 20% of their food budget on 'eating outside', the poorest spend comparatively more on readily available snacks, small dishes, and sweat tea from roadside shops and mobile street vendors (Keck et al. 2008; Etzold et al. 2009).

Table 19.4 Relative income, food expenditure, rice and fish consumption patterns in Dhaka's slums (based on a food consumption survey in nine slums in April–June 2009; $n = 205$)

Relative Income Group based on total HH income per month	Share of total income spent on food (%)	Rice expenditure p. pers. in HH per week (BDT)	Rice expen-diture as share of total food expend (%).	Rice amount consumed in HH (per week) (kg)	Share of HH eating lowest quality rice (coarse rice) (%)	Fish amount consumed in HH (/week) (kg)	Share of HH eating mainly dried fish (Choto Mach) (%)	Meat amount consumed in HH (/week) (kg)
Poorest quintile	127	65	56	11	90	1.6	85	0.25
Poor quintile	116	72	35	12	79	2.3	65	0.23
Middle quintile	96	73	40	13	74	2.7	54	0.66
'Rich' quintile	93	61	20	9	73	3.2	40	0.53
'Richest' quintile	69	78	35	13	60	2.5	59	0.94
Total/mean	**99**	**71**	**40**	**12**	**75**	**2.4**	**62**	**0.57**

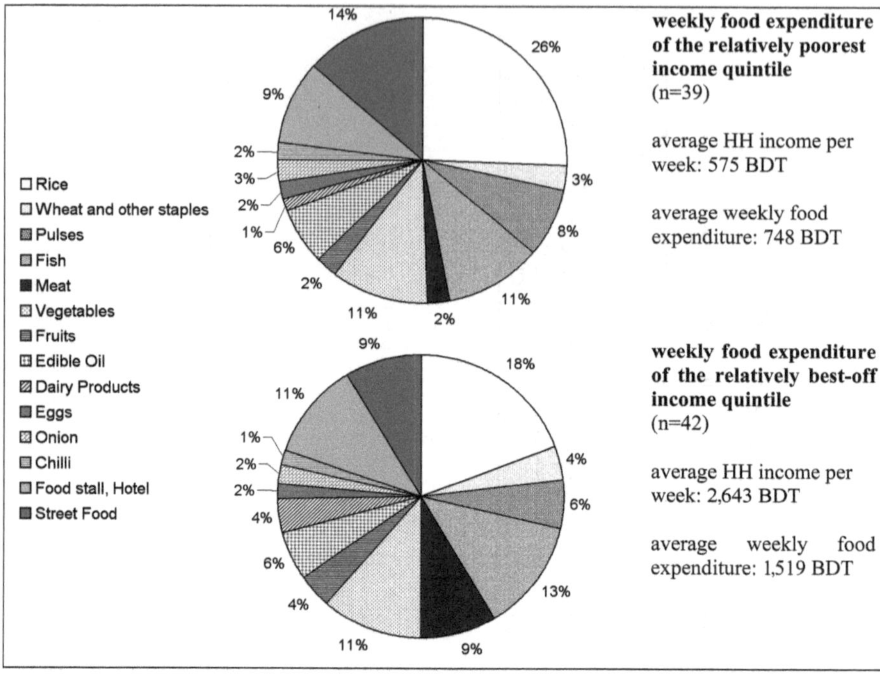

Fig. 19.2 'Food Basket' of slum households in Dhaka (based on Food Consumption Survey in nine slums in April–June 2009; $n = 205$)

As indicated in Tables 19.1 and 19.2 the health of slum dwellers varies strongly with occupation and income. The relation between food provision, the utilization of the nutrients within the body and health consequences are too complex to be dealt with in detail here. For the limited purpose of this study two simple measurements might suffice, i.e. the HH's health situation as perceived by the respondents themselves as a subjective and their Body Mass Index (BMI) as an objective one.

The high vulnerability to food insecurity of the occupational group of HH services (mean CSI-score 63) is reflected by the high share of underweight persons: 50% of respondents from this group have a BMI below 18.5, a clear indication of under-nourishment according to Pryer (2003: 149ff). The self-perceived health status of domestic servants and security guards is, however, not so bad: 40% said that they have a good health, 30% stated that their health is 'so so', while another 30% rated their health as 'fair' or 'poor'. In contrast, those HHs that depend on urban agriculture for their main income, which are the group with the highest average HH income per person, and which are among the least food insecure according to a mean CSI-score of 46, have the highest share of underweight persons with 57% and also a health status below average: 57% of them rate their own health as 'fair' or 'poor', while 43% stated that they have a good health. Interestingly, the healthiest occupational groups in our sample were those HHs that depend on construction work: only few of these families are in the lowest income quintile;

nevertheless their HH income and food expenditure are slightly below the total average, while their food security status (mean CSI-score 58) is average. With 23%, the share of underweight persons in this group is, however, the lowest. Twenty seven percent of construction workers HHs rate their own health as 'excellent' or 'good' and only 15% as 'fair' or 'poor'.

If only the HH income is taken as a basis for calculation (see Table 19.2) then it shows that the poorest income quintile is not only the most food insecure (lowest food expenditure per person and mean CSI-score of 60), but also the group with the highest share of underweight persons (46%); 18% are severely thin (BMI < 16), while 44% have normal weight. This goes in hand with this group's worst health status of all income groups (mean = 3.4): while 20% see themselves as being in good health (the same percentage as in the highest income quintile), 39% said their health was 'fair' or 'poor' (in contrast 20% said so in the highest income group).

If the aforementioned results are 'turned on their head', it shows that the BMI also has strong explanatory value for a HH's food security condition and its health status (Table 19.5). Households, in which the interviewed person was underweight (BMI < 18.5), disproportionally often had women as the main (and often also the sole) income earner. Moreover, most of the undernourished are in the lowest income quintile, their food insecurity status (mean CSI-score 62) is worse than average, and so is their health. If only the 9.5% of all interviewees that are severely thin (BMI < 16) are looked at, the data becomes even more contrasting: 21% of these HHs are headed by women, they are likely to be in the lowest income group, they have the lowest average HH income per person and also the lowest average food expenditure per person, with 65 their CSI-score is the highest of all, and with 3.5 they have the worst health status. While 28% of the underweight persons rated their health as 'fair' or 'poor', 37% of the severely thin people said so; in turn, only

Table 19.5 Body Mass Index, income, food expenditure, food security and health status in Dhaka's slums (based on a food consumption survey in nine slums in April–June 2009; $n = 205$)

Body Mass Index (kg/m^2) based on weight/ height of interviewees	% of all HH ($n = 200$)	% of HH headed by women	% of HH in lowest income quintile	HH income per pers. in HH per month (BDT)	HH food expenditure per pers. in HH per month (BDT)	CSI-Score	Health status
Underweight (BMI < 18.5)	38	13	24	1,458	1,377	62	3.3
Severe thinness (BMI < 16)	9.5	21	37	1,026	993	65	3.5
Moderate thinness (BMI 16–17)	8.5	12	18	1,463	1,264	61	3.1
Mild thinness (BMI 17–18.5)	19.5	10	21	1,660	1,617	61	3.3
Normal weight (BMI 18.5–25)	50	8	17	1,561	1,250	59	2.9
Overweight (BMI 25–30)	11	0	19	1,257	1,251	52	3.2
Obese (BMI > 30)	2	0	0	1,456	1,163	46	3.0
Total/mean	**100**	**9**	**17**	**1,488**	**1,298**	**59**	**3.1**

See FN19: BMI and Health Status data provided by the Bielefeld-Berlin INNOVATE consortium (cf. Khan et al. 2009)

11% of the severely thin and only 15% of the underweight persons said they were in good health (none said their health was excellent), for the people with normal weight the respective figure is 33%. This observation fits into the overall picture: All over South Asia a food related gender bias can be observed: South Asian women and girls are not only less well fed than South Asian men and boys, they are also less well fed than women and girls coming from similar income groups in Africa (Smith and Haddad 2000; Pryer 2003; Gragnolati et al. 2006).

19.5 Coping with the Food Price Hike in Dhaka

The prize hike of food in 2008 was a severe shock for the 'rice nation' of Bangladesh. The media reported extensively on price changes and how people from all classes had to change their food consumption patterns. In contrast to poor rural HHs, slum dwellers in cities have no direct access to rice and therefore depend on the market, on government (important especially for people in government service) and on NGOs. As the national government's efforts to curb the spiraling prices were not as effective as expected (or hoped), the urban poor were hit the hardest by the price hike. Within Dhaka's slums 77% of HHs in the poorest income quintile stated that the price hike of particularly rice affects them very severely. Additionally, also 45% of the comparatively most affluent slum HHs perceived the price hike to be 'very severe' (see Table 19.5).

Due to public pressure the government of Bangladesh took up public food distribution – if only half-heartedly – again that were almost abolished for several decades. The food procured by the government was sold through sales units set up by the Bangladesh Rifles (BDR), so called BDR markets, or through 'open market sale' (OMS) shops, i.e. licensed retail shops selling subsidized rice. Every person was allowed to buy up to 3 kg of rice at a rate of 25 BDT per kg when market prices were already at 30–35 BDT.[14]

The public food distribution schemes, however, did not reach all HHs affected by the price hike. Many slum and also pavement dwellers could not afford to stand in a queue for hours in order to get a few kilograms of rice as this time was lost for income generating activities. Our data shows that only a meagre portion of slum dwellers buys rice, pulses and/or vegetable oil from BDR shops; most people buy their food at normal market rates from the local bazaars and retail shops. Moreover, hardly any slum dweller benefits from group feeding programmes that operate through NGOs or community centres; in particular the neediest ones do not have access to such social charity schemes (see Table 19.6). Social networks are therefore highly important for the food provisioning of slum dwellers. In this regard, the family plays an important role. It seems that the more affluent slum dwellers are more in a position to maintain their ties to their families in the home districts, especially those who own some land in the village and can manage to go there

[14]Interview with Mr. Hiran Maya Barai, Chief Controller Dhaka Rationing, on 21 January, 2008

Table 19.6 Coping with the price hike: Relative income, coping strategies and food security in Dhaka's slums (based on a food consumption survey in nine slums in April–June 2009; $n = 205$)

Relative Income Group based on total HH income per month	Share of HHs that …						
	… perceive price hike of rice as very severe (%)	… never buy food from BDR shops (%)	… never get free food from NGOs, community centre, neighbours, etc. (%)	… sometimes bring food from their village (%)	… daily buy food from retailer on credit (%)	… never borrow food/money from relatives/ neighbours (%)	… never send children to eat with relatives/ neighbours (%)
Poorest quintile	77	87	97	0	21	31	69
Poor quintile	59	88	97	12	35	24	62
Middle quintile	54	89	94	8	38	33	82
'Rich' quintile	40	93	80	20	27	33	60
'Richest' quintile	45	81	93	24	33	43	76
Total	56	87	94	11	32	33	73

Relative Income Group based on total HH income per month	Share of HHs that …						
	… daily try to work and earn more than before (%)	… daily save less and send less money to family in rural areas (%)	… daily eat less preferred, but less expensive food (%)	… daily eat less in quantity (%)	… never skip meals during a day (%)	in which the mother often or daily eats less so that children can eat more (%)	… go to bed hungry a few times a week (%)
Poorest quintile	54	13	16	8	31	15	13
Poor quintile	35	6	27	0	47	6	9
Middle quintile	49	20	25	3	58	4	3
'Rich' quintile	60	20	20	0	60	0	7
'Richest' quintile	50	9	19	0	69	2	7
Total	49	14	22	3	54	6	7

Photo 19.1 People queue up at a public food distribution point of the Bangladesh Rifles (BDR) in Dhaka in order to get subsidised rice during the food price crisis in 2008 (Picture: M. Keck, February 2008)

regularly and bring food to the city. The poorest HHs, in contrast, cannot even afford a trip to their village (Photo 19.1).

The family also serves as the most important source of credit: On average two-thirds of the slum HHs borrow food or money from their family or neighbours; the poorest families particularly depend on this type of social capital. Thirty percent of them occasionally send their children to eat with other family members or neighbours. Another important group in this regard is local businessmen. Due to irregular and insecure income one-third of the interviewed families buy food items from 'their' local grocery shop or retailer on credit on a daily basis. Trust plays a pivotal role in these informal relations. Nevertheless, interest rates can be high, adding to a HH's overall debt. Seventy seven percent of the HHs in the poorest income quintile, for instance, stated that at the end of each week they have a severe financial gap to fill (in contrast to only 17% of the most affluent quintile). Fifty nine percent of them said that they are heavily indebted (in contrast to only 19% of the most affluent quintile).

The most important coping strategy that slum HHs pursue under conditions of a food price crisis is again related to their labour power. Almost half of all respondents stated that they are forced to work more, which is considered to be 'very severe' by 43% of the respondents. Permanent employment, in particular in public services, goes usually along with long-term security, better access to health and educational services, and thus a better food security and health status (Pryer 2003: 36ff). For the majority, though, this formal labour market stays barred. For them getting a job in the informal urban labour market again heavily depends on their social standing and their social capital.

Another coping strategy is to change the diet. In the wake of the food price crisis, 22% of the slum HHs studied in Dhaka ate cheaper food, which they preferred less. They bought cheaper and less valued rice or substituted expensive fresh with cheaper dried fish. Likewise, the size of portion had to be reduced, in the case of members of the poorest income quintile on a daily basis. Seventy percent of this group occasionally skipped meals in order to save money and 13% went to bed feeling hungry several times a week. Another coping strategy that is applied by 15% of the poorest households regularly, but far less common for the relatively more affluent slum dwellers, is food abstinence of mothers for the sake of their children. While this practice ensures the feeding of the children at the lowest nutritional level, it is highly detrimental for the food security and thus also for the health of the women.

19.6 Conclusion and Research Needs

In his book on "Globalisierte Nahrungskrisen. Bruchzone Kairo", Jörg Gertel (2010) has shown how global food crises pierce down to the local levels of Cairo in Egypt, and even to the bodies of the individuals. The same is true to our study of slum dwellers in the megacity of Dhaka: the global food crisis has had its impacts on food security and the health of poor slum populations. Slums in Dhaka, however, are by no means homogenous, as the disaggregation of our data in terms of employment, income and the health of slum families has proved. It appears that the lack of income, in the context of dramatically rising food prices, is the most serious threat not only to food security, but also to health. Access to employment and income is particularly limited for female-headed households, i.e. the families of divorced or widowed women.

The recent food crisis has also shown that the urban poor rely heavily on their informal networks. In the context of limited access to the formal social system, help and support from family members, from neighbours and sometimes also from traders becomes crucial. Data from our survey prove that government distribution and aid programmes, managed by national and international NGOs, had hardly any impact on the extreme poor.

Looking at both food insecurity and ill-health through a vulnerability lens has provided the opportunity in this chapter to broaden our perspective beyond exposure and stress for vulnerable populations. Actor-oriented perspectives on vulnerability, as employed in this chapter, can address the agency of actors who are at risk to food insecurity and ill-health. They can highlight the multiple ways how vulnerable people cope with and adapt to social and natural environments in crises situations (Bohle 2008).

Our chapter has also shown that a number of crucial questions regarding food and health of the poor in megacities still remain open. One aspect is that social and cultural factors that shape labour markets and income opportunities in the megacity have to be further scrutinized. A second aspect is that the linkages of slum dwellers to their former home villages need to be further investigated, since they seem to be of major importance to buffer food crisis and health risks. Moreover, a focus on the poor and vulnerable only which neglects middle and upper classes is too narrow to understand vulnerabilities in terms of food and health. As Lohr and Dittrich (2007) have demonstrated, it is frequently the higher classes that decide upon the poor's access to employment, food and health. The study of slum dwellers and poor sections of the megacity alone will not provide us the information that is necessary to critically address present-day challenges. Ideas on supporting systems of "adaptive food governance" (Bohle et al. 2009) may be regarded as one of these challenges.

References

Ahmed R (2001) Retrospects and Prospects of the Rice Economy of Bangladesh. Dhaka

BBS (2008) Statistical Pocket Book Bangladesh 2007. Dhaka

Bohle HG (2008) Food Vulnerability and Health Vulnerability: Convergence and Common Ground in Global Change Research. In: Exner, M. et al. (Ed.): Towards Sustainable Global Health. SOURCE 11/2008. Bonn, pp. 38–41

Bohle HG and Adhikari J (2002) The Metropolitan Food System of Kathmandu - Conceptual Considerations and Empirical Evidence. In: Die Erde 133, H. 4, pp. 401–421

Bohle HG and Sakdapolrak P (2009) Chennai. Alltagskampf um Wasser, Nahrung und Gesundheit. In: Geographie und Schule 31, H. 181, pp. 4–10

Bohle HG et al. (2009) Adaptive Food Governance. In: IHDP Update 3, Magazine of the International Human Dimensions Programme on Global Environmental Change, pp. 53–58

Chambers R (1989) Editorial introduction: vulnerability, coping and policy. In: IDS bulletin 20, H. 2, S. 1–7

Cohen M, Garrett JL (2009) The food price crisis and urban food (in)security. London / New York

Community-Studies-Team (2007) Food and Nutritional Security in the Slums of Hyderabad. Berlin

Dorosh PA et al (2004) Overview of the Bangladesh Foodgrain Economy. In: Dorosh, P. A. et al. (Ed.): The 1998 Floods and Beyond. Towards Comprehensive Food Security in Bangladesh. Dhaka, pp. 13–52

Dorosh PA, Murshid KAS (2004) Trade Liberalization and National Food Security: Rice Trade between Bangladesh and India. Dhaka

Etzold B et al (2009) Informality as Agency. Negotiating Food Security in Dhaka. In: Die Erde 140, H. 1, pp. 3–24

FAO (2004) The State of Food Insecurity in the World 2004. Monitoring progress towards the World Food Summit and the Millenium Development Goals. Rome

FAO (2008) Rice Market Monitor. April 2008. Washington D.C. (Volume XI - Issue 1)

FAO (2009a) Rice Market Monitor. February 2009. Washington D.C. (Volume XII - Issue No. 1)

FAO (2009b) Rice Market Monitor. December 2009. Washington D.C. (Volume 7 - Issue No. 2)

FAOSTAT (2010) Database on Food Production and Trade, Food and Agricultural Organisation of the United Nations (FAO), http://faostat.fao.org, accessed: 07.02.2010
FPMU (2009) Bangladesh Food Situation Report. Volume 78. Dhaka
Gertel J (2010) Globalisierte Nahrungskrisen. Bruchzone Kairo. Bielefeld
Government of Bangladesh und WFP (2004) The Food Security Atlas of Bangladesh. Towards a poverty and hunger free Bangladesh. Dhaka
Gragnolati M et al (2006) India's Undernourished Children. A Call for Reform and Action. Washington D.C.
Hackenbroch K et al (2009) Spatiality of Livelihoods: Urban Public Space and the Urban Poor in Dhaka. In: Die Erde 140, H. 1, pp. 47–68
Keck M et al (2008) Reis für die Megacity. Die Nahrungsversorgung von Dhaka zwischen globalen Risiken und lokalen Verwundbarkeiten. In: Geographische Rundschau 60, H. 11, pp. 28–37
Khan, MMH et al (2009) Comparison of Health-Related Outcomes between Urban Slums, Urban Affluent and Rural Areas in and around Dhaka Megacity, Bangladesh. In: Die Erde 140, H. 1, pp. 69–87
Kulke E, Staffeld R (2009) Informal Production Systems - The Role of the Informal Economy in the Plastic Recycling and Processing Industry in Dhaka. In: Die Erde 140, H. 1, pp. 25–43
Lam, T T-F (1982) Food for the City: The Role of the Informal Sector. In: Geo Journal Supplementary Issue Vol. 4, pp. 49–59
Lohr, K and Dittrich C (2007) Changing food purchasing and consumption habits among urban middle-class in Hyderabad. Research Reports for Analysis and Action for Sustainable Development of Hyderabad. Report 3. Berlin
Maxwell D (1996) Measuring Food Security: The Frequency and Severity of 'Coping Strategies'. In: Food Policy 21, H. 3, pp. 291–303
Maxwell D (1999) The Political Economy of Urban Food Security in Sub-Saharan Africa. In: World Development 27, H. 11, pp. 1939–1953
Pryer J, Crook N (1988) Cities of Hunger. Urban Malnutrition in Developing Countries. Oxford
Pryer J (2003) Poverty and Vulnerability in Dhaka Slums: The Urban Livelihood Study. Aldershot
Ruel MT et al (1998) Urban Challenges to Food and Nutrition Security: A Review of Food Security, Health, and Caregiving in the Cities. Washington D.C. (FCND Disccussion Paper No. 51.)
Sakdapolrak P (2011) Orte und Räume der Health Vulnerability. Bourdieus Theorie der Praxis für die Analyse von Krankheit und Gesundheit in megaurbanen Slums von Chennai, Südindien. Dissertation. Universität Bonn (Geographisches Institut). Bonn
Sen A (1981) Poverty and Famines: an Essay on Entitlement and Deprivation. Oxford
Smith L, Haddad L (2000) Overcoming Child Malnutrition in Developing Countries. Past Achievements and Future Choices. Washington D.C.
UN-Habitat (2006) State of the World Cities 2006/7. The Millenium Development Goals and Urban Sustainability. 30 Years of Shaping the Agenda. New York
UN (2008) World Urbanization Prospects: The 2007 Revision. Highlights. New York
USDA (2010) International Food Consumption Patterns Database, United States Department of Agriculture. Economic Research Service
Watts M (1983) Silent Violence: Food, Famine, and Peasantry in Northern Nigeria. Berkeley
Watts M, Bohle HG (1993) The space of vulnerability: the causal structure of hunger and famine. In: Progress in Human Geography 17, H. 1, pp. 43–67

FSC
www.fsc.org
MIX
Papier | Fördert
gute Waldnutzung
FSC® C083411

Kolibri 360 GmbH
Ferdinand-Jühlke-Straße 7
99095 Erfurt, Deutschland
produktsicherheit@kolibri360.de